Emerging Technologies in the Hydrometallurgical Recycling of Critical Metals

Emerging Technologies in the Hydrometallurgical Recycling of Critical Metals

Guest Editor

Ana Paula Paiva

MDPI

Basel • Beijing • Wuhan • Barcelona • Belgrade • Novi Sad • Cluj • Manchester

Guest Editor
Ana Paula Paiva
Faculdade de Ciências
Universidade de Lisboa
Lisboa
Portugal

Editorial Office
MDPI AG
Grosspeteranlage 5
4052 Basel, Switzerland

This is a reprint of the Special Issue, published open access by the journal *Recycling* (ISSN 2313-4321), freely accessible at: https://www.mdpi.com/journal/recycling/special_issues/1K6PDNL095.

For citation purposes, cite each article independently as indicated on the article page online and as indicated below:

Lastname, A.A.; Lastname, B.B. Article Title. *Journal Name* **Year**, *Volume Number*, Page Range.

ISBN 978-3-7258-6424-9 (Hbk)
ISBN 978-3-7258-6425-6 (PDF)
https://doi.org/10.3390/books978-3-7258-6425-6

Contents

About the Editor vii

Preface ix

Katarzyna Leszczyńska-Sejda, Arkadiusz Palmowski, Michał Ochmański, Grzegorz Benke, Alicja Grzybek, Szymon Orda, et al.
Recycling of Rhenium from Superalloys and Manganese from Spent Batteries to Produce Manganese(II) Perrhenate Dihydrate
Reprinted from: *Recycling* **2024**, *9*, 36, https://doi.org/10.3390/recycling9030036 1

Wojciech Hyk and Konrad Kitka
Selective Recovery of Tin from Electronic Waste Materials Completed with Carbothermic Reduction of Tin (IV) Oxide with Sodium Sulfite
Reprinted from: *Recycling* **2024**, *9*, 54, https://doi.org/10.3390/recycling9040054 16

Jian-Zhi Wang, Yi-Chin Tang and Yun-Hwei Shen
Experimental Study on the Separation of Selected Metal Elements (Sm, Co, Fe, and Cu) from Nitric Acid Leachate Using Specific Precipitants
Reprinted from: *Recycling* **2024**, *9*, 111, https://doi.org/10.3390/recycling9060111 25

Angeliki Lampou, Evgenios Kokkinos, Charikleia Prochaska, Theodosios Tsiogkas, Effrosyni Peleka, Anthimos Xenidis and Anastasios Zouboulis
A Hydrometallurgical Process for the Recovery of Noble Metals (Au, Pt, Ir, and Ta) from Pyrolyzed and Acid-Digested Solutions of Single-Use Medical Devices
Reprinted from: *Recycling* **2024**, *9*, 118, https://doi.org/10.3390/recycling9060118 39

K. Yamini, Laurence Dyer, Bona Lim and Richard Diaz Alorro
Technospheric Mining of Cobalt and Nickel from Waste Nickel Furnace Slag by Ascorbic Acid-Assisted Citric Acid Leaching
Reprinted from: *Recycling* **2025**, *10*, 43, https://doi.org/10.3390/recycling10020043 59

Isabel F. F. Neto, Márcia A. D. Silva and Helena M. V. M. Soares
Effective Recovery of Gold from Chloride Multi-Metal Solutions Through Anion Exchange
Reprinted from: *Recycling* **2025**, *10*, 64, https://doi.org/10.3390/recycling10020064 75

Ahmed Tarek Ismail Mohamed, Giuliana Schimperna, Gianluca Cantoni, Francesca Demichelis, Debora Fino, Sara Perucchini, et al.
Effect of Solvent Pre-Treatment on the Leaching of Copper During Printed Circuit Board Recycling
Reprinted from: *Recycling* **2025**, *10*, 80, https://doi.org/10.3390/recycling10030080 91

Monika Gajec, Anna Król, Jadwiga Holewa-Rataj, Ewa Kukulska-Zając and Tomasz Kuchta
Electrolytic Recovery of Indium from Copper Indium Gallium Selenide Photovoltaic Panels: Preliminary Investigation of Process Parameters
Reprinted from: *Recycling* **2025**, *10*, 86, https://doi.org/10.3390/recycling10030086 108

Evan Lelong, Julien Couturier, Clément Levard, Stéphane Pellet-Rostaing and Guilhem Arrachart
Extraction of Rare Earth Elements from Organic Acid Leachate Using Formo-Phenolic-like Resins
Reprinted from: *Recycling* **2025**, *10*, 165, https://doi.org/10.3390/recycling10040165 130

Ahmed H. Ibrahim, Xianjun Lyu, Hani E. Sharafeldin and Amr B. ElDeeb
Eco-Friendly and Complex Processing of Vanadium-Bearing Waste for Effective Extraction of Valuable Metals and Other By-Products: A Critical Review
Reprinted from: *Recycling* **2025**, *10*, 6, https://doi.org/10.3390/recycling10010006 . . . . **148**

Zahra Ilkhani and Farid Aiouache
Bioleaching of Gold from Printed Circuit Boards: Potential Sustainability of Thiosulphate
Reprinted from: *Recycling* **2025**, *10*, 87, https://doi.org/10.3390/recycling10030087 . . . . **201**

About the Editor

Ana Paula Paiva

Ana Paula Paiva is an Associate Professor with Habilitation at the University of Lisbon in Portugal. Her main research interests have been focused on the investigation of the hydrometallurgical recovery of platinum-group metals from end-of-life devices, particularly the development of novel solvent extraction systems with the aim to efficiently and selectively recover Pd(II) and Pt(IV) from spent automotive and petrochemical catalysts' aqueous leaching solutions. She is the author and co-author of about 75 articles, including original research and review articles in peer-reviewed journals and conference proceedings. In 2025, she was included in the World's Top 2% Scientists list (career award), recognizing the world's most highly cited scientists according to metrics defined by Stanford University and the Elsevier database.

Preface

As primary metal resources are finite, investment in successful routes to reutilization and recycling of metals from end-of-life devices, effluents, residues, dust, etc., are of crucial environmental and economic importance. This Reprint collects nine scientific articles reporting the use of hydrometallurgy to recover critical metals from different sorts of waste, ranging from permanent magnets, photovoltaic panels, electronic waste, various slag dusts, and single-use medical devices to superalloys and batteries. The hydrometallurgical approaches that were employed encompass a very broad range of methods, aiming to optimize their performance and, simultaneously, achieve the necessary sustainability. Two review articles are also included, critically discussing (1) the most relevant information relating to the potential sustainability of thiosulphate for the bioleaching of gold from printed circuit boards and (2) the processing of vanadium-bearing waste for effective extraction of valuable metals by eco-friendly routes. In summary, all readers who are interested in expanding their knowledge of hydrometallurgical techniques and, at the same time, keeping up to date with the latest developments in the field of recycling are invited to browse this Reprint.

Ana Paula Paiva
Guest Editor

recycling

MDPI

Article

Recycling of Rhenium from Superalloys and Manganese from Spent Batteries to Produce Manganese(II) Perrhenate Dihydrate

Katarzyna Leszczyńska-Sejda *, Arkadiusz Palmowski, Michał Ochmański, Grzegorz Benke, Alicja Grzybek, Szymon Orda, Karolina Goc, Joanna Malarz and Dorota Kopyto

Łukasiewicz Research Network—Institute of Non-Ferrous Metals, Sowińskiego 5, 44-100 Gliwice, Poland; arkadiusz.palmowski@imn.lukasiewicz.gov.pl (A.P.); michal.ochmanski@imn.lukasiewicz.gov.pl (M.O.); grzegorz.benke@imn.lukasiewicz.gov.pl (G.B.); alicja.grzybek@imn.lukasiewicz.gov.pl (A.G.); szymon.orda@imn.lukasiewicz.gov.pl (S.O.); karolina.goc@imn.lukasiewicz.gov.pl (K.G.); joanna.malarz@imn.lukasiewicz.gov.pl (J.M.); dorota.kopyto@imn.lukasiewicz.gov.pl (D.K.)
* Correspondence: katarzyna.leszczynska-sejda@imn.lukasiewicz.gov.pl

Abstract: This work presents the research results on the development of an innovative, hydrometallurgical technology for the production of manganese(II) perrhenate dihydrate from recycled waste. These wastes are scraps of Ni-based superalloys containing Re and scraps of Li–ion batteries containing Mn—specifically, solutions from the leaching of black mass. This work presents the conditions for the production of $Mn(ReO_4)_2 \cdot 2H_2O$. Thus, to obtain $Mn(ReO_4)_2 \cdot 2H_2O$, manganese(II) oxide was used, precipitated from the solutions obtained after the leaching of black mass from Li–ion batteries scrap and purified from Cu, Fe and Al (pH = 5.2). MnO_2 precipitation was carried out at a temperature < 50 °C for 30 min using a stoichiometric amount of $KMnO_4$ in the presence of H_2O_2. MnO_2 precipitated in this way was purified using a 20% H_2SO_4 solution and then H_2O. Purified MnO_2 was then added alternately with a 30% H_2O_2 solution to an aqueous $HReO_4$ solution. The reaction was conducted at room temperature for 30 min to obtain a pH of 6–7. $Mn(ReO_4)_2 \cdot 2H_2O$ precipitated by evaporating the solution to dryness was purified by recrystallization from H_2O with the addition of H_2O_2 at least twice. Purified $Mn(ReO_4)_2 \cdot 2H_2O$ was dried at a temperature of 100–110 °C. Using the described procedure, $Mn(ReO_4)_2 \cdot 2H_2O$ was obtained with a purity of >99.0%. This technology is an example of the green transformation method, taking into account the 6R principles.

Keywords: batteries; superalloys; rhenium; manganese; metal perrhenates

1. Introduction

Rhenium appears extremely rarely in nature, only in a dispersed state, mainly in molybdenite, columbite and copper-bearing shales [1]. Manganese, on the other hand, is the twelfth-most abundant metal in the Earth's crust, occurring mainly in the form of oxides, carbonates and silicates. It is the most common heavy metal after iron [2–4].

The chemical properties of rhenium determine its applications, mainly in the production of superalloys but also in catalysis [5,6]. Rhenium is a heavy metal with a high melting point (>3180 °C), significant density (21.0 g/cm^3), high hardness (Mohs 7), high strength, thermal and chemical resistance and can also be easily shaped by plastic processing. It is characterized by the highest modulus of elasticity of all refractory metals, −420 GPa [1,3,7].

Manganese, on the other hand, is a hard, silvery and brittle metal with a pink luster. It is widely used as an additive in steel, lowering its melting point and improving its mechanical properties. It is pyrophoric when crushed. Manganese is in huge demand around the world due to its numerous and diverse applications, mainly in the production of steel and high-capacity batteries [2,8,9].

The world's annual production of rhenium is up to 75 tons, the vast majority of which is recovered from so-called primary raw materials, and only ~8 tons are produced from recycling [1,10,11]. The main manufacturers of rhenium from recycling are countries such as the USA, Canada, Germany, Czechia and Poland [1,12]. Rhenium recycling was much greater in 2014–2015, generating 13 tons/year. Rhenium is manufactured as a side product during the production of other elements both from primary or secondary resources, for example, during the recovery of Co and Ni from superalloys scraps. Consequently, the production of rhenium is influenced by the demand for other metals like cobalt and nickel. Currently, on the metals market, cobalt is considered a critical material in most of the world [13–15]. In contrast, nickel is a strategic material and therefore sought after, mainly due to the development of the battery industry [16,17].

However, the world annual production of manganese is not comparable to rhenium, as it constitutes ~6.2 million tons. South Africa is the world's largest manufacturer of manganese, accounting for 33.5% of its global production. It should also be noted that most of the mining of manganese is concentrated in the Kalahari Desert, which is believed to contain over 70% of the world's resources of this metal. Manganese is recycled mainly from iron and steel scraps; a small amount of it is recycled from aluminum beverage cans. The recycling rate is 37%, and the efficiency is estimated at 53% [18–20].

This publication concerns the preparation of manganese(II) perrhenate dihydrate. Manganese and rhenium share many of the chemical characteristics of transition metals, including multiple valency, the ability to form stable complex ions, paramagnetism and catalytic properties. However, in many aspects, rhenium is chemically more similar to technetium than to manganese [21,22]. For this reason, $Mn(ReO_4)_2 \cdot 2H_2O$ has potential for use in the production of alloys, catalysts or in electrical components [23].

There are not many reports in the literature about the combination of rhenium and manganese and, consequently, about $Mn(ReO_4)_2 \cdot 2H_2O$. There are several reports on the preparation, properties and use of manganese(II) perrhenate and its hydrates.

In 1949, W. T. Smith and G. E. Maxwell described the synthesis of manganese(II) perrhenate dihydrate, which was obtained as a result of the reaction of manganese(II) carbonate or manganese(II) hydroxide with perrhenic acid. This publication also specified the physicochemical properties of the above-mentioned compound, such as solubility, density and freezing point [24].

In 1969, H. G. Mayfield, Jr. and W. E. Bull described the process of forming the complexes of manganese(II) perrhenate with pyridine [25].

However, the synthesis of manganese(II) perrhenate from manganese(IV) oxide, metallic manganese and rhenium(VI) oxide has been patented and described in patent no. US 4027004A. It also explains the properties of the obtained manganese(II) perrhenate, which allows it to be used in electrical components [23].

In 1981, K.V. Ovchinnikov and his team investigated the thermal decomposition of manganese(II) perrhenate in vacuum [26].

In 1997, Charles Torardi and his team presented the magnetic properties of anhydrous manganese(II) perrhenate in a publication. This compound was found to have antiferromagnetic properties. It was also determined that anhydrous manganese(II) perrhenate has trigonal symmetry [27].

A. Butz, G. Miehle, H. Paulus, P. Strauss and H. Fuess, in 1998, researched the crystallographic structure of manganese(II) perrhenate in dihydrate and anhydrous forms and published the results of the dihydrate dehydration process. This compound was obtained as a result of the reaction of manganese(II) carbonate with perrhenic acid. The process was carried out in a temperature range of 50–60 °C in the presence of carbon dioxide [28].

In *Z. Für Krist. -New Cryst. Struct.*, in the same year, an article was published presenting the results of research on the crystallographic structure of $Mn(ReO_4)_2(H_2O)_2$. The compound was obtained as a result of the reaction of manganese(II) carbonate with perrhenic acid [29].

A work about $[Mn(H_2O)_2](ReO_4)_2$ was also published by J. Hetmańczyk and Ł. Hetmańczyk These studies were conducted using spectroscopic methods. The authors presented the dynamics of H_2O ligands and a perrhenate anion in the $[Mn(H_2O)_2](ReO_4)_2$ molecule at the moment of phase transition, as well as the crystallographic structure and thermal properties of the above-mentioned compounds. These compounds were obtained as a result of the reaction of manganese(II) carbonate with perrhenic acid [30].

In 2021, B.C. Gong, H.C. Yang, J.F. Zhang, K. Liu and Z.Y. Lu conducted research in which they found that manganese(II) perrhenate, which has a wavy layered structure, doped with electrons exhibits ferromagnetic properties. They also determined that this compound can be a so-called van der Walls magnetic material, which means that it can be used for research on magnetism and in spintronics [31].

In 2020, in *OSTI.GOV*, a detailed crystallographic analysis of manganese(II) perrhenate was described. It showed that $Mn(ReO_4)_2$ crystallizes in a trigonal space group and its structure is two-dimensional [32].

The objective of this study was to produce manganese(II) perrhenate dihydrate exclusively from two waste streams that, to the knowledge of the authors, have not been previously reported in the literature. Furthermore, the innovative method involves handling all waste materials, solid and liquid, to eliminate any loss of components and the need for further processing. This waste-free technology enables the production of high-purity manganese(II) perrhenate dihydrate, a valuable material for future novel applications in alloys, catalysts and electrical components. It is also in agreement with the principles of sustainable development and green transformation, and importantly, it meets the 6R principles in all six areas (rethink, refuse, reduce, reuse, repair and recycle) [33].

2. Results and Discussion

2.1. Recovery of Manganese from Post-Leaching Solutions of Li–ion Battery Scrap and Analysis of the Obtained MnO_2

Manganese precipitation tests were carried out using the post-leaching solutions of Li–ion battery scrap. Before the main precipitation of manganese, the solution was purified from Cu and then from Al and Fe. Copper was precipitated using 1 mole of iron in the form of metallic Fe powder for each mole of copper in the solution, which takes around 45 min with filtration. After filtering the precipitate, 30% H_2O_2 solution and then aqueous 20%NaOH solution were added over 30 min until the pH 5.2 was obtained. Neutralization was carried out at a temperature not exceeding 50 °C, and after obtaining the desired pH, the reaction was conducted for 30 min in the obtained conditions. After that time, the precipitated aluminum–iron sludge was filtered to obtain a solution that was sent for the manganese recovery tests. Table 1 lists the compositions of all the solutions, and Figure 1 shows (a) losses of valuable metals in the purification processes (Li, Mn, Co and Ni) and (b) efficiency of the precipitation of pollutants (Cu, Fe and Al).

Table 1. Composition of the solutions obtained in the purification tests.

Type of Solution	Volume, dm^3	Concentration, g/dm^3						
		Ni	Co	Al	Mn	Li	Fe	Cu
after leaching	10.0	8.20	25.60	2.50	6.54	2.56	1.80	2.50
after Cu precipitation (E)	9.65	8.42	26.32	2.59	6.78	2.65	4.14	<0.01
after Fe and Al precipitation (F)	10.12	7.91	24.74	<0.01	6.45	2.51	<0.01	<0.01

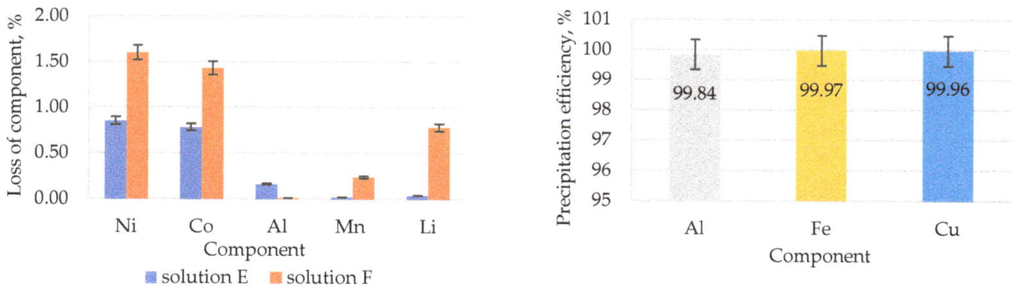

(**a**) losses of valuable metal

(**b**) efficiency of the precipitation

Figure 1. Results of the solution purification tests from Cu, Fe and Al.

As can be seen in Figure 1, the losses of valuable metals (Li, Mn, Co and Ni) did not exceed 2% for each component, while the precipitation efficiencies of impurities, such as Cu, Fe and Al, were 99.84%, 99.97% and 99.96%, respectively.

In the next stage, the purified solution was sent for the manganese precipitation tests. Precipitation was carried out using a stoichiometric amount of $KMnO_4$, and the resulting mixture was intensively stirred for 60 min at room temperature. After this time, the precipitate was vacuum-filtered, washed (with solutions of sulfuric acid and water) and analyzed. The solution from the filtration was directed to the recovery of other metals (Ni, Co and Li).

In this way, manganese(II) oxides were obtained, with the compositions depending on the purification stage listed in Table 2. No significant losses of Ni, Li and Co were observed during the process. Cobalt and nickel losses, associated with MnO_2 precipitation, were reduced by proposing a recycling of the solutions resulting from the purification. For the selected manganese(II) oxide, after all purification steps, XRD analysis was performed (Figure 2). During the research, a waste management method was developed (Figure 3). SEM analysis was performed for the purified with 20% H_2SO_4 manganese(II) oxide—Figure 4.

Table 2. Compositions of the obtained manganese(II) oxides.

Type of MnO_2	Composition, wt%			
	Mn	Co	Ni	Fe
crude	56.10	4.14	0.90	<0.01
purified with 1% H_2SO_4	53.30	3.55	0.67	<0.01
purified with 20% H_2SO_4	55.90	0.20	0.16	<0.01

The purification of MnO_2 with 20% H_2SO_4 solution allowed to reduce the level of impurities below 0.20 wt% of each metal. Thus, MnO_2 was obtained, in which the amount of individual impurities such as Co, Ni, and Fe was (wt%): 0.20, 0.16 and <0.01, respectively (Table 2).

As can be seen in Figure 2, XRD analysis confirmed the possibility of producing MnO_2 using the described conditions, as the obtained diffractogram shows only the crystalline form of MnO_2. The form of the diffractogram indicates a large amount of the amorphous phase.

The obtained MnO_2 precipitate was contaminated with cobalt and nickel but did not contain iron. After purification with a 20% H_2SO_4 solution, a product free of the above-mentioned elements was obtained. It was the substrate for the production of manganese(II) perrhenate.

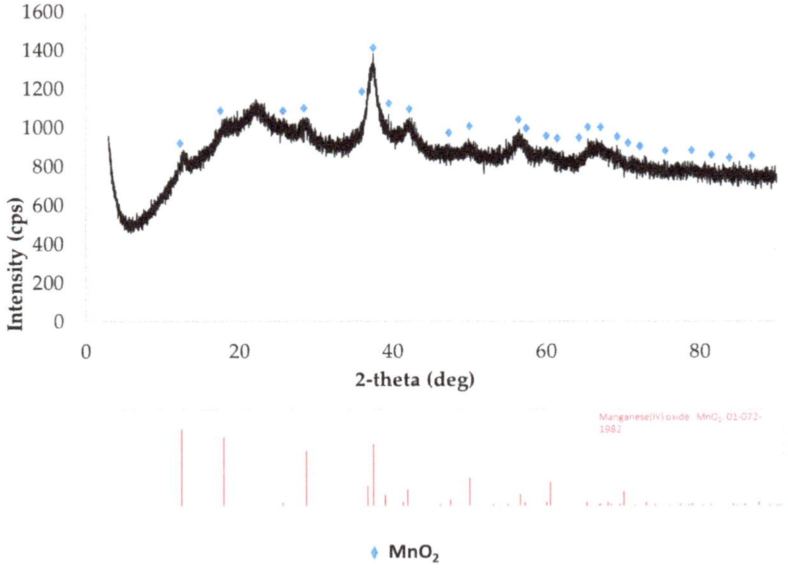

Figure 2. XRD pattern of the purified MnO_2 (purified with 20% H_2SO_4).

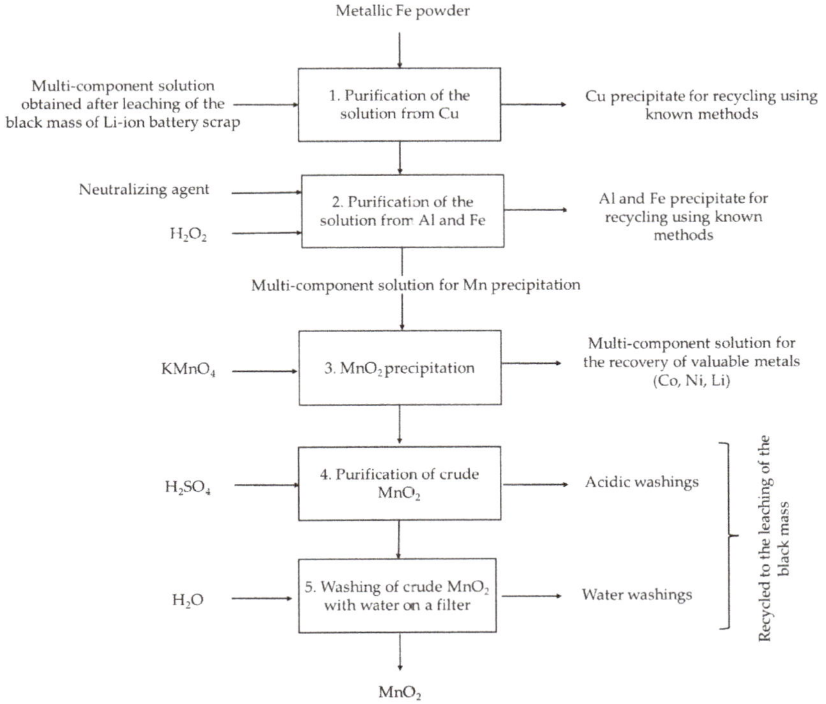

Figure 3. Scheme of MnO_2 precipitation from the post-leaching solutions of the black mass from Li–ion batteries.

Figure 4. SEM analysis of the purified MnO_2.

2.2. Precipitation and Characterization of the Obtained Manganese(II)Perrhenate Dihydrate

The production of manganese(II) perrhenate dihydrate was carried out according to the reaction (1)

$$MnO_2 + H_2O_2 + 2HReO_4 = Mn(ReO_4)_2 + 2H_2O + O_2 \quad (1)$$

Purified MnO_2 (weighing 10 g) was added alternately with a 30% H_2O_2 solution (50 cm^3 of hydrogen peroxide for each 10 g of manganese(II) oxide) to the aqueous solutions of perrhenic acid with the concentrations of 18.0 or 295.0 g/dm^3 of Re. The tests were carried out at room temperature. In this way, a solution with the pH ranging from 6 to 7 was obtained. After reaching the pH, the solution was stirred for 90 min and successively filtered from the unreacted residue. The resulting solutions were directed to the evaporation to dryness, which was carried out at a temperature < 50 °C with an addition of 30% H_2O_2 solution using 15 cm^3 of hydrogen peroxide for every 50 g of rhenium. Evaporation tests were also carried out to obtain the first crystals. The crude wet sludge precipitated in this way was combined and sent to the cyclic purification stage. A 30 g portion of crude manganese(II) perrhenate dihydrate was used in the tests. Purification was carried out using recrystallization from water with an addition of 15 cm^3 of 30% aqueous H_2O_2 solution for every 50 g of Re in the precipitate and successively evaporated to obtain the first crystals. Four purification cycles were carried out. The results of the obtained tests are presented in Tables 3 and 4. Figures 5 and 6 illustrate the influence of the used rhenium concentration on the precipitation efficiency of the obtained manganese(II) perrhenate dihydrate and the influence of purification cycles on the purification efficiency of the obtained compound.

With the use of perrhenic acid with the rhenium concentrations of 18 and 295 g/dm^3 both, similar $Mn(ReO_4)_2{\cdot}2H_2O$ precipitation efficiencies were obtained using both precipitation methods. The efficiencies were high, for example, over 92%, while using the evaporation to dryness method. Therefore, the choice of acid concentration for the technology in industry will result from the scale of the production, the amount of waste solutions and the price of energy and will be analyzed at each stage of the technology implementation.

Table 3. Results of $Mn(ReO_4)_2{\cdot}2H_2O$ precipitation.

Concentration of Re in $HReO_4$, g/dm³	Volume of $HReO_4$, dm³	Efficiency of $Mn(ReO_4)_2{\cdot}2H_2O$ Precipitation, %	Composition in Crude $Mn(ReO_4)_2{\cdot}2H_2O$, wt%				
			Mn	Re	Co	Ni	Fe
18.0	2.38	92.51	9.31	62.30	0.55	0.12	0.02
		70.56 *	9.22	62.30	0.54	0.08	<0.01
295.0	0.15	96.52	9.54	64.30	0.51	0.13	<0.01
		75.25 *	9.58	64.10	0.55	0.13	0.02

* Research conducted to the precipitation of the first crystals.

Table 4. $Mn(ReO_4)_2{\cdot}2H_2O$ purification results.

Number of the Cycle	Mass of $Mn(ReO_4)_2{\cdot}2H_2O$, g	Precipitation Efficiency of $Mn(ReO_4)_2{\cdot}2H_2O$ *, %	Composition in Crude $Mn(ReO_4)_2{\cdot}2H_2O$, wt%				
			Mn	Re	Co	Ni	Fe
0	30.0	-	9.41	63.25	0.54	0.16	<0.01
I	24.0	80.0	9.29	62.95	0.15	0.10	<0.01
II	19.6	65.3	9.29	62.95	0.10	0.10	<0.01
III	17.5	58.3	9.29	62.95	0.10	0.10	<0.01
IV	15.2	50.7	9.29	62.95	0.10	0.10	<0.01

* Efficiency calculated in relation to the initial mass of crude $Mn(ReO_4)_2{\cdot}2H_2O$.

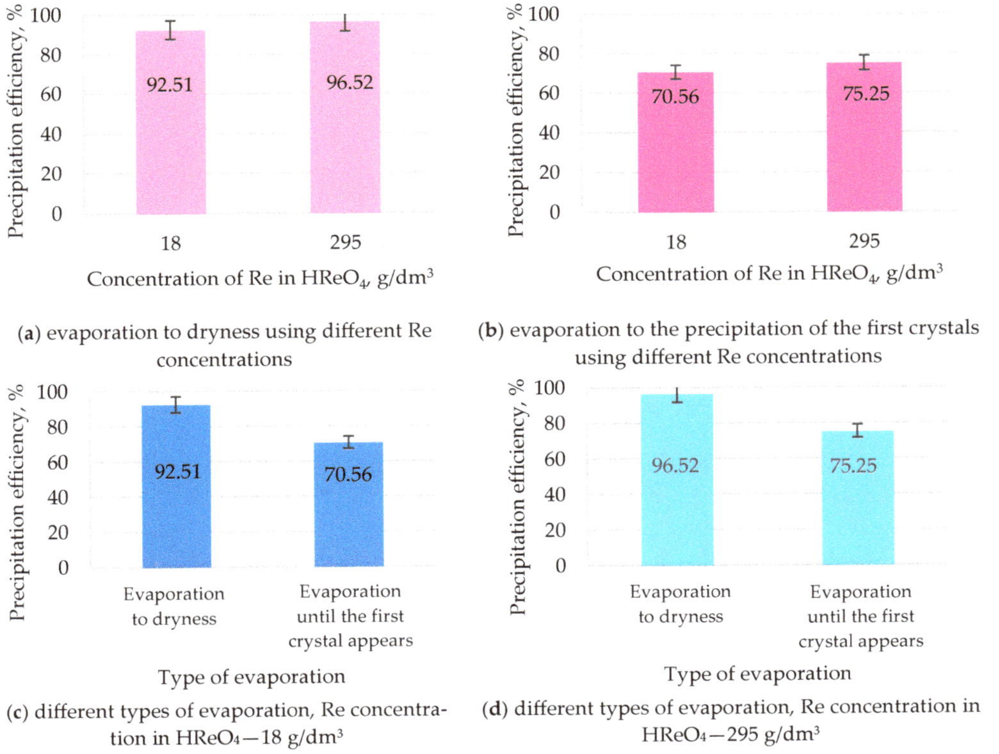

Figure 5. Results of the influence of the Re concentration in $HReO_4$ and the evaporation method on the efficiency of $Mn(ReO_4)_2{\cdot}2H_2O$ precipitation.

The use of the second stage of purification allows to obtain $Mn(ReO_4)_2{\cdot}2H_2O$ of satisfactory purity, i.e., 9.29% of Mn, 62.9% of Re, 0.10% of Co and Ni and <0.01% of Fe.

In this way, 15.20 g of wet manganese(II) perrhenate dihydrate was obtained, which was dried at 100–110 °C until a constant weight was reached. Thus, manganese(II) perrhenate dihydrate weighing 14.54 g with a purity > 99% was obtained. XRD analysis was performed for this compound—Figure 7.

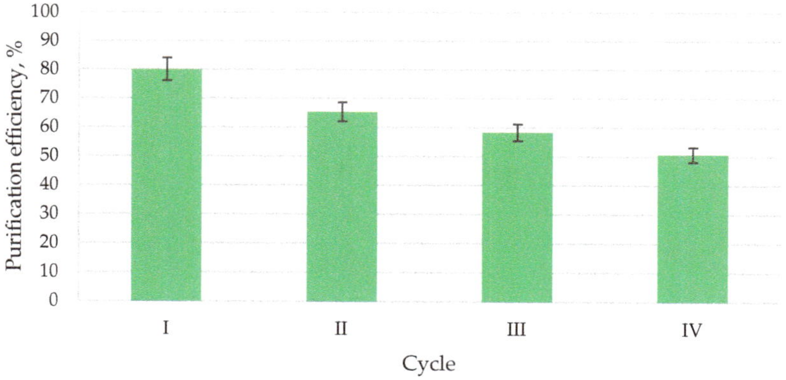

Figure 6. Results of the influence of the cyclic purification of $Mn(ReO_4)_2{\cdot}2H_2O$ on the efficiency of the purification of this compound.

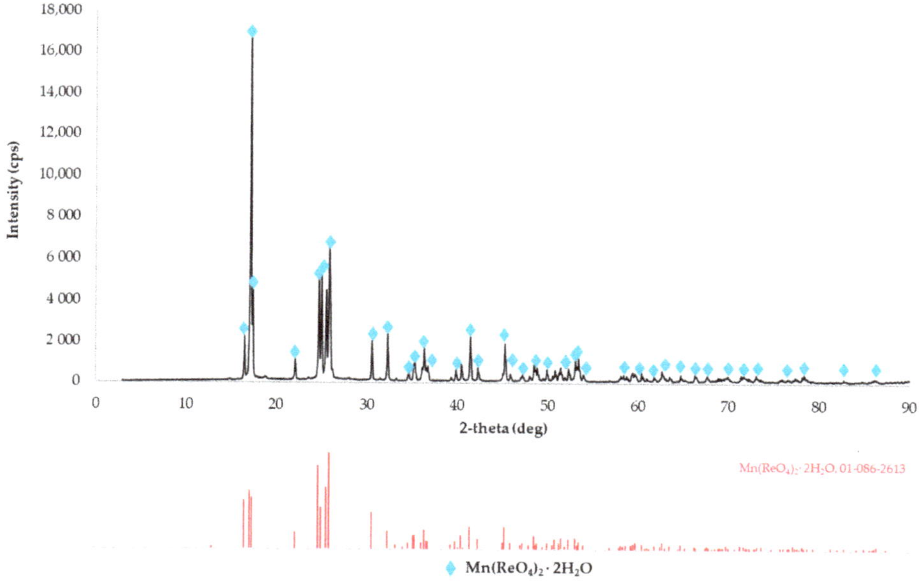

Figure 7. XRD pattern of $Mn(ReO_4)_2{\cdot}2H_2O$, after four purification cycles.

As can be seen in the diffractogram (Figure 7), a crystalline form of $Mn(ReO_4)_2{\cdot}2H_2O$ was obtained using the described conditions. This diffraction pattern also does not show any amorphous phases and substances.

For this $Mn(ReO_4)_2{\cdot}2H_2O$, its solubility was also determined. The test results are presented in Table 5.

Table 5. $Mn(ReO_4)_2 \cdot 2H_2O$ solubility test results.

Temperature, °C	Solubility of $Mn(ReO_4)_2 \cdot 2H_2O$, %	Standard Deviation
0	31.5	0.09
10	28.2	0.10
20	22.7	0.22
30	20.5	0.16

The determined solubilities of $Mn(ReO_4)_2 \cdot 2H_2O$ obtained at various temperature values are consistent with the literature data [24,34]. Each test using the gravimetric method was repeated three times, giving consistent results on the solubility of manganese(II) perrhenate dihydrate. Standard deviations were calculated and are included in Table 5.

The analysis of the conducted research on the precipitation of $Mn(ReO_4)_2 \cdot 2H_2O$ showed that this compound can be obtained using purified MnO_2, adding it alternately with the 30% H_2O_2 solution (50 cm^3 of hydrogen peroxide for each 10 g of manganese(II) oxide) to the aqueous solution of perrhenic acid. The reaction should be carried out at room temperature for 30 min until a pH 6–7 is obtained. $Mn(ReO_4)_2 \cdot 2H_2O$, precipitated by evaporating the solution to dryness, should be purified by recrystallization from water with the addition of H_2O_2 at least twice. The purified $Mn(ReO_4)_2 \cdot 2H_2O$ should be dried at 100–110 °C.

Based on the described research, this new, innovative, hydrometallurgical technology was developed for obtaining manganese(II) perrhenate dihydrate that was produced entirely from waste, i.e., superalloy scrap and Li–ion batteries. The next stage of the research will be to find a real application of manganese(II) perrhenate dihydrate in the catalytic, battery and defense industries. Research will be carried out on the possibility of obtaining manganese(II) perrhenate dihydrate with dedicated physicochemical properties, e.g., grain size below 100 nm.

3. Materials and Methods

3.1. Materials

Perrhenic acid, used in this research, was obtained by leaching the superalloy scrap (Figure 8) with a mixture of acids with an addition of oxidants. The compositions of the materials are shown in Table 6 [35,36]. In this way, a solution was obtained containing 1.1 g/dm^3 of Re and mainly nickel but also cobalt, chromium and aluminum. This solution was directed to the sorption of rhenium using a weakly basic ion exchange resin A170 (Purolite, King of Prussia, PA, USA, hydroxide form). The solution obtained after the sorption of rhenium was directed to the recovery of valuable metal components (Ni and Co), while rhenium absorbed in the ion exchange resin was eluted with an aqueous ammonia solution (25%, Chempur, Piekary Śląskie, Poland, p.a.). Ammonium perrhenate was crystallized from the obtained ammonia eluate, which was then dissolved in water and directed to the sorption of ammonium ions using a strongly acidic cation exchange resin C160 (Purolite, USA, hydrogen form) [37,38]. The post-sorption solution, containing perrhenic acid, was sent to the concentration stage, or manganese(II) perrhenate was obtained directly from it. Figure 9 shows the laboratory equipment used in the preparation of perrhenic acid using the method described above. Perrhenic acid used in the conducted research differed in rhenium concentration. The first type came directly from the ion exchange process and was not concentrated. The second type of perrhenic acid was first concentrated and then used to obtain manganese(II) perrhenate dihydrate.

Figure 8. Pictures of the crushed superalloy scrap used in the research.

Table 6. Compositions of the superalloy scrap used in the research.

Material	Composition, wt%					
	Re	Ni	Co	Cr	Al	Fe
A	1.78	43.32	6.50	4.70	14.50	1.30
B	1.10	37.90	19.50	9.10	13.90	0.08

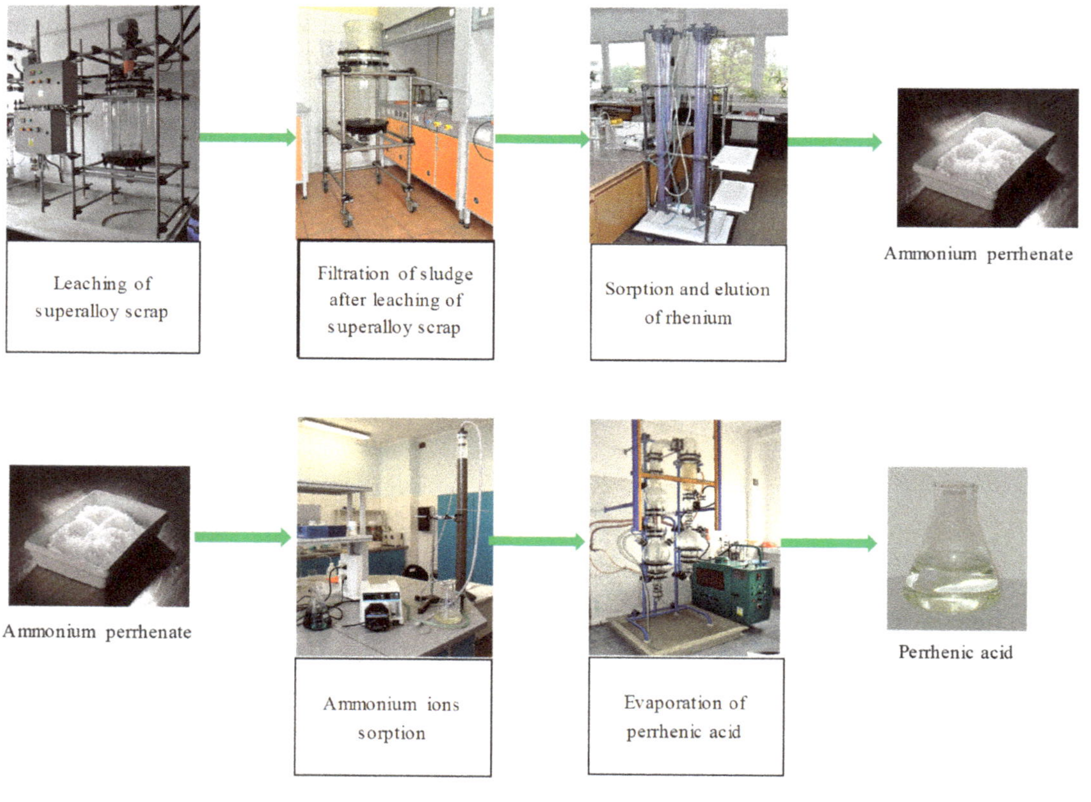

Figure 9. Scheme of obtaining perrhenic acid from the superalloy scrap.

In this way, two types of perrhenic acid were obtained, containing (1) 18.0 g/dm^3 of Re and <0.0001% of Ca, K, Mg, Cu, Na, Mo, Ni, Pb, Fe, NH_4^+, Bi, Zn, W, As and Al each and (2) 295.0 g/dm^3 of Re; <0.0005% of K; <0.0002% of NH_4^+; and <0.0001% of Ca, Mg, Cu, Na, Mo, Ni, Pb, Fe, Bi, Zn, W, As and Al each [38].

The solutions obtained after the leaching of the black mass from Li–ion batteries were used as the source of manganese—Table 7. Battery masses for the leaching process were obtained as a result of using the technology developed and patented by Łukasiewicz-IMN [39,40].

Table 7. Compositions of the black mass from Li–ion batteries used in the research.

Material	Composition, wt%						
	Ni	Co	Al	Mn	Li	Fe	Cu
C	6.50	32.00	2.00	5.45	2.50	1.90	2.03
D	12.70	17.00	2.79	10.02	2.70	2.10	1.78

After leaching, a solution with the composition of 2.50 g/dm^3 of Al, 25.60 g/dm^3 of Co, 2.56 g/dm^3 of Li, 6.54 g/dm^3 of Mn, 2.50 g/dm^3 of Cu, 8.20 g/dm^3 of Ni and 1.80 g/dm^3 of Fe was obtained, which was directed to the MnO_2 precipitation tests.

The following materials and reagents were used in this research: sulfuric acid (95%, Chempur, Piekary Śląskie, Poland, p.a.), aqueous solution of hydrogen peroxide (30%, P.P.H. Stanlab, Lublin, Poland, p.a.), demineralized water (<2 μS/cm, Łukasiewicz-IMN, Gliwice, Poland), sodium hydroxide (>98% Stanlab, Poland, p.a.), $KMnO_4$ (99% P.P.H. Stanlab, Lublin, Poland, p.a.) and Fe dust (>98%, Chempur, Piekary Śląskie, Poland, p.a.).

3.2. Analyses

All analyses were performed at the Łukasiewicz Research Network—Institute of Non-Ferrous Metals, mainly at the Centre for Analytical Chemistry (Gliwice, Poland).

The rhenium content in manganese(II) perrhenate and perrhenic acid was determined by thin-layer X-ray fluorescence spectrometry using an X-ray fluorescence spectrometer (ZSX Primus, Rigaku, Tokyo, Japan).

Ammonium ions in the aqueous solutions of perrhenic acid were determined by a distillation method with titration after distillation—the Nessler method.

The following instrumental techniques were used to determine the concentrations of Mn, Li, Cu, Mo, Pb, Bi, Zn, W, As, Al, Mg, Ca, K, Mg, Na, Fe, Co and Ni: GFAAS (graphite furnace atomic absorption spectroscopy; Z-2000, HITACHI, Tokyo, Japan), ICP-OES (inductively coupled plasma-optical emission spectroscopy; ULTI-MA 2, HORIBA Jobin-Ivon, Kyoto, Japan), ICP-MS (inductively coupled plasma-mass spectroscopy; Nexion, PerkinElmer, Waltham, MA, USA), FAAS, (flame atomic absorption spectrometry; THERMO SOLAAR S4, Thermo Fisher Scientific, Waltham, MA, USA, equipped with a flame module—deuterium background and correction) and ICP-OES (Optima 5300V, Perkin Elmer, Waltham, MA, USA).

XRD analyses were also performed at the Łukasiewicz Research Network—Institute of Non-Ferrous Metals, Centre of Functional Materials (Gliwice, Poland). The diffraction patterns were prepared using a Rigaku MiniFlex 600 XRD diffractometer equipped with an X-ray tube with a wavelength of 1.5406 Å, a D/TeX silicon strip detector and a Soller slit with a high resolution of 2.5″ on the primary and scattered beam, calibrated using NIST SRM (Standard Reference Material) 640d Si powder.

SEM analyses were also performed at the Łukasiewicz Research Network—Institute of Non-Ferrous Metals, Centre of Advanced Materials Technologies, using a high-resolution Zeiss Gemini 1525 scanning electron microscope, equipped with a Quantax xFlash®6 Bruker Nano X-ray spectrometer.

For the precipitated and purified manganese(II) perrhenate dihydrate, its solubility was measured using a gravimetric method. Figure 10 shows the used laboratory equipment. The excess amount of salt was dissolved in water with a conductivity of 0.06 μS in a thermostated reactor with a heating jacket with a working capacity of 0.5×10^{-3} m^3. The saturated solution was stirred for 4 h using a magnetic stirrer at a constant temperature. The temperature was controlled using a Huber Ministat CC-K6 thermostat (accuracy +/−0.02 °C). After stabilization, a sample of the saturated solution was taken using a syringe equipped with a filter with a pore diameter of 0.22 μm. In order to avoid crystallization during sample collection, the filter and syringe were heated to a temperature 5 °C higher than the solution temperature. The sample was dried in a laboratory dryer at 30 °C. The solubility of manganese(II) perrhenate was calculated based on the mass difference. The experiments were carried out for four temperature values, i.e., 0 °C, 10 °C, 20 °C and 30 °C. Each measurement was repeated three times.

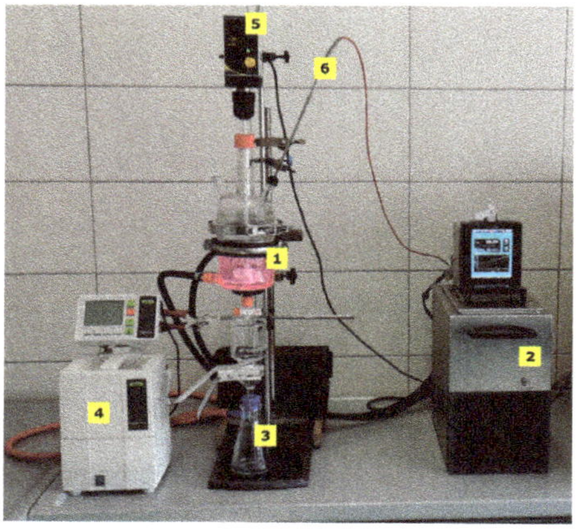

Figure 10. The equipment for determining the solubility of $Mn(ReO_4)_2{\cdot}2H_2O$: (1) reactor with the heating jacket; (2) heating and cooling thermostat; (3) vacuum filtration kit; (4) vacuum pump; (5) mechanical stirrer; (6) thermocouple.

4. Conclusions

This work showed that with the use of manganese(II) oxide, precipitated from the post-leaching solutions of the black mass from Li–ion batteries, after prior purification from Cu, Fe, and Al (pH = 5.2)—respectively, steps 1 and 2—$Mn(ReO_4)_2{\cdot}2H_2O$ can be obtained. MnO_2 precipitation should be carried out at a temperature not exceeding 50 °C for 30 min using a stoichiometric amount of $KMnO_4$—step 3. The precipitated MnO_2 should be washed using a 20% solution of sulfuric acid and then water—steps 4 and 5. Purified MnO_2 should be added alternately with a 30% H_2O_2 solution (50 cm^3 of hydrogen peroxide for each 10 g of manganese(II) oxide) to an aqueous solution of perrhenic acid. The reaction should be carried out at room temperature for 30 min to obtain a pH of 6–7—step 6. $Mn(ReO_4)_2{\cdot}2H_2O$ precipitated by evaporating the solution to dryness—step 7—should be purified by recrystallization from water with the addition of H_2O_2 a minimum of two times—step 8. The purified Mn described $(ReO_4)_2{\cdot}2H_2O$ should be dried at a temperature of 100–110 °C—step 9. Using the described procedure, $Mn(ReO_4)_2{\cdot}2H_2O$ with a purity > 99.0% and with a good precipitation efficiency can be obtained. A complete diagram of the entire developed technology is shown in Figure 11.

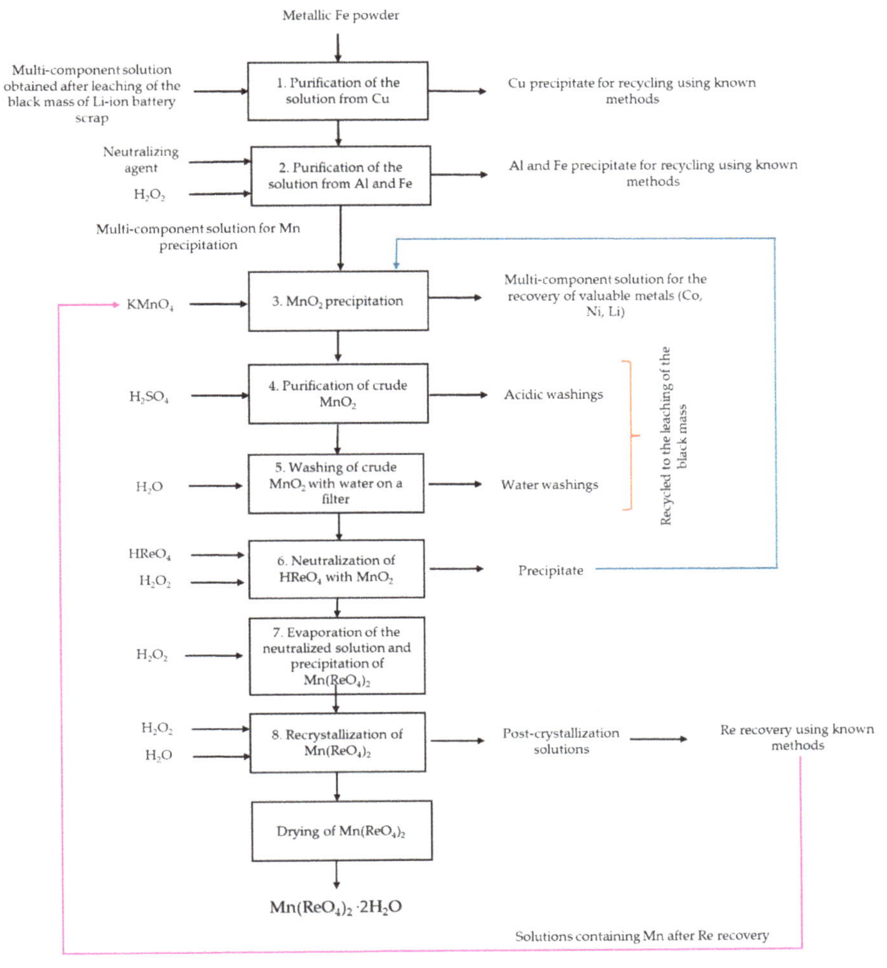

Figure 11. Diagram of the entire $Mn(ReO_4)_2{\cdot}2H_2O$ production technology from waste.

Figure 11 shows the diagram of the developed method in its entirety, with all the recirculation possibilities and indication of the so-called auxiliary operations. The blue line shows the recycling of sludge formed after the neutralization of perrhenic acid to prevent the loss of manganese and rhenium. The purple line represents the recirculation of the solutions resulting from the purification of $Mn(ReO_4)_2{\cdot}2H_2O$ to minimize the losses of rhenium and manganese. The orange line shows the recycling of the solutions from MnO_2 purification to minimize the manganese loss.

5. Patents

Part of the results of the work presented in this publication is the material submitted for patenting in the Patent Office of the Republic of Poland on 22 March 2024 entitled Sposób otrzymywania dwuwodnego renianu(VII) manganu(II) z roztworów pochodzących z ługowania masy czarnej baterii Li–ion (English title: Method of obtaining manganese(II) perrhenate dihydrate from the post-leaching solutions of the black mass from Li–ion batteries).

Author Contributions: Conceptualization, K.L.-S., M.O., A.P., S.O. and G.B.; methodology, K.L.-S., M.O., G.B. and D.K.; validation, K.L.-S., D.K., J.M. and M.O.; investigation, J.M., M.O., A.P. and S.O.; resources, K.L.-S., G.B., D.K., M.O. and A.P.; data curation, K.L.-S., A.G., G.B., M.O., A.P. and D.K.; writing—original draft preparation, K.L.-S., S.O. and A.G.; writing—review and editing, K.L.-S., K.G. and A.G.; visualization, K.L.-S., J.M., K.G. and M.O.; supervision, G.B., D.K. and K.L.-S.; project administration, K.L.-S.; funding acquisition, K.L.-S. All authors have read and agreed to the published version of the manuscript.

Funding: This work was funded by the Norwegian Financial Mechanism 2014–2021—Small Grant 2020 NOR/SGS//RenMet/0049/2020-00 (11/PE/0146/21), entitled Innovative hydrometallurgical technologies for the production of rhenium compounds from recycled waste materials for catalysis, electromobility, aviation and defense industry.

Data Availability Statement: Data available on request due to restrictions of privacy. The data presented in this study are available on request from the corresponding author. The data are not publicly available due to patent application and project contract.

Acknowledgments: The authors would like to express their thanks for the paid quantitative chemical analyses carried out in the Łukasiewicz Research Network—Institute of Non-Ferrous Metals, Centre of Analytical Chemistry and for the XRD qualitative analyses conducted in the Łukasiewicz Research Network—Institute of Non-Ferrous Metals, Centre of Functional Materials (especially Łukasz Hawełek) and Centre of Advanced Materials Technologies.

Conflicts of Interest: The authors declare no conflicts of interest. The funders had no role in the design of the study; in the collection, analyses or interpretation of the data; in the writing of the manuscript or in the decision to publish the results.

References

1. Roskill. *Rhenium: Outlook to 2029*, 11th ed.; Roskill Information Services Ltd.: London, UK, 2019; ISBN 978-1-910-92279-8.
2. Howe, P.; Dobson, S. *Manganese and Its Compounds: Environmental Aspects*; World Health Organization: Geneva, Switzerland, 2005.
3. Kemmitt, R.D.W.; Peacock, R.D. *The Chemistry of Manganese, Technetium and Rhenium*; Pergamon Press: Oxford, UK, 1975; ISBN 0-08-018870-2.
4. Leszczyńska-Sejda, K.; Majewski, T.; Benke, G.; Piętaszewski, J.; Anyszkiewicz, K.; Michałowski, J.; Chmielarz, A. Production of High-Purity Ammonium Perrhenate for W-Re-Ni-Fe Heavy Alloys. *J. Alloys Compd.* **2012**, *513*, 347–352. [CrossRef]
5. Huang, M.; Zhu, J. An Overview of Rhenium Effect in Single-Crystal Superalloys. *Rare Met.* **2016**, *35*, 127–139. [CrossRef]
6. Leszczynska-Sejda, K.; Benke, G.; Malarz, J.; Ciszewski, M.; Kopyto, D.; Piatek, J.; Drzazga, M.; Kowalik, P.; Zemlak, K.; Kula, B. Rhenium(VII) Compounds as Inorganic Precursors for the Synthesis of Organic Reaction Catalysts. *Molecules* **2019**, *24*, 1451. [CrossRef] [PubMed]
7. Shabalin, I.L. *Ultra-High Temperature Materials I*; Springer: Dordrecht, The Netherlands, 2014; ISBN 978-94-007-7587-9.
8. Lee, Y.K.; Han, J. Current Opinion in Medium Manganese Steel. *Mater. Sci. Technol.* **2015**, *31*, 843–856. [CrossRef]
9. Xie, C.; Li, T.; Deng, C.; Song, Y.; Zhang, H.; Li, X. A Highly Reversible Neutral Zinc/Manganese Battery for Stationary Energy Storage. *Energy Environ. Sci.* **2020**, *13*, 135–143. [CrossRef]
10. Shen, L.; Tesfaye, F.; Li, X.; Lindberg, D.; Taskinen, P. Review of Rhenium Extraction and Recycling Technologies from Primary and Secondary Resources. *Miner. Eng.* **2021**, *161*, 106719. [CrossRef]
11. Xu, D.; Zheng, S.; Chen, P.; Wei, B.; Zhang, J.; Cheng, J. Recycling of Rhenium from W–Re-Alloyed Scraps by a Pyrometallurgical Method. *J. Sustain. Metall.* **2022**, *8*, 148–155. [CrossRef]
12. Yagi, R.; Okabe, T.H. Review: Rhenium and Its Smelting and Recycling Technologies. *Int. Mater. Rev.* **2024**, *69*, 142–177. [CrossRef]
13. Nansai, K.; Nakajima, K.; Kagawa, S.; Kondo, Y.; Shigetomi, Y.; Suh, S. Global Mining Risk Footprint of Critical Metals Necessary for Low-Carbon Technologies: The Case of Neodymium, Cobalt, and Platinum in Japan. *Environ. Sci. Technol.* **2015**, *49*, 2022–2031. [CrossRef]
14. Li, B.M.; Lu, J. Cobalt in Lithium-Ion Batteries. *Science* **2020**, *367*, 979–980. [CrossRef]
15. Botelho Junior, A.B.; Stopic, S.; Friedrich, B.; Tenório, J.A.S.; Espinosa, D.C.R. Cobalt Recovery from Li-ion Battery Recycling: A Critical Review. *Metals* **2021**, *11*, 1999. [CrossRef]
16. Xu, J.; Lin, F.; Doeff, M.M.; Tong, W. A Review of Ni-Based Layered Oxides for Rechargeable Li-Ion Batteries. *J. Mater. Chem. A* **2017**, *5*, 874–901. [CrossRef]
17. Cui, Z.; Manthiram, A. Thermal Stability and Outgassing Behaviors of High-Nickel Cathodes in Lithium-Ion Batteries. *Angew. Chem. -Int. Ed.* **2023**, *62*, e202307243. [CrossRef] [PubMed]
18. Beukes, N.J.; Swindell, E.P.W.; Wabo, H. Manganese Deposits of Africa. *Episodes* **2016**, *39*, 285–317. [CrossRef]
19. Gutzmer, J.; Du Plooy, A.P.; Beukes, N.J. Timing of Supergene Enrichment of Low-Grade Sedimentary Manganese Ores in the Kalahari Manganese Field, South Africa. *Ore Geol. Rev.* **2012**, *47*, 136–153. [CrossRef]

20. Keller, A.; Hlawitschka, M.W.; Bart, H.J. Manganese Recycling of Spent Lithium Ion Batteries via Solvent Extraction. *Sep. Purif. Technol.* **2021**, *275*, 119166. [CrossRef]
21. Rouschias, G. Recent Advances in the Chemistry of Rhenium. *Chem. Rev.* **1974**, *74*, 531–566. [CrossRef]
22. Darab, J.G.; Smith, P.A. Chemistry of Technetium and Rhenium Species during Low-Level Radioactive Waste Vitrification. *Chem. Mater.* **1996**, *8*, 1004–1021. [CrossRef]
23. Sleight, A.W. Rhenium Oxides of Types MReO4 and M'2ReO6. U.S. Patent 40,270,04A, 31 May 1977.
24. Smith, W.T.; Maxwell, G.E. The Salts of Perrhenic Acid. II. The Iron Family and Manganese. *J. Am. Chem. Soc.* **1949**, *71*, 578–580. [CrossRef]
25. Mayfield, H.G., Jr.; Bull, W.E. Perrhenato Complexes of Bivalent Cations. *Inorganica Chim. Acta* **1969**, *3*, 676–680. [CrossRef]
26. Ovchinnikov, K.V.; Nikolaev, E.N.; Semenov, G.A. Thermal Decomposition of Perrhenates of Cr(3), Mn, Fe(3), Co and Ni in Vacuum. *Zh. Obs. Khim.* **1981**, *50*, 485–488.
27. Torardi, C.; Reiff, W.M.; Dodrill, B.C.; Vogt, T. Layered 3-D Ferromagnets AND Antiferromagnets, $M^{2+}(ReO_4)_2$ (M=Mn, Fe, Co, Ni, Cu): Importance of Dipolar Interactions. *Solid-State Chem. Inorg. Mater.* **1997**, *453*, 399–403. [CrossRef]
28. Butz, A.; Miehe, G.; Paulus, H.; Strauss, P.; Fuess, H. The Crystal Structures of $Mn(ReO_4)_2{\cdot}2H_2O$ and of the Anhydrous Perrhenates$M(ReO_4)_2$ of Divalent Manganese, Cobalt, Nickel, and Zinc. *J. Solid State Chem.* **1998**, *138*, 232–237. [CrossRef]
29. Mujica, C. Crystal Structure of Díaquabis (Perrhenato) Manganese(II). *Z. Für Krist. -New Cryst. Struct.* **1998**, *213*, 229.
30. Hetmańczyk, J.; Hetmańczyk, Ł. Dynamics of H_2O Ligands and ReO^{4-} Anions at the Phase Transition in $[Mn(H_2O)_2](ReO_4)_2$ Studied by Complementary Spectroscopic Methods. *J. Raman Spectrosc.* **2018**, *49*, 298–311. [CrossRef]
31. Gong, B.C.; Yang, H.C.; Zhang, J.F.; Liu, K.; Lu, Z.Y. Inducing High-Tc Ferromagnetism in the van DerWaals Crystal $Mn(ReO_4)_2$ via Charge Doping: A First-Principles Study. *Phys. Rev. B* **2021**, *104*, 075133. [CrossRef]
32. osti.gov. Available online: https://www.osti.gov/biblio/1205969 (accessed on 21 August 2023).
33. Maťová, H.; Kaputa, V.; Triznová, M. *Forestry and Forestry Based Industry Implications Digitalisation and Circular Economy: Forestry and Forestry Based Industry Implications*; WoodEMA, i.a, Zagreb; Union of Scientists of Bulgaria: Sofia, Bulgaria, 2019; ISBN 9789543970421.
34. Borisova, L.V.; Ermakov, A.N. *Analytical Chemistry of Elements—Analytical Chemistry of Rhenium*; Publishing House Nauka: Moscow, Russia, 1974.
35. Leszczyńska-Sejda, K.; Benke, G.; Kopyto, D.; Majewski, T.; Drzazga, M. Production of High-Purity Anhydrous Nickel(II) Perrhenate for Tungsten-Based Sintered Heavy Alloys. *Materials* **2017**, *10*, 448. [CrossRef] [PubMed]
36. Leszczyńska-Sejda, K.; Benke, G.; Kopyto, D.; Drzazga, M.; Ciszewski, M. Application of Ion Exchange for Preparation of Selected Metal Perrhenates—Precursors for Superalloy Production. *Metals* **2019**, *9*, 201. [CrossRef]
37. Benke, G.; Anyszkiewicz, K.; Hac, D.; Litwinionek, K.; Leszczyńska-Sejda, K. Progress in the Methods of Recovering Rhenium from Copper Metallurgy. *Przem. Chem.* **2006**, *85*, 793–797.
38. Leszczyńska-Sejda, K.; Benke, G.; Chmielarz, A.; Krompiec, S.; Michalik, S.; Krompiec, M. Synthesis of Perrhenic Acid Using Ion Exchange Method. *Hydrometallurgy* **2007**, *89*, 289–296. [CrossRef]
39. Lewandowski, D.; Kozłowski, J.; Mikłasz, W. Sposób Recyklingu Baterii Wtórnych. PL 240110 B1, 2017. Available online: https://worldwide.espacenet.com/patent/search/family/063854944/publication/PL240110B1?q=PL%20240110%20B1 (accessed on 29 January 2024).
40. Lewandowski, D.; Kozłowski, J.; Potempa, M. Recycling of Li-Ion Batteries with the Application of Mechanical and Pyrometallurgical Methods. *World Metall. -Erzmetall* **2020**, *73*, 123–127.

recycling

MDPI

Article

Selective Recovery of Tin from Electronic Waste Materials Completed with Carbothermic Reduction of Tin (IV) Oxide with Sodium Sulfite

Wojciech Hyk [1,2,*] and Konrad Kitka [1]

1 Faculty of Chemistry, University of Warsaw, Pasteura 1, PL-02-093 Warsaw, Poland
2 Biological and Chemical Research Center, University of Warsaw, Żwirki i Wigury 101, PL-02-089 Warsaw, Poland
* Correspondence: wojhyk@chem.uw.edu.pl; Tel.: +48-22-552-6359

Abstract: A new approach for the thermal reduction of tin dioxide (SnO_2) in the carbon/sodium sulfite (Na_2SO_3) system is demonstrated. The process of tin smelting was experimentally optimized by adjusting the smelting temperature and amounts of the chemical components used for the thermal reduction of SnO_2. The numbers obtained are consistent with the thermodynamic characteristics of the system and molar fractions of reactants derived from the proposed mechanism of the SnO_2 thermal reduction process. They reveal that the maximum yield of tin is obtained if masses of C, Na_2SO_3 and SnO_2 are approximately in the ratio 1:2:3 and the temperature is set to 1050 °C. The key role in the suggested mechanism is the thermal decomposition of Na_2SO_3. It was deduced from the available experimental data that the produced sulfur dioxide undergoes carbothermic reduction to carbonyl sulfide—an intermediate product involved in the bulk reduction of SnO_2. Replacing sodium sulfite with sodium sulfate, sodium sulfide and even elemental sulfur practically terminated the production of metallic tin. The kinetic analysis was focused on the determination of the reaction orders for the two crucial reactants involved in the smelting process.

Keywords: tin; tin dioxide; carbothermic reduction; sodium sulfite; smelting

1. Introduction

Tin is one of the most important metals commonly used in various industrial areas. It is the main component of low-melting alloys used as solders in the electronic industry. For the last two decades, the electronic industry has been the major consumer of this metal.

Cassiterite, a naturally occurring oxide of tin (SnO_2) containing about 79% tin, is the principal primary source of tin. It is the only natural mineral of tin from which extraction of tin is economically justified. The production of tin from its ore involves roasting and smelting processes [1]. In the first stage, the concentrates of cassiterite are roasted with specific reagents (including calcium chloride, calcium oxide and pyrite) to eliminate the sulfur at temperatures between 550 and 650 °C. The type of the roasting chemical depends on the type and amount of impurities. After the preparation (roasting) step, the material for smelting comprises tin dioxide and some impurities (mainly iron oxides). The smelting process carried out in either reverberatory, blast or electric furnaces is based on the reduction of tin dioxide by heating to 1300–1400 °C with carbon (in the form of anthracite coal or coke) and limestone (slag-producing agent) to produce tin and carbon dioxide [2]. The product of thermal reduction is then electrolytically refined and, as a result, up to 99.8% pure tin is obtained.

The tin-bearing natural sources will be exhausted in several years; therefore, currently, an increased effort is directed to the recovery of tin from tin-bearing secondary resources, including tin anode slime, tin slag, tin-bearing tailings and waste electronic and electrotechnical devices. A rapid growth in the amount of e-waste observed recently makes this class of tin secondary sources of special importance. Many scientific papers deal with the methods for the extraction of tin from these materials. In general, the procedures proposed are based on the chemical leaching of metallic tin, followed by its hydrometallurgical or pyrometallurgical conversion back to its metallic form [3]. It is desired that leaching agents selectively extract tin from its alloys (which contain either lead or silver and bismuth and are usually deposited on copper substrates) to generate tin compounds in either ionic or oxide forms. In a large number of examples, the cheap and effective reagents for the preparation of a water-soluble tin compound involve solutions of either sodium hydroxide [4–6], hydrochloric acid [6–8] or iron (III) chloride [9]. A strong base reacts with metallic tin to produce well-soluble stannate, while acidic solutions of chlorides lead to the formation of chloride complexes of tin. Both stannate and tin chloride hydrolize to the oxide form of tin or can be easily transformed into tin dioxide. By applying appropriate conditions, the soluble tin compounds can be electrolyzed. Alternatively, they can be quantitatively transformed into tin (IV) oxide and thermally reduced using various reducing agents. These include carbon/carbon monoxide [10], hydrogen [11], methane [12], carbon/molten $Na_2CO_3/NaNO_3$ [13] and Ar-balanced CH_4 (decomposed partially to H_2 and C) [14] systems. The optimal method for tin recovery is usually a combination of both hydrometallurgical and pyrometallurgical processes.

Recently, we have developed and implemented a method for selective recovery of tin from e-waste materials by employing the leaching process with diluted nitric (V) acid [15]. Under the specified conditions, most metals from e-scrap are leached into the aqueous solution in the form of cations (Pb^{2+}, Cu^{2+}, Ag^{+}, etc.). Tin, in turn, is transformed into a solid oxide phase, contaminated, to some extent, with lead oxides.

This paper presents a new approach for the final step in the tin recovery process. It utilizes a carbothermic reduction of tin dioxide with an addition of sodium sulfite (Na_2SO_3). The efficiency of the smelting process was studied as a function of temperature and the composition of the smelting mixture. Of special interest was the dependence of the amount of sulfites added to the system on the temperature at which tin dioxide is quantitatively reduced to metallic tin. Discussion on a possible mechanism of the SnO_2 thermal reduction is supported by the chemical, kinetic and thermodynamic analyses. The results allowed us to establish the optimal composition of the reaction mixture that can produce a maximum yield of tin in the smelting process. The possibility of Na_2SO_3 regeneration from SO_2 generated during the smelting process makes the whole process economically advantageous.

2. Results and Discussion

2.1. *The Effect of Temperature and the Amount of Sodium Sulfite*

The smelting process was examined under varying experimental conditions using fixed amounts of SnO_2 and charcoal. The effect of temperature and varying concentration of Na_2SO_3 on the efficiency of the thermal reduction of SnO_2 was investigated. It is worth noting that under the conditions employed for the experiments, no quantifiable amounts of metallic tin were obtained for the system, with no deliberately added sodium sulfite. The experimental results are presented in Figure 1 (in the 3D and 2D forms) where the percent yield of metallic tin is plotted against the temperature and the amount of Na_2SO_3 added to the thermally treated system. Each set of the results (represented by empty circles, grey squares, dark grey triangles and filled circles) corresponds to the selected temperature and represents a collection of averaged data obtained for melting crucibles made of different materials. The temperatures selected for the examination were set to 950, 1000, 1050 and 1140 °C, while the amount of sulfite ranged from 7.5 to 15.5 g. The experimental results clearly indicate that the best performance of the reduction process is achieved at 1050 °C for samples that comprise approximately 5 g of charcoal and not less than 10 g of Na_2SO_3 (i.e., 1/3 and 2/3 of the mass of tin dioxide in the sample, respectively). The quality of the smelting products was inspected using the elemental mapping mode of scanning electron microscopy. Two extreme sets of reduction conditions were selected for the comparison.

One (Figure 2A) was related to the conditions where the best performance of the process is achieved (i.e., at 1050 °C for the mixture composed of 15.0 g of SnO_2, 10.1 g of Na_2SO_3, 5.0 g of charcoal), while the other one (Figure 2B) represented the effect of the reduction under the unfavorable conditions (i.e., at 950 °C for the mixture composed of 15.0 g of SnO_2, 5.2 g of Na_2SO_3, and 5.5 of g charcoal). In the latter case, unreacted particles of sodium sulfite were clearly seen while the former case yielded a homogeneous metallic product covered with a very thin film of unreacted components. The differences observed may be due to the insufficient amount of carbon, which accelerates the Na_2SO_3 decomposition. Additionally, at lower temperatures, the reduction process involving sulfur-bearing species is not thermodynamically favorable. At this temperature, carbon is the only component that may reduce tin dioxide to the metallic phase.

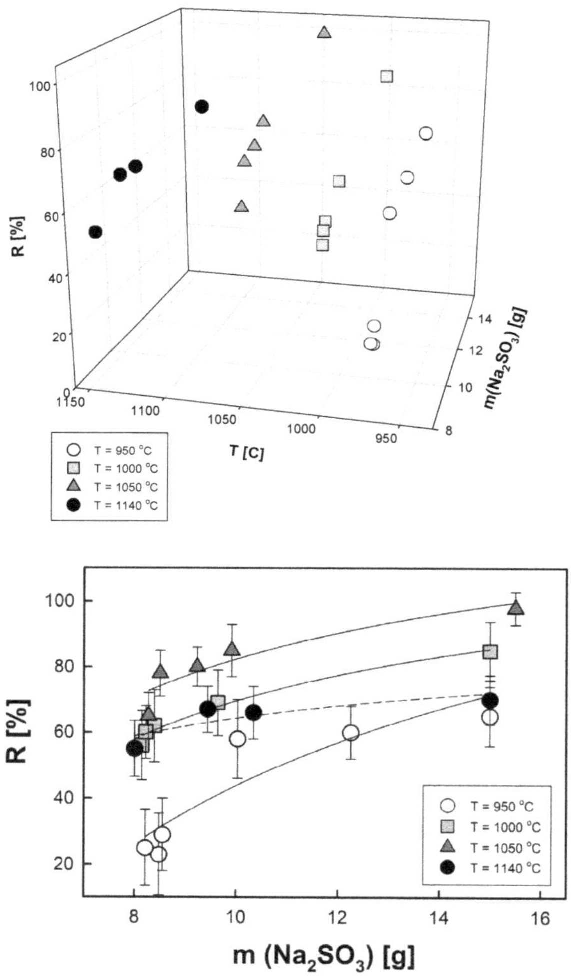

Figure 1. Percent yield of tin versus temperature and mass of sodium sulfite added to the system. Amount of charcoal: ~5 g. Each set of results corresponds to the selected temperature and represents a collection of averaged data obtained for melting crucibles made of different materials. Error bars represent expanded uncertainties of the determined quantities (i.e., their 95% confidence intervals).

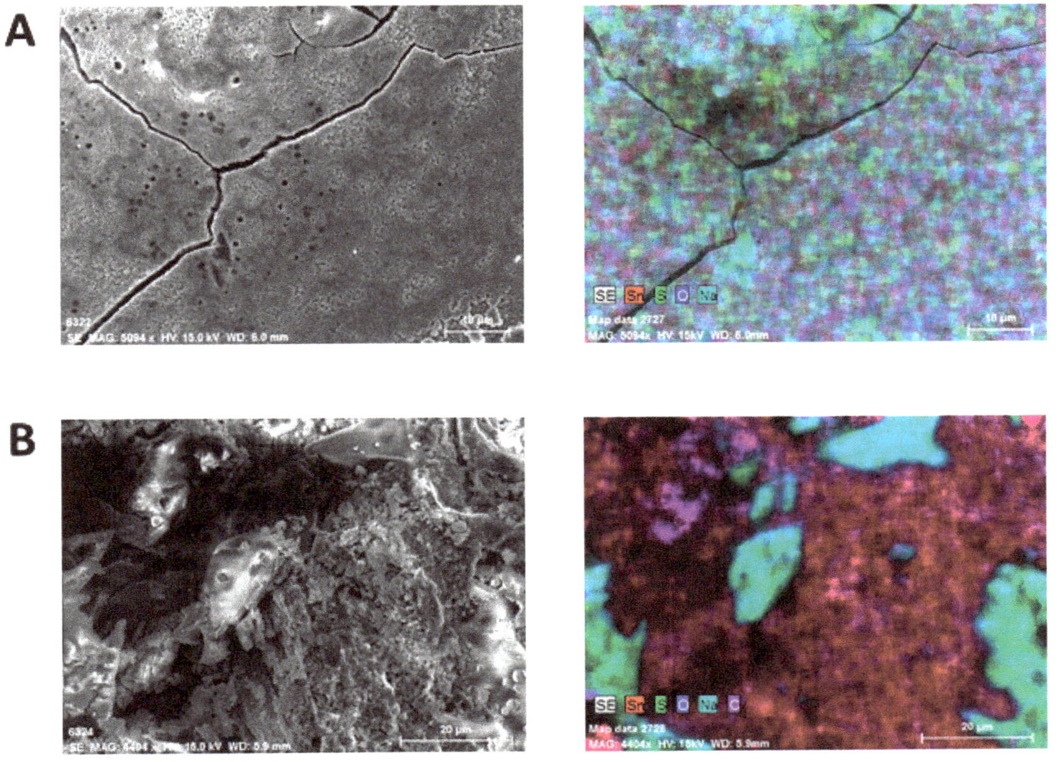

Figure 2. SEM images and elemental mapping of surfaces of tin samples obtained for the systems composed of 15.0 g of SnO_2, 10.1 g of Na_2SO_3, 5.0 of g charcoal at 1050 °C (**A**) and 15.0 g of SnO_2, 5.2 g of Na_2SO_3, 5.5 g of charcoal at 950 °C (**B**).

2.2. Kinetic Studies

To validate the observations, a more detailed kinetic analysis was performed for the process at 1050 °C. The partial orders of the reaction with respect to sodium sulfite and carbon were determined via the unweighted linear regression analysis of the dependencies between logarithms of the normalized average reaction rates and logarithms of normalized molarities of either sodium sulfite or carbon (Figure 3). The average reaction rate is directly proportional to $R/\Delta t$, where Δt is the duration of the smelting process at the given temperature. The numerical results obtained are as follows: 0.81 ± 0.32 for sodium sulfite and -1.05 ± 0.82 for carbon. The numbers after the $\pm$ sign represent expanded uncertainties (with a coverage factor of 2) of the determined kinetic parameters. The negative partial order of the reaction indicates that the concentration of that species inversely controls the overall reaction rate and suggests a complex mechanism of the reduction process. The mechanism may involve a reaction step that proceeds at the phase boundaries limited by the adsorption/desorption rates. It is also possible that one of the final products of the reduction process is involved in another step of the reaction mechanism.

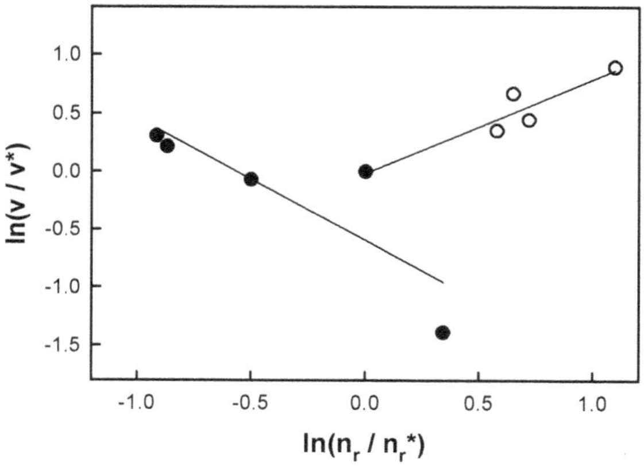

Figure 3. Logarithmic dependencies of normalized reaction rates and normalized numbers of moles of reactants examined: sodium sulfite (empty circles) and charcoal (filled circles). The rates and numbers of moles were related to the selected points that correspond to the following compositions of the system: 15.60 g of SnO_2, 5.00 g of charcoal and 5.18 g of sodium sulfite (empty circles), and 15.06 g of SnO_2, 10.90 g of charcoal and 10.00 g of sodium sulfite (filled circles). Smelting temperature: 1050 °C.

2.3. *The Proposed Mechanism of the Smelting Process*

All the information obtained from the kinetic studies and the qualitative chemical analysis (Section 3.4) allowed us to establish a possible mechanism for the thermal reduction of SnO_2 with Na_2SO_3. The mechanism can be summarized by three key reactions represented by the following equations:

(A) $2\ Na_2SO_3 \rightarrow 2\ Na_2O + 2\ SO_2$;

(B) $3\ C + 2\ SO_2 \rightarrow 2\ COS + CO_2$;

(C) $3\ SnO_2 + 2\ COS \rightarrow 3\ Sn + 2\ CO_2 + 2\ SO_2$.

The smelting process can be thus represented by the following reaction scheme:

$$SnO_2 + C \xrightarrow{Na_2SO_3, T} Sn + CO_2$$

First of all, the molar fractions of the crucial reagents (SnO_2: C: Na_2SO_3 = 1 mole: 1 mole: 2/3 mole, i.e., 150.7 g: 12 g: 84.0 g) confirm roughly the empirically determined mass ratio of the reactants (i.e., 150 g: (0.6 × 50 g): 100 g, where the number 0.6 corresponds to the fraction of carbon in charcoal) that generates the maximum percent yield of tin for temperatures ranging from 950 to 1140 °C. Secondly, it suggests the presence of an intermediate sulfide product—carbonyl sulfide (COS)—in the reaction pathway that may speed up the reduction of tin dioxide. Depending on the temperature and proportions of the reactants used, it may act as a reducing agent (more effectively than carbon) or decompose to produce another effective reducing agent—CO. COS is a product of the carbothermic reduction of SO_2. The latter substance can be easily produced by the thermal decomposition of sodium sulfite, which occurs above 500 °C [16]. This property justifies the application of sodium sulfite and excludes other sulfur-bearing compounds. Additional experiments performed with other sulfur-bearing additives (sodium sulfate, sodium sulfide and even elemental sulfur) supported this conclusion. Replacing sodium sulfite with the above-mentioned alternative additives practically terminated the production of metallic tin (the percent yield of tin was close to 0).

At temperatures above 700 °C, SO_2 undergoes carbothermic reduction to produce carbonyl sulfide. The formation of the latter substance can be catalyzed by some portion of SnO_2 as shown by Han et al. [17]. This stage of the process can be affected to some extent by another reaction path:

(B′) $C + SO_2 \rightarrow CO_2 + S$

However, at the boiling point of sulfur, the reaction B′ also leads to the formation of carbonyl sulfide [18]. At higher temperatures (above 650 °C), the rate of COS decomposition increases leading to the formation of CO and elemental sulfur [19].

In step (C), involving COS, SnO_2 is reduced to metallic tin by either COS or CO. At this stage, SnO_2 may also be partially transformed to SnS_2 by COS [17], which will lower, to some extent, the reaction yield. An alternative reaction path can be represented by the following equation:

(C′) $SnO_2 + 2\ COS \rightarrow SnS_2 + 2\ CO_2$ (and parallelly $Na_2O + COS \rightarrow Na_2S + CO_2$)

The proposed mechanism and the experimental conditions employed for the studies were also characterized thermodynamically. The expression for the standard free enthalpy for the reduction process is summarized by the net chemical reaction ($SnO_2(s) + C(s) + (2/3)\ Na_2SO_3(s) \rightarrow Sn(s) + CO_2(g) + (2/3)\ SO_2(g) + (2/3)\ Na_2O(s)$), which is as follows:

$$\Delta G^\circ = 444 - 0.328\ T\ [kJ] \quad (1)$$

where T is the smelting temperature.

The expression for ΔG° was derived using the tabulated values of the standard molar enthalpies of formation (in kJ/mol) and standard molar entropies (in J/mol/K) of the reactants of Equation (1), i.e., −577.6 and 49, 0 and 5.7, −1100.8 and 145.9, 0 and 51.2, −393.5 and 213.6, −296.8 and 248.2 and −414.2 and 75.1 for $SnO_2(s)$, C(s), $Na_2SO_3(s)$, Sn(s), $CO_2(g)$, $SO_2(g)$ and $Na_2O(s)$, respectively.

According to Equation (1), at temperatures higher than 1080 °C, the value of ΔG° is negative, and the reaction may proceed in the indicated direction.

2.4. The Effect of the Presence of Lead

The interfering effect of lead or copper is related to the fact that lead and copper (e.g., present in the oxide forms) react preferably with sulfur-bearing fluxes (carbonyl sulfide, elemental sulfur) at higher temperatures, producing lead or copper sulfides. These reactions would consume a significant part of the reducing agent and result in the decrease or even termination of the tin smelting process. The results of experiments performed for SnO_2 and PbO_2 mixed at various mass ratios, presented in Figure 4, led to the conclusion that the use of the method results in an acceptable yield of tin (more than 70%) if the level of lead impurities does not exceed 5%.

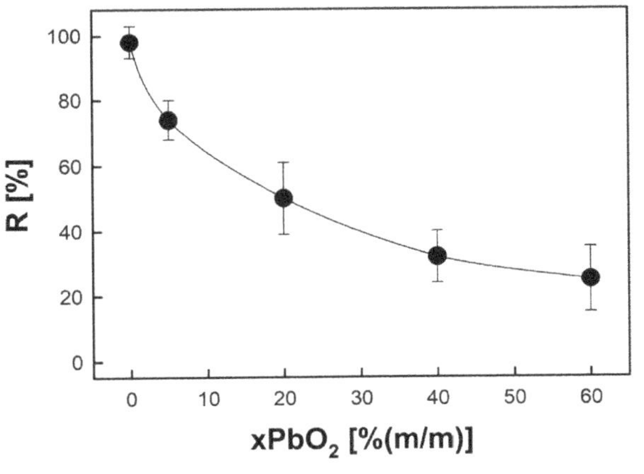

Figure 4. Percent yield of tin versus mass fraction of PbO_2 in the SnO_2/PbO_2 mixture. Masses of mixture of tin/lead oxides, charcoal and sodium sulfite: 15, 5 and 10 g, respectively. Smelting temperature: 1050 °C. Error bars represent expanded uncertainties of the determined quantities (i.e., their 95% confidence intervals).

3. Materials and Methods

3.1. Chemicals

Tin oxide (IV) (98%, SnO_2), charcoal (containing approximately 60% of carbon), sodium sulfate (IV) (99.8%, Na_2SO_3), lead (II) nitrate (V) (99,9%, $Pb(NO_3)_2$) and sodium hydroxide (99%, NaOH) were purchased from Keramos (Poland) and Sigma-Aldrich (Poland).

3.2. Instrumentation

The smelting process was conducted in a laboratory electric furnace (Argenta, Poland).

An X-ray Fluorescence Spectrometer (XRF) (model S1 Titan, Bruker, Billerica, MA, USA) was used for the determination of the chemical composition of the solid materials.

An Inductively Coupled Plasma Mass Spectrometer (ICP-MS) (NexION 300D, PerkinElmer, Poland) was used for the validation of the results produced by the XRF method.

The inspection of the surfaces of the smelting products was performed using a scanning electron microscope, model LEO 435VP.

3.3. Smelting Process

In general, the smelting process was examined on batches containing approximately 15 g of tin dioxide, 5 g of charcoal and varying content of sodium sulfite. Kinetic studies were performed for two sets of samples containing approximately 15 g of tin dioxide and either a fixed amount of charcoal (5 g) and varying content of sodium sulfite (ranging from 5.2 to 15.6 g) or a fixed amount of sodium sulfite (10 g) and varying content of charcoal (ranging from 4 to 15.5 g). All reactants were mixed thoroughly in a crucible and ground into fine powder. The electric furnace with a maximum temperature of 1150 °C was preheated to the desired temperature. Then, the content of the crucible was transferred to a melting pot, which was then inserted into the furnace. The smelting process was carried out for 25–30 min with access to air. After completing the reduction reaction for each run, the pot was removed, liquid metal was poured out and the slag was collected. The obtained metallic tin and slag were weighed and analyzed. The smelting process was examined in graphite, stainless steel and fireclay crucibles. Each run was repeated 3–5 times.

3.4. Chemical Analysis

To determine qualitatively the products of the reduction process, the reaction solid residue and gaseous products were chemically tested.

The XRF analysis of the solid residue revealed the presence of tin, sodium and sulfur. This implied the presence of unreacted tin compounds, sodium oxide/carbonate, small amounts of sulfides and elemental sulfur in the residue. The speciation process was performed via qualitative chemical analysis using selective precipitation. The residue was dissolved in ultrapure water and the resulting suspension was filtered. The filtrate pH was significantly greater than 7. The basic character of the solution confirmed the presence of sodium oxide and probably sodium carbonate. The resulting solution was then treated with lead (II) nitrate (V). The formation of black precipitate suggested the presence of sulfide anions. The presence of sulfides in the smelting residue implied the formation of sulfide by-products during the smelting process.

The gaseous products of the smelting process were directed to an aqueous solution of sodium hydroxide and the resulting solution was treated with lead (II) nitrate (V). The formation of a white precipitate suggested the presence of sulfite and carbonate anions—the products of the neutralization of SO_2 and CO_2 by the strong base.

The allotropic form of the metallic product was determined from the examination of its density. The obtained value, 7.1 g/cm^3, confirmed that SnO_2 was reduced to crystalline β-Sn.

3.5. Experimental Data Analysis

The obtained metallic tin and the slag were analyzed chemically using a handheld XRF spectrometer (S1 Titan, Bruker) using the tin alloy calibration mode. Based on the XRF measurement results, the grade of tin metal and the composition of the slag were determined. The efficiency of the smelting process was quantified by the percent yield of tin given by the following formula:

$$R = 100 \times m_{Sn} \times P_{Sn}/(0.788 \times m_{SnO2} \times P_{SnO2}) \quad (2)$$

where R is the percent yield of tin metal, m_{Sn} and P_{Sn} are the mass and purity of tin metal produced, respectively, m_{SnO2} and P_{SnO2} are the mass and purity of tin dioxide in the charge, respectively, and 0.788 is the fraction of tin in tin dioxide.

4. Conclusions

The thermodynamic characteristics (in the form of the temperature dependence of the standard free enthalpy) of the reduction process validate the optimum smelting temperature determined experimentally. It confirms that SnO_2 in the C/Na_2SO_3 system can be converted to metallic tin at a temperature greater than 1080 °C. The significant increase in temperature above the equilibrium level leads to a drop in the percent yield of tin, probably due to the more effective release of the gaseous intermediate products from the system.

The smelting process was tested in crucibles made of various materials: graphite, stainless steel and fireclay. Although we have not observed any relation between the efficiency of SnO_2 reduction and the type of the crucible material, it is worth noting that the porous structure of the graphite material was easily penetrated by gaseous products, leading to the intensification of its corrosion.

The data on the experimental conditions and the tin recovery levels extracted from the related works in the literature, along with the details from our work, are summarized in Table 1.

Table 1. The experimental systems employed for the thermal reduction of SnO_2.

Reducing Agent (Temperature [°C])	Tin Recovery [%]	Reference
C graphite/CO (1000)	100	10
H_2 (1300)	100	11
CH_4 (1473)	100	12
C charcoal/Na_2CO_3-$NaNO_3$ (1000)	95	13
Ar-balanced CH_4 (1000)	80	14
C charcoal/Na_2SO_3 (1050)	98	This work

Among the limitations of the method developed, one may list the concurrent reactions involving possible contaminates of tin dioxide. This is especially vital for tin dioxide obtained from the secondary sources of tin (mainly e-scrap materials) containing lead additives. The method produces an acceptable yield of tin (more than 70%) if the level of lead impurities does not exceed 5%.

Author Contributions: Conceptualization, K.K. and W.H.; methodology, W.H.; validation, K.K. and W.H.; formal analysis, W.H. and K.K.; investigation, W.H. and K.K.; writing—original draft preparation, W.H. and K.K.; writing—review and editing, W.H.; visualization, W.H.; supervision, W.H.; project administration, W.H. All authors have read and agreed to the published version of the manuscript.

Funding: This research was funded by the National Centre for Research and Development under the Smart Growth Operational Programme, grant number POIR.04.01.04-00-0094/18.

Data Availability Statement: No other data are available.

Conflicts of Interest: The authors declare no conflicts of interest.

References

1. Angadi, S.I.; Sreenivas, T.; Jeon, H.-S.; Baek, S.-H.; Mishra, B.K. A review of cassiterite beneficiation fundamentals and plant practices. *Miner. Eng.* **2015**, *70*, 178. [CrossRef]
2. Barry, B.T.K. *Tin Processing in Encyclopedia Britannica*; Encyclopaedia Britannica, Inc.: Chicago, IL, USA, 2017.

3. Su, Z.; Zhang, Y.; Liu, B.; Lu, M. Extraction and Separation of Tin from Tin-Bearing Secondary Resources: A Review. *JOM* **2017**, *69*, 2364. [CrossRef]
4. Jun, W.S.; Yun, P.S.; Lee, E.C. Leaching behavior of tin from Sn–Fe alloys in sodium hydroxide solutions. *Hydrometallurgy* **2004**, *73*, 71. [CrossRef]
5. Choi, Y.I.; Salman, S.; Kuroda, K.; Okido, M. Synergistic corrosion protection for AZ31 Mg alloy by anodizing and stannate post-sealing treatments. *Electrochim. Acta* **2013**, *97*, 313. [CrossRef]
6. Ranitović, M.; Kamberović, Z.; Korać, M.; Jovanović, N.; Mihjalović, A. Hydrometallurgical recovery of tin and lead from waste printed circuit boards (WPCBs): Limitations and opportunities. *Metalurgija* **2016**, *55*, 153.
7. Barakat, M.A.; Koike, K. Acid leaching of indium–lead–tin alloy wire scrap. *J. Soc. Mater. Eng. Resour. Jpn.* **1997**, *10*, 36. [CrossRef]
8. Castro, L.A.; Martins, A.H. Recovery of tin and copper by recycling of printed circuit boards from obsolete computers. *Braz. J. Chem. Eng.* **2009**, *26*, 649. [CrossRef]
9. Kim, S.K.; Lee, J.C.; Yoo, K.K. Leaching of tin from waste Pb-free solder in hydrochloric acid solution with stannic chloride. *Hydrometallurgy* **2016**, *165*, 143. [CrossRef]
10. Leveque, G.; Abanades, S. Thermodynamic and Kinetic Study of the Carbothermal Reduction of SnO_2 for Solar Thermochemical Fuel Generation. *Energy Fuels* **2014**, *28*, 1396. [CrossRef]
11. Kim, B.-S.; Lee, J.-C.; Yoon, H.-S.; Kim, S.-K. Reduction of SnO_2 with Hydrogen. *Mater. Trans.* **2011**, *52*, 1814. [CrossRef]
12. Cetinkaya, S.; Eroglu, S. Thermodynamic analysis and reduction of tin oxide with methane. *Int. J. Miner. Process.* **2012**, *110–111*, 71. [CrossRef]
13. El Deeb, A.B.; Morsi, I.M.; Atlam, A.A.; Omar, A.A.; Fathy, W.M. Pyrometallurgical Extraction of Tin Metal from the Egyptian Cassiterite Concentrate. *Int. J. Sci. Eng. Res.* **2015**, *6*, 54.
14. Ha, H.; Yoo, M.; An, H.; Shin, K.; Han, T.; Sohn, Y.; Kim, S.; Lee, S.-R.; Han, J.H.; Kim, H.Y. Design of Reduction Process of SnO_2 by CH_4 for Efficient Sn Recovery. *Sci. Rep.* **2017**, *7*, 14427. [CrossRef] [PubMed]
15. Hyk, W.; Kitka, K.; Rudnicki, D. Method for the Selective Recovery of Tin and a Reactor for Use in Said Method. PCT/IB2019/052273, 26 September 2019.
16. Lewis, R.J., Sr. (Ed.) *Hawley's Condensed Chemical Dictionary*, 13th ed.; John Wiley & Sons, Inc.: New York, NY, USA, 1997; p. 1029.
17. Han, G.B.; Park, N.-K.; Yoon, S.H.; Lee, T.J. Catalytic reduction of sulfur dioxide with carbon monoxide over tin dioxide for direct sulfur recovery process. *Chemosphere* **2008**, *72*, 1744. [CrossRef] [PubMed]
18. Stacy, W.O.; Vastola, F.J.; Walker, P.L., Jr. Interaction of sulfur dioxide with active carbon. *Carbon* **1968**, *6*, 917. [CrossRef]
19. Karan, K.; Mehrotra, A.K.; Behie, L.A. Thermal decomposition of carbonyl sulfide at temperatures encountered in the front end of modified claus plants. *Chem. Eng. Comm.* **2005**, *192*, 370. [CrossRef]

recycling

MDPI

Article

Experimental Study on the Separation of Selected Metal Elements (Sm, Co, Fe, and Cu) from Nitric Acid Leachate Using Specific Precipitants

Jian-Zhi Wang [1,2,*], Yi-Chin Tang [1,3] and Yun-Hwei Shen [1]

1 Department of Resources Engineering, National Cheng Kung University, Tainan 70101, Taiwan; n48091500@gs.ncku.edu.tw (Y.-C.T.); yhshen@mail.ncku.edu.tw (Y.-H.S.)
2 Department of Mechanical Engineering, National Kaohsiung University Science and Technology, No. 415, Jiangong Rd, Kaohsiung 80778, Taiwan
3 Green Energy and System Integration Research and Development Department, China Steel Corporation, Kaohsiung 81233, Taiwan
* Correspondence: n48091518@gs.ncku.edu.tw; Tel.: +886-979884436

Abstract: As more countries emphasize the importance of the circular economy, recycling resources from waste has become increasingly crucial. This study proposes a novel separation process for SmCo magnets, which can separate and recover metals by precipitation, thus reducing the amount of solvent used. The precipitation process involved the use of Na_2SO_4, NH_4OH, and $H_2C_2O_4$ to separate Sm, Fe, Cu, and Co, resulting in high precipitation efficiencies of 96.11%, 99.97%, 93.81%, and 98.15%, respectively. Moreover, the recovered metals can be directly used to create magnets after calcination, making this process a step towards achieving a circular economy.

Keywords: SmCo magnet; precipitation; resource recovery; recycling

1. Introduction

Rare earth elements include light rare earth elements (LREEs) such as Lanthanum (La), Cerium (Ce), Praseodymium (Pr), Neodymium (Nd), Promethium (Pm), Samarium (Sm), and Europium (Eu), and heavy rare earth elements (HREEs) such as Gadolinium (Gd), Terbium (Tb), Dysprosium (Dy), Holmium (Ho), Erbium (Er), Thulium (Tm), Ytterbium (Yb), and Lutetium (Lu). Yttrium (Y) and Scandium (Sc) are two other elements excluded from the LREEs and HREEs classifications [1,2]. The Sm in SmCo magnets belongs to the light rare earth elements.

In 2022, the global rare earth reserves amounted to approximately 130 million metric tons. Among them, China, Vietnam, Brazil, and Russia hold 44 million, 22 million, 21 million, and 21 million metric tons of rare earth reserves, respectively [3]. These four countries together possess more than 80% of the world's rare earth reserves. Rare earth reserves are excessively concentrated in specific countries. In addition, the production of rare earth is also concentrated in specific countries. China's rare earth production in 2022 was 210,000 metric tons, which represented about 70% of the total global rare earth production [3]. Furthermore, in 2010, China issued the "The Opinion on Promoting Mergers and Acquisitions of Enterprises" [4], which listed rare earth elements as a key industry for mergers, acquisitions, and reorganization, in an attempt to reduce the export of rare earths. This led to a historical high in the price of rare earths [5]. The most effective method currently to reduce the risk of rare earth supply is to recycle metals from waste [6,7].

Rare earth elements have become a significant topic of discussion among researchers, entrepreneurs, and politicians in recent years due to their crucial role in current technological advancements. Rare earths are utilized in a broad range of applications [8–10], including wind turbines, electric vehicles [11], mobile phones, hard disk drives, fluorescent and LED lamps, defense applications, catalysts, pharmaceuticals, and medicine [12]. Of these, over a quarter of rare earths are utilized in magnet manufacturing [13], which is the field that consumes the most rare earths. Consequently, if a large number of magnets can be recycled, rare earth elements can be effectively recovered.

Among the rare earth elements, Nd and Sm are commonly used in magnet production. Nd is primarily used to create NdFeB magnets, which are widely applied across industries due to their high energy density. These industries include acoustic transducers, wind turbines, electric vehicles, industrial motors, missiles, tanks, warplanes, submarines, microwaves, computers, printers, and hard disk drives [14–16]. In comparison, SmCo magnets exhibit unique properties, including higher coercivity and superior temperature resistance, making them ideal for specialized applications like aero engines [17–19]. While NdFeB magnets currently account for over 90% of annual rare earth magnet production, many experts predict an increase in demand for SmCo magnets in the coming years [20]. This shift is largely due to the high cost of dysprosium (Dy), a key component in NdFeB magnets. Consequently, innovative direct and indirect recycling methods for SmCo magnets must be developed.

Currently, methods for recovering rare earth from NdFeB magnets include hydrometallurgical processes [21–24] and pyrometallurgical processes [25–28]. However, there have been few studies on the recovery of rare earths from SmCo magnets. In the literature [29–31] on the pre-treatment of SmCo magnets, waste magnets were first demagnetized in a high-temperature furnace. Sinha et al. (2017) [29] proposed demagnetizing them at a temperature of 850 °C for 6 h. Then, SmCo magnets should be crushed and ground using a hydraulic press and grinder, which is beneficial for follow-up research. Sinha et al. (2017) [29] also suggested mechanically crushing, grinding, and screening the SmCo magnet ring with a 150 μm sieve. After completing the pretreatment process, most studies [29,31–33] use hydrometallurgical methods to purify and separate valuable metals. First, the powder is leached with either an inorganic acid (such as sulfuric acid, hydrochloric acid, or nitric acid) [29,31,32] or an organic acid (such as citric acid) [34]. However, Zhou et al. [33] (2017) proposed using sulfuric acid to leach waste SmCo magnets. Unfortunately, Sm easily reacts with sulfuric acid to form samarium sulfate precipitation, which results in poor leaching efficiency. Therefore, most studies [29,32] have used hydrochloric acid and nitric acid for leaching. After the leaching experiment was completed, the study used solvent extraction [17,29,32,35], ion exchange [36], and precipitation [31,33,37,38] methods for metal separation. Finally, the recovered powder was calcined to meet the raw materials for making magnets. Sahoo et al. proposed that the separated Sm and Co solutions be precipitated with oxalic acid, and then the samarium oxalate and cobalt oxalate were calcined at 800 °C and 450 °C, respectively, to form metal oxides.

One of the primary strategies for circulating magnet materials throughout the supply chain involves proposing the most effective approach to refine valuable materials from discarded SmCo magnets. However, the above studies [17,29,31–33,35,37,38] only focused on the recovery of Sm and Co, and did not purify and separate all metals (Sm, Co, Fe, Cu, and Zr) in the waste. Therefore, this study hopes to propose a complete recovery method to deal with real waste. In addition, in the separation method, after the metal is separated by the solvent extraction method, metal precipitation is required, and the process is relatively complicated. The ion exchange method, on the other hand, requires diluting the aqueous solution, resulting in increased wastewater volumes. In this study, a simpler process was developed, using the precipitation method to directly separate the metal by precipitation, so the metal could be reprocessed into a magnet.

2. Results and Discussion

2.1. Selective Precipitation of Sm by Na_2SO_4

To facilitate subsequent research, precipitation experiments were carried out using a $L_9(3^4)$ orthogonal table to determine the precipitation efficiency of Sm. The results of these experiments, showing the precipitation efficiencies of Sm under different parameters, are presented in Table 1. The precipitation efficiency was observed to be affected by the temperature, addition of Na_2SO_4, and time [33].

Table 1. The orthogonal array experiment results of Sm precipitation.

No.	Temp.	Time	Na_2SO_4	Sm	Co	Fe	Cu
Unit	°C	min	(w/v)%	%	%	%	%
1	70	20	6	17.69	0.09	0.64	0
2	70	40	9	90.88	2.05	3.51	1.13
3	70	60	12	99.19	2.31	6.06	0.88
4	80	20	9	90.22	1.78	5.74	0.84
5	80	40	12	99.31	2.85	4.31	0.84
6	80	60	6	44.63	1.25	2.71	0
7	90	20	12	99.40	3.03	4.78	1.34
8	90	40	6	48.59	0	0	0
9	90	60	9	95.60	0	3.19	0

In this study, the priority of each factor on Sm precipitation efficiency was analyzed through factor effect analysis using results from orthogonal table experiments. The precipitation efficiency at the addition of Na_2SO_4 of K3 (as shown in Table 2) represents the average precipitation efficiency with an addition of Na_2SO_4 of 12 (w/v)% (as shown in Table 1). The extreme deviation is the difference between the best and worst precipitation rates of Sm under K1 to K3 conditions. The order of priority for factors that affect precipitation efficiency was determined using extreme deviation [39,40].

Table 2. Factor response table for Sm precipitation.

	Effect Factor	Temperature	Time	Na_2SO_4
Sm	K1	69.25%	69.10%	36.97%
	K2	78.05%	79.60%	92.23%
	K3	81.20%	79.81%	99.30%
	Extreme Deviation	11.94%	10.70%	62.33%
	Priority Order	Na_2SO_4 > Temperature > Time		

Table 2 shows the factor response table for the precipitation rate, which revealed that the order of influence on precipitation efficiency was Na_2SO_4 > temperature > time. The preliminary optimal parameters determined were 12 (w/v)% Na_2SO_4 addition, 80 °C temperature, and 40 min time, which resulted in the highest precipitation efficiency. As the precipitation efficiency at 40 and 60 min was found to be equivalent, 40 min was chosen as the initial optimum parameter. Similarly, as the precipitation efficiency at 80 and 90 °C was also found to be equivalent, 80 °C was selected as the initial optimum parameter time.

Figure 1 illustrates the results of the Sm precipitation confirmation experiment. As shown in Figure 1a, Sm precipitation efficiency was less than 50% when 6 (w/v)% of sodium sulfate was added. However, when the amount of sodium sulfate was increased to 10 (w/v)%, the precipitation efficiency of Sm significantly improved, reaching 96.25%. Therefore, the amount of sodium sulfate added was identified as the most crucial parameter, as it had a significant impact on precipitation effectiveness.

Compared with the amount of sodium sulfate added, the temperature had a negligible effect on Sm precipitation efficiency, as depicted in Figure 1b. When the Sm temperature was 60 °C, the precipitation efficiency reached 93.44%, and increasing the temperature did not result in a significant improvement in precipitation efficiency. However, at a temperature of 60 °C, the co-precipitation efficiency of Fe was 18.56%. Thus, 60 °C was not the optimal precipitation parameter. The primary reason for Fe co-precipitation was that Fe readily forms ferric sulfate precipitation at lower temperatures [41–43], resulting in increased co-precipitation rates of Fe. Therefore, the optimal temperature was 70 °C.

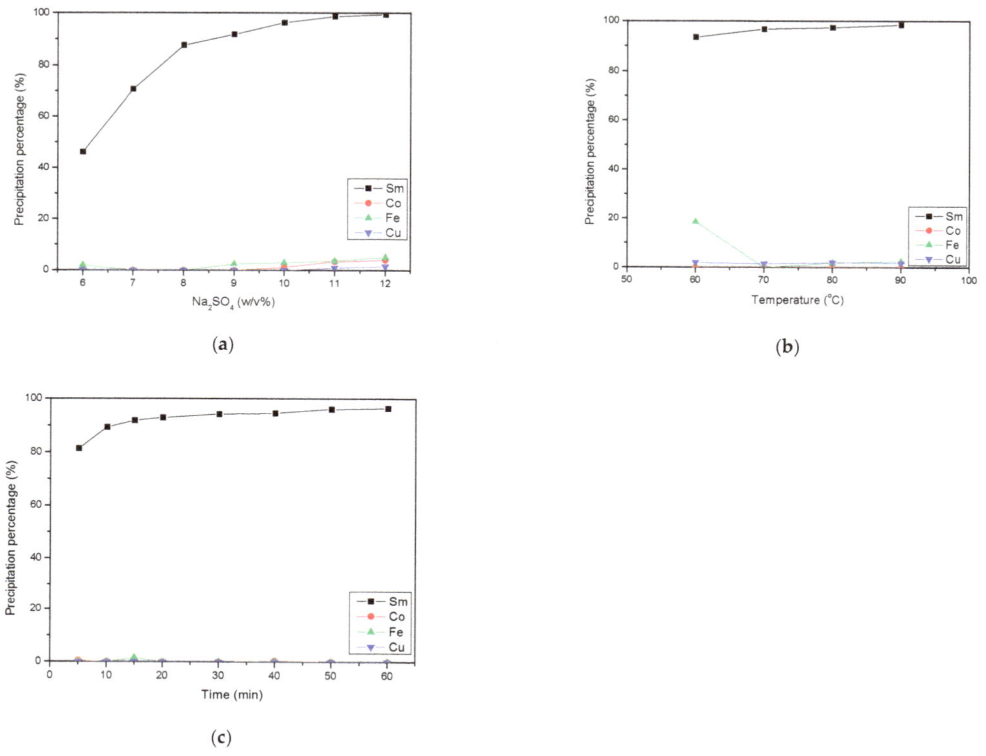

Figure 1. Effects of (**a**) addition of sodium sulfate (temperature = 80 °C, time = 40 min), (**b**) temperature ([Na_2SO_4] = 10 (w/v)%, time = 40 min), and (**c**) time ([Na_2SO_4] = 10 (w/v)%, temperature = 70 °C) on Sm precipitation.

Figure 1c shows that the Sm precipitation reaction was relatively slow, and it took 50 min to achieve a Sm precipitation rate higher than 95%.

Combining the above results, the optimal conditions for precipitation were determined to be the addition of 10 (w/v)% of sodium sulfate, a temperature of 70 °C, and a precipitation time of 50 min. Under these conditions, the precipitation rates of Sm, Co, Fe, and Cu were 96.11%, 0.05%, 0%, and 0%, respectively. This successful separation of Sm resulted in a co-precipitation rate of other metals that were less than 0.1%.

The findings of this study were compared to those of Zhou et al. [33] (Table 3). Zhou et al. [33] utilized sulfuric acid for leaching, resulting in a solution with a high concentration of sulfate ions, which promoted the precipitation of sodium samarium bisulfate. Consequently, less sodium sulfate was added in their study. The precipitation of sodium samarium bisulfate is demonstrated in Formula (1). Despite the higher molar ratio used in this study, a greater amount of Sm was successfully precipitated at lower temperature and time conditions. Moreover, this study effectively optimized the parameters for Sm precipitation.

$$Sm_2(SO_4)_3 + Na_2SO_4 + 2H_2O \leftrightarrow 2NaSm(SO_4)_2 \cdot H_2O \quad (1)$$

Table 3. Comparison of results for Sm precipitation.

No.	This Study	Zhou et al. [33]
Temp.	70 °C	80 °C
Time	50 min	60 min
Molar ratio	6.69:1	4:1
Leachate	HNO_3	H_2SO_4
Ions in the leachate	Sm, Co, Fe, and Cu	Sm and Co
The precipitation rate of Sm	96.11%	93.4%

2.2. Selective Precipitation of Fe by NaOH

This section analyzes the impact of pH and time on Fe precipitation. Figure 2a shows the effect of pH on Fe precipitation, revealing that the efficiency of Fe precipitation was less than 10% when the pH was below 2.5, and the co-precipitation rate of Cu exceeded 50% when the pH was above 4. Thus, pH was a crucial factor in Fe precipitation. The optimal Fe precipitation rate of 98.28% was achieved at a pH of 3.5, with less than 1% co-precipitation of other metals. Therefore, a pH of 3.5 was selected as the optimal parameter.

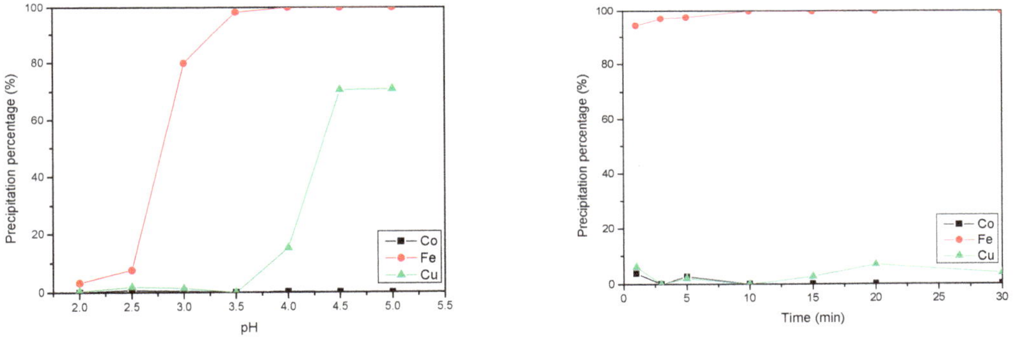

(a) (b)

Figure 2. Effects of (**a**) pH (temperature = 25 °C, time = 60 min) and (**b**) time (pH = 3.5, temperature = 25 °C) on Fe precipitation.

The results regarding the effect of time on Fe precipitation are shown in Figure 2b. The efficiency of Fe precipitation was faster compared with that of Sm precipitation. At a time of 10 min, the precipitation rate of Fe reached 99.97%.

Combining the aforementioned results, it can be concluded that the precipitation rates of Fe, Cu, and Co were 99.97%, 0%, and 0%, respectively, when the pH was 3.5 and the time was 10 min. Fe was successfully separated, and the co-precipitation rate of other metals was less than 0.1%.

The study of SmCo magnets has not yielded any other studies on Fe precipitation separation so far. Therefore, no comparison of experimental results was made, highlighting the novelty of this study.

2.3. Selective Precipitation of Cu by NaOH

This section discusses the impact of pH and time on the efficiency of Cu precipitation. The effect of pH on the Cu precipitation efficiency is demonstrated in Figure 3a. The pH level played a crucial role in Cu precipitation as it significantly influenced precipitation efficiency. The optimal precipitation pH value was found to be 6.5, resulting in a Cu precipitation efficiency of 94.93%.

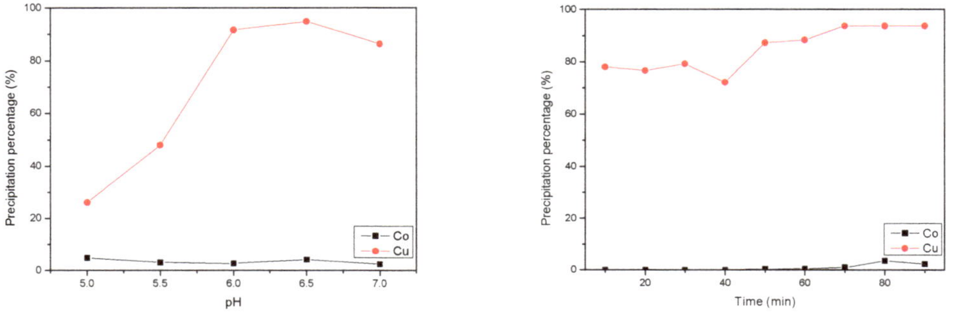

(**a**) (**b**)

Figure 3. Effects of (**a**) pH (temperature = 25 °C, time = 60 min) and (**b**) time (pH = 6.5, temperature = 25 °C) on Cu precipitation.

Figure 3b shows the effect of time on Cu precipitation efficiency. The Cu precipitation rate was slow, taking more than 70 min to achieve over 90% efficiency.

By combining the above findings, it was determined that a pH of 6.5 and a precipitation time of 70 min resulted in Cu and Co precipitation rates of 93.81% and 0.95%, respectively, indicating successful Cu separation with less than 1% Co co-precipitation.

The study of SmCo magnets has not yielded any other studies on Cu precipitation separation so far. Therefore, no comparison of experimental results was made, highlighting the novelty of this study.

2.4. Selective Precipitation of Co by Oxalic Acid

This part discusses the influence of molar ratio and time on the Co precipitation effect, as shown in Figure 4. As the molar ratio increased from 0.5 to 1.5, the Co precipitation efficiency increased from 76.76% to 96.84%. However, increasing the precipitation time did not significantly increase the precipitation efficiency. Therefore, the molar ratio had a greater influence on Co precipitation.

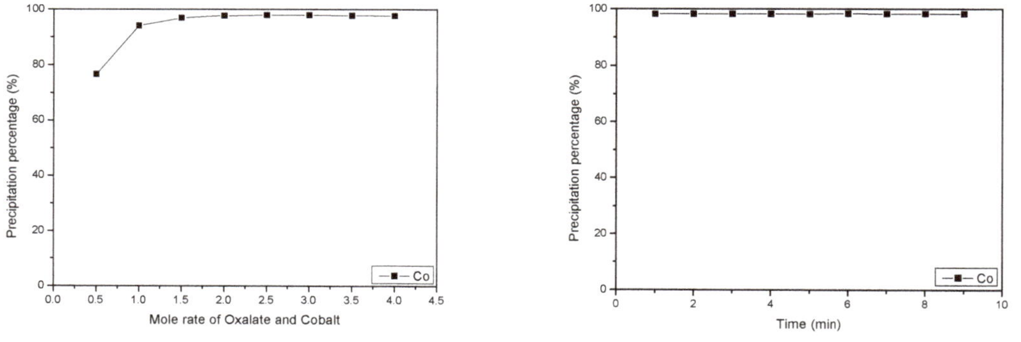

(**a**) (**b**)

Figure 4. Effects of (**a**) molar ratio (temperature = 25 °C, time = 60 min) and (**b**) time (molar ratio = 1.5, temperature = 25 °C) on Co precipitation.

In the study of the influence of time on Co precipitation, it was found that the Co precipitation rate was very fast. The Co precipitation efficiency reached 98.15% in 1 min.

Based on the above results, the precipitation rate of Co reached 98.15% under the condition of a molar ratio of 1.5 and a time of 1 min. These results were compared with those of Wang et al. [36] (Table 4). Despite the higher molar ratio used in this study, more Co was successfully precipitated under lower temperature and time conditions. Moreover, this study effectively optimized the parameters for Co precipitation.

Table 4. Comparison of results for Co precipitation.

No.	This Study	Wang et al. [36]
Temp.	25 °C	60 °C
Time	1 min	60 min
Molar ratio	1.5:1	1:1
Leachate	HNO_3	H_2SO_4
The precipitation rate of Co	98.15%	96.20%

2.5. Characterization of $NaSm(SO_4)_2$

As $NaSm(SO_4)_2$ releases sulfur oxides during TGA analysis, which can damage the instrument, the oxidation temperature was not determined. However, Denisenko et al. [44] proposed that heating $NaSm(SO_4)_2$ to 1187 °C would decompose it into liquid sodium sulfate and solid samarium oxide. In this study, $NaSm(SO_4)_2$ can be used to obtain samarium oxide, which can be reused to make magnets.

The Sm precipitate obtained was analyzed by XRD, and the results are displayed in Figure 5. The major peaks of $NaSm(SO_4)_2$ H_2O at 25.699°, 29.736°, 42.325°, and 49.278° were observed, confirming the presence of crystalline $NaSm(SO_4)_2$ H_2O as the dominant phase.

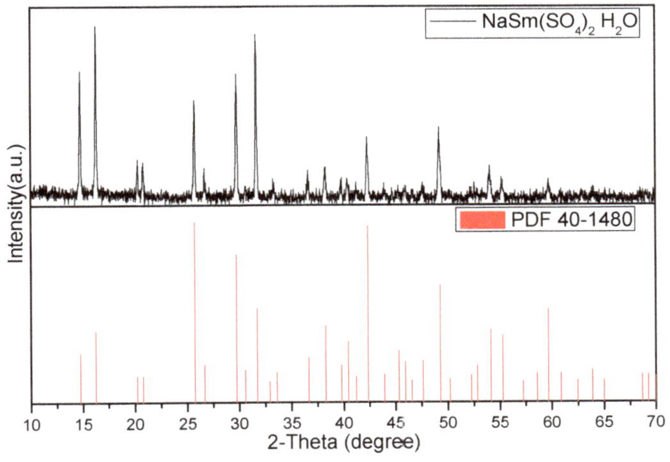

Figure 5. The XRD analysis of $NaSm(SO_4)_2$.

2.6. Preparation of Iron Oxide and Characterization

The TGA diagram and oxidation formula of $Fe(OH)_3$ are shown in Figure 6a and Formula (2), respectively. It can be observed that there is no significant weight loss when the temperature reached 500 °C, and the weight remained at about 59.40% of the original weight. Based on the theory of weight loss after oxidation, it can be concluded that the final oxidation temperature of $Fe(OH)_3$ was 500 °C.

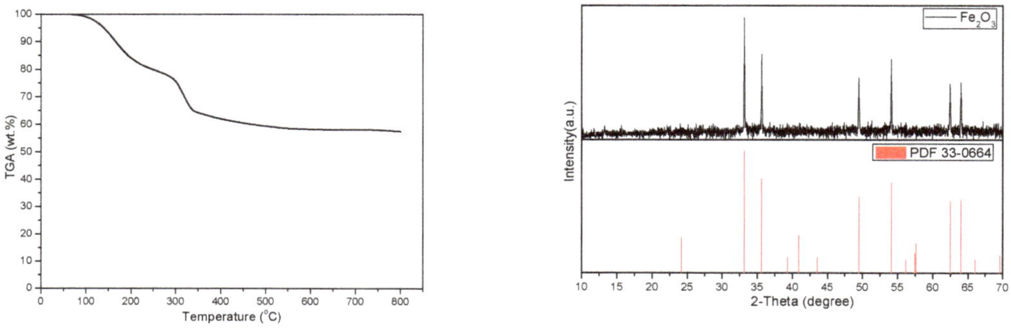

(**a**) (**b**)

Figure 6. The (**a**) TGA analysis of $Fe(OH)_3$ and the (**b**) XRD analysis of Fe_2O_3.

$Fe(OH)_3$ was calcined at 500 °C for 4 h, and the XRD analysis was performed (Figure 6b). The results showed that after calcination, the sample exhibited the main peak of Fe_2O_3 at 33.152°, confirming that $Fe(OH)_3$ had transformed into Fe_2O_3.

$$2Fe(OH)_3 \rightarrow Fe_2O_3 + 3H_2O \quad (2)$$

2.7. Preparation of Copper Oxide and Characterization

Figure 7a and Formula (3) depict the TGA diagram and oxidation formula of $Cu(OH)_2$, respectively. Upon analysis, it was observed that there was no significant weight loss when the temperature reached 500 °C. At this temperature, the weight was approximately 80.36% of the original weight, which was consistent with the theoretical value of weight loss after oxidation. Thus, 500 °C was determined to be the final oxidation temperature of $Cu(OH)_2$.

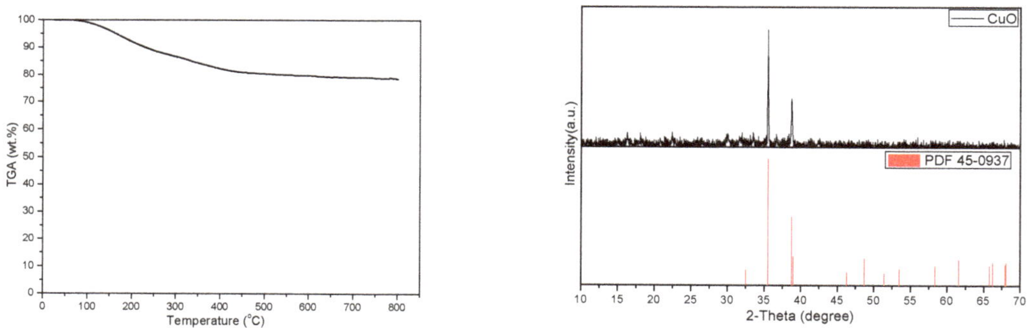

(**a**) (**b**)

Figure 7. The (**a**) TGA analysis of $Cu(OH)_2$ and the (**b**) XRD analysis of CuO.

After calcining $Cu(OH)_2$ at 500 °C for 4 h, the XRD analysis was performed, and the results are shown in Figure 7b. The XRD pattern of the calcined $Cu(OH)_2$ matches with the main peaks of CuO at 35.495° and 38.730°, confirming the transformation of $Cu(OH)_2$ to CuO upon calcination.

$$Cu(OH)_2 \rightarrow CuO + H_2O \quad (3)$$

2.8. Preparation of Cobalt Oxide and Characterization

The TGA diagram and oxidation formula of CoC_2O_4 are shown in Figure 8a and Formula (4), respectively. Upon analysis, it was observed that there was no significant weight loss when the temperature reached 470 °C. At this temperature, the weight was approximately 38.98% of the original weight, which was consistent with the theoretical value of weight loss after oxidation. Therefore, 470 °C was determined to be the oxidation temperature of CoC_2O_4.

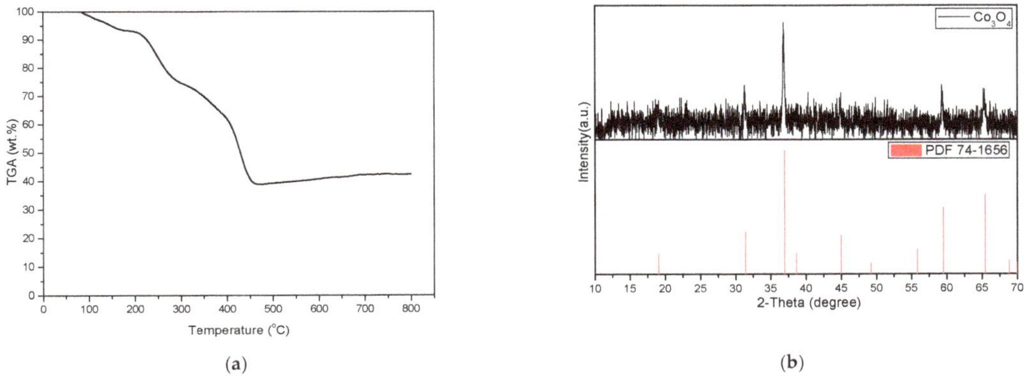

Figure 8. The (**a**) TGA analysis of CoC_2O_4 and the (**b**) XRD analysis of Co_3O_4.

After calcination of CoC_2O_4 at 500 °C for 4 h, the XRD analysis was performed, and the results are shown in Figure 8b. The XRD pattern of the calcined CoC_2O_4 corresponds to the main peaks of Co_3O_4 at 36.935°, 59.508°, and 65.406°, confirming the transformation of CoC_2O_4 to Co_3O_4 upon calcination.

$$3CoC_2O_4 + 2O_2 \rightarrow Co_3O_4 + 6CO_2 \quad (4)$$

3. Materials and Methods

3.1. Materials

Tai Chuan Metal Co., Ltd. (Taipei, Taiwan) provided the SmCo magnet scraps used in this study, which were abundant in Sm, Co, Fe, Cu, and Zr. Table 5 displays the composition of the SmCo magnet scrap after digestive analysis and the concentrations of Sm, Co, Fe, and Cu in the leachate, which only contains these metals since Zr, which cannot be leached [45], was initially separated during the leaching process. The leachate was produced at a temperature of 25 °C, a nitric acid concentration of 2.5 mol L^{-1}, a solid-to-liquid ratio of 65 g L^{-1}, and a time of 1 min, resulting in a leaching efficiency of 99.52%, 92.45%, 92.84%, and 94.30% for Sm, Co, Fe, and Cu, respectively.

Table 5. The content of the SmCo magnet and the concentration in the leachate.

Element	Sm	Co	Fe	Cu	Zr
The weight percent of element (wt.%)	22.70	51.08	14.51	5.12	4.31
Metal concentration in leachate (ppm)	15,756	31,088	8762	3174	-

Nitric acid (HNO_3, 70%) and hydrochloric acid (HCl, 37%) for aqua regia digestion, nitric acid (HNO_3, 70%) for leaching, sodium sulfate (Na_2SO_4, 99.5%) for Sm precipitation, ammonia (NH_4OH, 29%) for precipitation of Fe and Cu, and oxalic acid ($H_2C_2O_4$, 99%)

for Co precipitation were purchased from Echo Chemical Co., Ltd. (Miaoli, Taiwan). All aqueous solutions were prepared in deionized water.

3.2. Equipment

The pH of the solution was monitored by a pH meter (PL-700PVS, Dogger Science, Taipei, Taiwan). In order to control the precipitation and leaching temperature, the precipitation and leaching process was carried out in a thermostatic bath with magnetic stirring (Shin-Kwang Precision Industry Ltd., New Taipei, Taiwan). Subsequently, the solid–liquid separation was carried out through a filter device (Chemker 300, Rone Scientific Co., Ltd., New Taipei, Taiwan). An atomic absorption spectrometer (AAS, PinAAcle 900F AA Spectrometer, PerkinElmer Inc., Waltham, MA, USA) was used to analyze the concentration of metals in the aqueous solution. The oxidation temperature of $Fe(OH)_3$, $Cu(OH)_2$, and CoC_2O_4 in the precipitation were analyzed by thermogravimetry analysis (TGA, TGA-51, Shimadzu, Japan). The crystal structure of the product ($NaSm(SO_4)_2$, Fe_2O_3, CuO, and Co_3O_4) were analyzed via X-ray diffractometry (XRD, DX-2600, Dandong, China), with Cu Kα radiation ranging from 2θ = 10–70°, to identify the phases present. The precipitate was calcined using a high-temperature furnace (Thermo Scientific Lindberg/Blue M BF51866C Muffle Furnace, Hogentogler & Co., Inc., Howard County, Columbia, MD, USA).

3.3. Experimental Method

3.3.1. The Composition of the Magnet

To confirm the powder composition, aqua regia digestion was utilized. In detail, 1 g of SmCo magnet powder was added to a 250 mL conical flask containing 10 mL of nitric acid and 30 mL of hydrochloric acid. The flask was then placed in a thermostatic bath at 90 °C for one day to guarantee complete metal dissolution. The metal concentration in the resulting aqueous solution was analyzed using AAS. The contents of Sm, Co, Fe, Cu, and Zr in the magnet are displayed in Table 5.

3.3.2. Leaching of Magnets

The hydrometallurgical method was employed in this study to recover the metals from the magnet. For leaching, SmCo magnet powder and 100 mL of an acid solution (a mixture of deionized water and the required amount of acid) were added to a 250 mL conical flask, and the experimental temperature was regulated using a thermostatic bath. The resulting solution was filtered, and the metal concentration was analyzed using AAS. The leaching efficiency (L%) was determined using Formula (5) [33].

$$L(\%) = \frac{V_1(L) \times C_1(g\,L^{-1})}{M_1(g) \times W_1} \times 100\% \quad (5)$$

where M_1 is the mass of the alloy sample, W_1 is the metal content of the alloy sample, V_1 is the volume of the leach solution, and C_1 is the mass concentration of the metal in the leach solution.

3.3.3. Precipitation Method

For the precipitation experiment, this study added 100 mL of leachate and precipitant to a 250 mL Erlenmeyer flask. The experimental temperature and stirring rate were controlled using a thermostatic bath. The resulting solution was filtered, and the metal concentration was analyzed using AAS. The precipitation efficiency (P%) was determined using Formula (6) [33].

$$P\% = \frac{V_2 \times C_2 - V_3 \times C_3}{V_2 \times C_3} \times 100\% \quad (6)$$

where P is the precipitation percentage of metal, V_2 and V_3 are the volume of the leachate and the filtrate, respectively, and C_2 and C_3 are the mass concentrations of metal in the leachate and in the filtrate, respectively.

3.3.4. Sm Precipitation

The previous study [33] revealed that sodium sulfate exhibited selective separation characteristics for Sm precipitation under low pH conditions. Therefore, this study first employed sodium samarium sulfate to investigate Sm separation. In contrast to the study by Zhou et al. [33], which focused solely on the separation of Sm and Co, this study aims to investigate the separation of Sm, Co, Fe, and Cu with more complex components. Furthermore, this study employed the Taguchi method to determine the priority order of each parameter on precipitation efficiency and determine the optimal precipitation parameters. In comparison to the one-factor-at-a-time experimental method and the full-factor experiment method, the Taguchi method requires fewer experiments to obtain useful statistical information. Therefore, the Taguchi method was adopted for this study.

The efficiency of Sm precipitation is influenced by various factors, such as temperature, the addition of Na_2SO_4, and time [33]. To determine the priority order of these three factors on precipitation efficiency, this study used the Taguchi method. First, the orthogonal array experiment was implemented. This study used a $L_9(3^4)$ orthogonal table with selected control factors and levels of temperature (70–90 °C), an addition of Na_2SO_4 (6–12 g 100 mL^{-1}), and time (20–60 min), as shown in Table 6. After completing the orthogonal array experiment, this study performed a factor effect analysis to understand the order of priority of each factor affecting the Sm precipitation efficiency and found the preliminary optimal experimental parameters. Finally, this study used the priority ranking of precipitation by each factor to conduct confirmation experiments and obtained the best precipitation parameters.

Table 6. A $L_9(3^4)$ orthogonal table of Sm precipitation.

No.	Temp.	Time	Na_2SO_4
Unit	°C	min	$(w/v)\%$
1	70	20	6
2	80	40	9
3	90	60	12

3.3.5. Fe and Cu Precipitation

This study examines the precipitation efficiency of Fe, Cu, and Co at different pH values. The results [38] show that Fe precipitation occurs at a pH range of 2–2.7, while Cu precipitation occurs at pH 4.5–6.5, and Co precipitation at pH 7.5–9. Based on these findings, it is believed that adjusting the pH value is an effective method for separating these metals. Furthermore, the research findings [38] indicate that precipitation of each metal can be achieved at room temperature. As a result, this study opted to investigate the precipitation of Fe and Cu at 25 °C.

Given that the parameters that need to be adjusted for Fe and Cu precipitation are pH and time, and based on previous studies [38], it is clear that pH affects metal precipitation more than time. Therefore, the optimal precipitation parameters for Fe and Cu were determined without evaluating the relative importance of each parameter. For Fe precipitation parameter adjustments, the pH range was from 2 to 5 and the time range was from 1 to 30 min. For Cu precipitation parameter adjustments, the pH range was from 5 to 7 and the time range was from 10 to 90 min.

3.3.6. Co Precipitation

To produce a reusable product, this study employed oxalic acid for Co precipitation. For Co precipitation parameter adjustments, the molar ratio ranged from 0.5 to 4, and the time ranged from 1 to 9 min.

3.3.7. Product Preparation and Characterization

To obtain raw materials suitable for magnet production, this study analyzed the oxidation temperature of the precipitate using TGA to confirm the oxidation temperature of each metal. However, sodium samarium bisulfate was not analyzed using TGA due to the generation of sulfur oxides during the analysis process, which could cause damage to the instrument.

After confirming the oxidation temperature of each metal, this study proceeded with the oxidation of the precipitate. Specifically, 5 g of the precipitate was placed into a ceramic vessel, which was put into a high-temperature furnace. The temperature was raised to the required level in order to obtain the oxides of each metal. Finally, the XRD [46] was used to confirm the composition of the product ($NaSm(SO_4)_2$, Fe_2O_3, CuO, and Co_3O_4).

3.3.8. Separation Process

Figure 9 shows the research process. First, this study added sodium sulfate to precipitate Sm from the leachate. The Sm formed a precipitate of $NaSm(SO_4)_2$ which was separated from the leachate. Secondly, this study added ammonia to precipitate Fe from the leachate by adjusting the pH, and the Fe formed a $Fe(OH)_3$ precipitate. Thirdly, this study added ammonia to precipitate Cu by adjusting the pH and obtained a $Cu(OH)_2$ precipitate. Fourth, the Co was precipitated as CoC_2O_4 by the addition of oxalic acid. Fifth, this study carried out a TGA analysis and calcination experiments to obtain the metal oxides of the precipitates. Finally, this study carried out XRD analysis to confirm the composition of the precipitates.

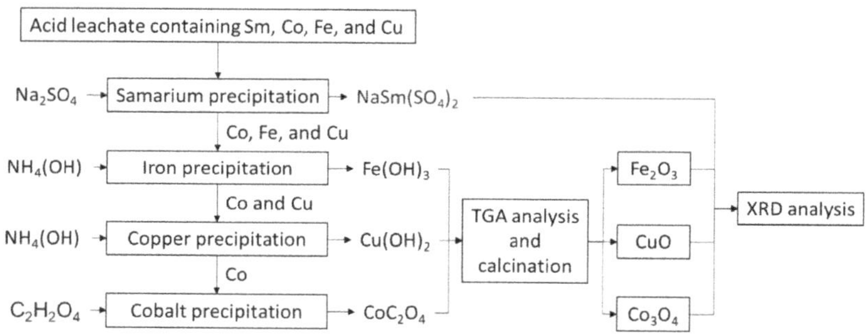

Figure 9. Proposed methodology for recycling Sm, Co, Fe, and Cu.

4. Conclusions

In this study, Na_2SO_4, NH_4OH, and $H_2C_2O_4$ were used to precipitate and separate Sm, Fe, Cu, and Co. The precipitation efficiencies for Sm, Fe, Cu, and Co were 96.11%, 99.97%, 93.81%, and 98.15%, respectively, with co-precipitation efficiency below 1%. The XRD analysis indicated that the precipitates were $NaSm(SO_4)_2$, Fe_2O_3, CuO, and Co_3O_4, respectively. This study proposes a new method for recycling SmCo magnets, enabling the recovered metals to be converted back into magnets to support a circular economy.

Compared with solvent extraction and ion exchange techniques, the precipitation method proposed in this study significantly reduces wastewater output and enables a one-step recovery process, thereby streamlining the recovery procedure. Nevertheless, the approach to wastewater treatment in this study warrants further investigation. Given the advanced state of current industrial wastewater treatment technologies, it is anticipated that effective solutions can be implemented to address this aspect with clarity.

Author Contributions: Conceptualization, J.-Z.W.; methodology, J.-Z.W.; validation, J.-Z.W.; investigation, J.-Z.W. and Y.-C.T.; resources, J.-Z.W.; data curation, J.-Z.W.; writing—original draft preparation, J.-Z.W.; writing—review and editing, J.-Z.W. and Y.-C.T.; supervision, Y.-H.S.; project administration, Y.-H.S. All authors have read and agreed to the published version of the manuscript.

Funding: This research received no external funding.

Institutional Review Board Statement: Not applicable.

Informed Consent Statement: Not applicable.

Data Availability Statement: Data is contained within the article.

Acknowledgments: The authors wish to acknowledge the support of the Ministry of Science and Technology, R.O.C.

Conflicts of Interest: Author Yi-Chin Tang was employed by the company China Steel Corporation. The remaining authors declare that the research was conducted in the absence of any commercial or financial relationships that could be construed as a potential conflict of interest.

References

1. Golev, A.; Scott, M.; Erskine, P.D.; Ali, S.H.; Ballantyne, G.R. Rare earths supply chains: Current status, constraints and opportunities. *Resour. Policy* **2014**, *41*, 52–59. [CrossRef]
2. Massari, S.; Ruberti, M. Rare earth elements as critical raw materials: Focus on international markets and future strategies. *Resour. Policy* **2013**, *38*, 36–43. [CrossRef]
3. USGS. *Mineral Commodity Summaries 2023*; U.S. Geological Survey: Reston, VA, USA, 2023. [CrossRef]
4. Zhang, L.; Guo, Q.; Zhang, J.; Huang, Y.; Xiong, T. Did China's rare earth export policies work?—Empirical evidence from USA and Japan. *Resour. Policy* **2015**, *43*, 82–90. [CrossRef]
5. Fan, J.H.; Omura, A.; Roca, E. Geopolitics and rare earth metals. *Eur. J. Political Econ.* **2022**, *78*, 102356. [CrossRef]
6. Du, X.; Graedel, T.E. Uncovering the end uses of the rare earth elements. *Sci. Total Environ.* **2013**, *461–462*, 781–784. [CrossRef]
7. Salim, H.; Sahin, O.; Elsawah, S.; Turan, H.; Stewart, R.A. A critical review on tackling complex rare earth supply security problem. *Resour. Policy* **2022**, *77*, 102697. [CrossRef]
8. Wübbeke, J. Rare earth elements in China: Policies and narratives of reinventing an industry. *Resour. Policy* **2013**, *38*, 384–394. [CrossRef]
9. Wang, J.; Guo, M.; Liu, M.; Wei, X. Long-term outlook for global rare earth production. *Resour. Policy* **2020**, *65*, 101569. [CrossRef]
10. Paulick, H.; Machacek, E. The global rare earth element exploration boom: An analysis of resources outside of China and discussion of development perspectives. *Resour. Policy* **2017**, *52*, 134–153. [CrossRef]
11. Li, X.Y.; Ge, J.P.; Chen, W.Q.; Wang, P. Scenarios of rare earth elements demand driven by automotive electrification in China: 2018–2030. *Resour. Conserv. Recycl.* **2019**, *145*, 322–331. [CrossRef]
12. Ascenzi, P.; Bettinelli, M.; Boffi, A.; Botta, M.; Simone, G.D.; Luchinat, C.; Marengo, E.; Mei, H.; Aime, S. Rare earth elements (REE) in biology and medicine. *Rend. Lincei Sci. Fis. E Nat.* **2020**, *31*, 821–833. [CrossRef]
13. Penchoff, D.A.; Sims, C.B.; Windus, T.L. Rare Earth Elements and Critical Materials: Uses and Availability. In *Rare Earth Elements and Actinides: Progress in Computational Science Applications*; Penchoff, D.A., Windus, T.L., Peterson, C.C., Eds.; American Chemical Society: Washington, WA, USA, 2021; pp. 63–74. [CrossRef]
14. Yang, Y.; Lan, C.; Wang, Y.; Zhao, Z.; Li, B. Recycling of ultrafine NdFeB waste by the selective precipitation of rare earth and the electrodeposition of iron in hydrofluoric acid. *Sep. Purif. Technol.* **2020**, *230*, 115870. [CrossRef]
15. Padhan, E.; Nayak, A.K.; Sarangi, K. Recovery of neodymium and dysprosium from NdFeB magnet swarf. *Hydrometallurgy* **2017**, *74*, 210–215. [CrossRef]
16. Padhan, E.; Sarangi, K. Recovery of Nd and Pr from NdFeB magnet leachates with bi-functional ionic liquids based on Aliquat 336 and Cyanex 272. *Hydrometallurgy* **2017**, *167*, 134–140. [CrossRef]
17. Swain, N.; Mishra, S.; Acharya, M.R. Hydrometallurgical route for recovery and separation of samarium (III) and cobalt (II) from simulated waste solution using tri-n-octyl phosphine oxide—A novel pathway for synthesis of samarium and cobalt oxides nanoparticles. *J. Alloys Compd.* **2020**, *815*, 152423. [CrossRef]
18. Pragnell, W.M.; Williams, A.J.; Evans, H.E. The oxidation morphology of SmCo alloys. *J. Alloys Compd.* **2009**, *487*, 69–75. [CrossRef]
19. Su, X.; Wang, Y.; Guo, X.; Dong, Y.; Gao, Y.; Sun, X. Recovery of Sm(III), Co(II) and Cu(II) from waste SmCo magnet by ionic liquid-based selective precipitation process. *Waste Manag.* **2018**, *78*, 992–1000. [CrossRef]
20. Eldosouky, A.; Skulj, I. Recycling of SmCo5 magnets by HD process. *J. Magn. Magn. Mater.* **2018**, *454*, 249–253. [CrossRef]
21. Zheng, X.; Zhang, F.; Liu, E.; Xu, X.; Yan, Y. Efficient Recovery of Neodymium in Acidic System by Free-Standing Dual-Template Docking Oriented Ionic Imprinted Mesoporous Films. *ACS Appl. Mater. Interfaces* **2017**, *9*, 730–739. [CrossRef]
22. Maat, N.; Nachbaur, V.; Larde, R.; Juraszek, J.; Breton, J.M.L. An Innovative Process Using Only Water and Sodium Chloride for Recovering Rare Earth Elements from Nd–Fe–B Permanent Magnets Found in the Waste of Electrical and Electronic Equipment. *ACS Sustain. Chem. Eng.* **2016**, *4*, 6455–6462. [CrossRef]
23. Jha, M.K.; Kumari, A.; Panda, R.; Kumar, J.R.; Yoo, K.; Lee, J.Y. Review on hydrometallurgical recovery of rare earth metals. *Hydrometallurgy* **2016**, *165*, 2–26. [CrossRef]
24. Bogart, J.A.; Cole, B.E.; Boreen, M.A.; Schelter, E.J. Accomplishing simple, solubility-based separations of rare earth elements with complexes bearing size-sensitive molecular apertures. *Proc. Natl. Acad. Sci. USA* **2016**, *113*, 14887–14892. [CrossRef] [PubMed]

25. Walton, A.; Yi, H.; Rowson, N.A.; Speight, J.D.; Mann, V.S.J.; Sheridan, R.S.; Bradshaw, A.; Harris, I.R.; Williams, A.J. The use of hydrogen to separate and recycle neodymium–iron–boron-type magnets from electronic waste. *J. Clean. Prod.* **2015**, *104*, 236–241. [CrossRef]
26. Moore, M.; Gebert, A.; Stoica, M.; Uhlemann, M.; Loser, W. A route for recycling Nd from Nd-Fe-B magnets using Cu melts. *J. Alloys Compd.* **2015**, *647*, 997–1006. [CrossRef]
27. Maroufi, S.; Nekouei, R.K.; Sahajwalla, V. Thermal Isolation of Rare Earth Oxides from Nd–Fe–B Magnets Using Carbon from Waste Tyres. *ACS Sustain. Chem. Eng.* **2017**, *5*, 6201–6208. [CrossRef]
28. Bian, Y.; Guo, S.; Jiang, L.; Liu, J.; Tang, K.; Ding, W. Recovery of Rare Earth Elements from NdFeB Magnet by VIM-HMS Method. *ACS Sustain. Chem. Eng.* **2016**, *4*, 810–818. [CrossRef]
29. Sinha, M.K.; Pramanik, S.; Kumari, A.; Sahu, S.K.; Prasad, L.B.; Jha, M.K.; Yoo, K.; Pandey, B.D. Recovery of value added products of Sm and Co from waste SmCo magnet by hydrometallurgical route. *Sep. Purif. Technol.* **2017**, *179*, 1–12. [CrossRef]
30. Orefice, M.; Audoor, H.; Li, Z.; Binnemans, K. Solvometallurgical route for the recovery of Sm, Co, Cu and Fe from SmCo permanent magnets. *Sep. Purif. Technol.* **2019**, *219*, 281–289. [CrossRef]
31. Xu, T.; Zhang, X.; Lin, Z.; Lu, B.; Ma, C.; Gao, X. Recovery of rare earth and cobalt from Co-based magnetic scraps. *J. Rare Earths* **2010**, *28*, 485–488. [CrossRef]
32. Sahoo, K.; Nayak, A.K.; Ghosh, M.K.; Sarangi, K. Preparation of Sm_2O_3 and Co_3O_4 from SmCo magnet swarf by hydrometallurgical processing in chloride media. *J. Rare Earths* **2018**, *36*, 725–732. [CrossRef]
33. Zhou, K.; Wang, A.; Zhang, D.; Zhang, X.; Yang, T. Sulfuric acid leaching of SmCo alloy waste and separation of samarium from cobalt. *Hydrometallurgy* **2017**, *174*, 66–70. [CrossRef]
34. Wang, J.Z.; Tang, Y.C.; Shen, Y.H. Leaching of Sm, Co, Fe, and Cu from Spent SmCo Magnets Using Organic Acid. *Metals* **2023**, *13*, 233. [CrossRef]
35. Torkaman, R.; Moosavian, M.A.; Mostaedi, M.T.; Safdari, J. Solvent extraction of samarium from aqueous nitrate solution by Cyanex301 and D2EHPA. *Hydrometallurgy* **2013**, *137*, 101–107. [CrossRef]
36. Wang, J.Z.; Hsieh, Y.H.; Tang, Y.C.; Shen, Y.H. Separation of Cobalt, Samarium, Iron, and Copper in the Leaching Solution of Scrap Magnets. *Metals* **2023**, *13*, 90. [CrossRef]
37. Onoda, H.; Kurioka, Y. Selective removal and recovery of samarium from mixed transition metal solution using phosphoric acid. *J. Environ. Chem. Eng.* **2016**, *4*, 4536–4539. [CrossRef]
38. Monhemius, J. Precipitation diagrams for metal hydroxides, sulphides, arsenates and phosphates. *Trans. Inst. Min. Metall.* **1977**, *86*, C202–C206. Available online: https://www.researchgate.net/publication/266137129_Precipitation_diagrams_for_metal_hydroxides_sulphides_arsenates_and_phosphates (accessed on 28 August 2023).
39. Fan, E.; Yang, J.; Huang, Y.; Lin, J.; Arshad, F.; Wu, F.; Li, L.; Chen, R. Leaching Mechanisms of Recycling Valuable Metals from Spent Lithium-Ion Batteries by a Malonic Acid-Based Leaching System. *ACS Appl. Energy Mater.* **2020**, *3*, 8532–8542. [CrossRef]
40. Li, L.; Fan, E.; Guan, Y.; Zhang, X.; Xue, Q.; Wei, L.; Wu, F.; Chen, R. Sustainable Recovery of Cathode Materials from Spent Lithium-Ion Batteries Using Lactic Acid Leaching System. *ACS Sustain. Chem. Eng.* **2017**, *5*, 5224–5233. [CrossRef]
41. Ackermann, S.; Lazic, B.; Armbruster, T.; Doyle, S.; Grevel, K.D.; Majzlan, J. Thermodynamic and crystallographic properties of kornelite [$Fe_2(SO_4)_3 \cdot \sim 7.75H_2O$] and paracoquimbite [$Fe_2(SO_4)_3 \cdot 9H_2O$]. *Am. Mineral.* **2009**, *94*, 1620–1628. [CrossRef]
42. Majzlan, J.; Botez, C.; Stephens, P.W. The crystal structures of synthetic $Fe_2(SO_4)_3(H_2O)_5$ and the type specimen of lausenite. *Am. Mineral.* **2005**, *90*, 411–416. [CrossRef]
43. Majzlan, J.; Navrotsky, A.; Mccleskey, R.B.; Alpers, C.N. Thermodynamic properties and crystal structure refinement of ferricopiapite, coquimbite, rhomboclase, and $Fe_2(SO_4)_3(H_2O)_5$. *Eur. J. Mineral.* **2006**, *18*, 175–186. [CrossRef]
44. Denisenko, Y.G.; Sedykh, A.E.; Basova, S.A.; Atuchin, V.V.; Molokeev, M.S.; Aleksandrovsky, A.S.; Krylov, A.S.; Oreshonkov, A.S.; Khritokhin, N.A.; Salnikova, E.I.; et al. Exploration of the structural, spectroscopic and thermal properties of double sulfate monohydrate $NaSm(SO_4)_2 \cdot H_2O$ and its thermal decomposition product $NaSm(SO_4)_2$. *Adv. Powder Technol.* **2021**, *32*, 3943–3953. [CrossRef]
45. Deng, Y.; Zhang, Y.; Ding, Y. Recovery of rare earths in different media with novel dicarboxylate based ionic liquid and application to recycle SmCo magnets. *Hydrometallurgy* **2022**, *210*, 105844. [CrossRef]
46. Wang, J.Z.; Lin, H.H.; Tang, Y.C.; Shen, Y.H. Recovery of Calcium from Reaction Fly Ash. *Sustainability* **2023**, *15*, 2428. [CrossRef]

recycling

MDPI

Article

A Hydrometallurgical Process for the Recovery of Noble Metals (Au, Pt, Ir, and Ta) from Pyrolyzed and Acid-Digested Solutions of Single-Use Medical Devices

Angeliki Lampou [1], Evgenios Kokkinos [1,*], Charikleia Prochaska [1,*], Theodosios Tsiogkas [1], Effrosyni Peleka [1], Anthimos Xenidis [2] and Anastasios Zouboulis [1,*]

1 Laboratory of Chemical & Environmental Technology, Department of Chemistry, Aristotle University of Thessaloniki, 54124 Thessaloniki, Greece; angelina.lampou@gmail.com (A.L.); grtsiogkas@chem.auth.gr (T.T.); peleka@chem.auth.gr (E.P.)

2 School of Mining and Metallurgical Engineering, National Technical University of Athens, 15780 Athens, Greece; axen@mail.ntua.gr

* Correspondence: evgenios@chem.auth.gr (E.K.); prohaska@chem.auth.gr (C.P.); zoubouli@chem.auth.gr (A.Z.)

Abstract: Developing an efficient recycling route for spent single-use medical devices is essential for recovering precious metals. The proposed complete hydrometallurgical route goes through the initial pyrolysis and acid digestion steps, expanding upon our previous relevant work in the field, followed by solvent extraction, stripping, and precipitation procedures. In this study, a complete hydrometallurgical process was developed for the recovery of gold, platinum, iridium, and tantalum, separating them from other metals, i.e., from iron, chromium, and nickel, also present in the examined medical devices, i.e., (i) diagnostic electrophysiology catheters, containing gold, (ii) diagnostic guide wires, containing platinum and iridium alloys, and (iii) self-expanding stents, containing tantalum. This study reports the experimental results of selecting an efficient extractant, stripping, and precipitation agent, along with the effects of key factors that influence each consecutive step of the process, i.e., agent concentration, aqueous to organic phase ratio, contact time, and pH, using simulated metal solutions and also applying the obtained optimal conditions to the treatment of real sample solutions. For the selective separation of gold, Aliquat 336 was used to extract it in the organic phase; it was then stripped using a thiourea solution and precipitated by utilizing an iron sulfate (II) solution and proper pH adjustment. The selective separation of platinum was achieved by using Aliquat 336 for the organic phase extraction and a perchlorate acid solution for stripping it back into the aqueous solution and adding a sodium bromate solution to precipitate it. Due to the similar chemical behavior, the selective recovery of iridium followed the same processes as that of platinum, and the separation between them was achieved through selective precipitation, as heating the solution and adjusting the pH value resulted in the selective precipitation of iridium. Lastly, the selective recovery of tantalum consists of extraction by using Alamine 336, then stripping it back to the aqueous phase by using sodium chloride, and precipitation by using potassium salt solution and proper pH adjustment. A total recovery of 88% for Au, 86% for Pt, 84% for Ir, and 80% for Ta was obtained, thus achieving a high uptake of precious metals from the examined real spent/waste samples.

Keywords: medical waste; solvent extraction; stripping; precipitation; separation; hydrometallurgy; gold; platinum; iridium; tantalum

1. Introduction

Noble metals are crucial for manufacturing several different advanced technological applications, including (but not limited to) the automobile, chemical, electronic, and medical industries [1,2]. Furthermore, new technological developments require an increase in the market's needs regarding the availability of noble metals [3]. For instance, 5G networks and electric cars (current technological trends) are among the many applications requiring the presence of noble metals, such as tantalum [4].

Noble metals are often used during diagnostic or other medical procedures due to their inertness and visibility through fluoroscopy [5]. Many single-use medical devices usually contain noble metal parts, such as diagnostic electrophysiology catheters that contain Pt/Ir alloy marker bands [6], diagnostic guide wires with Au coating media [7], and self-expanding stents with Ta marker dots at each edge [8]. Due to health concerns, the reuse or recycling of such products are avoided [9]. Therefore, the waste generated from such products is conventionally sterilized and disposed of in landfills or incinerated after their first use [10].

With the primary sources of noble metals being mining, located mainly outside Europe, as Europe provides only around 3% of the respective demand/needs, their recovery from secondary sources can (at least partly) replace the need for imports and guarantee their supply [11]. Consequently, the European Union's directives call for considering secondary sources of noble metals as primary ones [12]. Hence, many studies focus on the recovery of noble metals from industrial slimes [13], spent catalysts [13,14], and electronic wastes [15–17]. The recovery of noble metals from spent/used medical devices has not been thoroughly investigated and is, to the best of our knowledge, limited to the recovery of gold from implantable devices containing printed circuit boards (e.g., from pacemakers) [18]. According to recent surveys, more than 1 million cardiac catheter ablation procedures [19] and more than 3 million cardiac catheterization procedures are performed worldwide annually [20]. Also, more than 2 million stents are being implanted annually [21]. Even though stents are implanted permanently inside patients' bodies, approximately 10% of hospitals' stent stock is estimated to be discarded due to product expiry [22]. Such products are sold at approximately 50–90% below the Manufacturer's Suggested Retail Price (MSRP) (Medical Materials, n.d.) for training and non-clinical research purposes or, most commonly, without further use. Considering these facts, the recovery of noble metals from medical devices is selected to be examined in this study as it poses great potential for enhancing their recycling potential.

Several processes have been proposed that can be used to recover noble metals from secondary sources [16]. The most widespread methods for recovering gold, platinum, iridium, and tantalum consist of pyrometallurgy and hydrometallurgy processes [23–25]. Hydrometallurgy presents more advantages than pyrometallurgy, including higher purification yields and lower temperatures during its implementation. This means that it consumes less energy on average and has promising scalability and recovery yields due to better process control [15]. The hydrometallurgical recovery of these noble metals focuses on their selective separation from and among other metals (impurities). In general, the hydrometallurgical route for the recovery of any metal entails the dissolution of the secondary source sample and then the application of a separation method, usually followed by precipitation.

Especially for the recovery of gold, the gold-containing sample is dissolved by using aqua regia [26], a solution of HCl with H_2O_2 as the oxidizing agent [27], or cyanide [16]. The separation of gold from the other metals that may have been co-dissolved along with it can be achieved in various ways, such as by the use of activated carbon [28], ion exchange resins [29], or polymeric membranes [30], or by solvent extraction and precipitation [26]. Only a few solvents have been used so far for the separation of gold, and the most effective of them proved to be Aliquat 336, previously used for the extraction of gold from an aqueous solution to the organic phase and, afterward, a thiourea solution utilized for the stripping of gold from the organic phase back to the aqueous one [26]. A similar system of Cyanex 272 and NH_4Cl, respectively, has also been applied for the selective recovery of gold [2]. The precipitation of gold can be subsequently achieved either with the addition of metal shavings or powders, such as iron, to the resulting solution [25] or by the use of proper salt solutions, such as Na_2SO_3 [17] and $FeSO_4$ [27].

Noble metals, such as platinum and iridium, are significantly harder to separate from each other due to their similar chemical behavior. The dissolution of platinum-containing samples was achieved by using aqua regia [31,32] or NaCN [16]. The dissolution of iridium in iridium–platinum alloys was successful with aqua regia treatment under heat. [33,34]. The selective recovery of platinum was achieved mainly through solvent extraction by using Aliquat 336 [32], LIX 63 [34], or Alamine 300 [35], and then, for the stripping of platinum from these solutions, $HClO_4$ [32,34] and NaSCN [35] were used. The selective recovery of iridium was also achieved with solvent extraction by applying organic reagents, such as Aliquat 336 [2,34], LIX 63 [34], or Cyanex 471X [33]. Subsequently, the iridium was stripped from these solutions by using $HClO_4$ [34] or thiourea [33]. Both platinum and iridium cementation can also be achieved by adding NH_4Cl [31,36] or $NaBrO_3$ [37].

Tantalum-bearing samples were typically dissolved using a concentrated HF and H_2SO_4 mixture [38–41]. The separation of tantalum was achieved with solvent extraction using Cyanex 923 [40] or Alamine 336 [24,38,39] and subsequently, NH_3 [40], H_2O [24], or NaCl [39] were examined to strip it from the organic solvents. Tantalum cementation is also achieved by adding NH_3 to the tantalum solution [39], or by the addition of KF or KCl in alkaline pH [24,40].

In our previous work, a process of recovery, consisting of the application of initial pyrolysis for the elimination of polymer content and from the selective dissolution of noble metals existing in diagnostic electrophysiology catheters (Pt and Ir), diagnostic guide wires (Au), and self-expanding stents (Ta), considered as a single waste stream, was proposed [1]. By using the corresponding results as our guiding points, in this study, we aim to develop an integrated hydrometallurgical process for the recovery of gold, platinum, iridium, and tantalum, separating them selectively from the other common co-existing metals (i.e., iron, chromium, and nickel) present in medical devices. The process involves three main treatment steps, applied consecutively, i.e., solvent extraction, stripping, and precipitation of the previously leached metals. The experimental optimization of the extractant, stripping, and precipitation agent selection, along with the effects of the critical factors that influence each consecutive step of the integrated process, i.e., agent concentration, aqueous to organic phase ratio, contact time, and pH, are investigated in this research.

2. Results and Discussion

2.1. Extraction Performance

2.1.1. Selection of Extraction Reagents

The results of the preliminary experiments, using single-metal synthetic aqueous solutions to determine the optimal reagent for each (selective) extraction, are presented in Figure 1. Au^{3+}, Ir^{3+}, and Pt^{4+} were successfully extracted (>85%) by Aliquat 336, as they present similar chemical behavior [33]. This extraction is mainly based on an ion exchange mechanism, in which the negatively charged chloro-complexes of these metals are exchanged with the Cl^- anions of the ammonium functional group of Aliquat 336, resulting in their transfer into the organic phase [42]. The extraction of Au, Pt, and Ir follows the reactions 1–3 [26,36,43,44]:

$$[AuCl_4]^-{}_{(aq)} + (R_3R'N^+Cl^-)_{(org)} \rightleftarrows [(R_3R'N^+)AuCl_4{}^-]_{(org)} + 2Cl^- \quad (1)$$

$$[PtCl_6]^{2-}{}_{(aq)} + 2(R_3R'N^+Cl^-)_{(org)} \rightleftarrows \left[(R_3R'N^+)_2PtCl_6{}^{2-}\right]_{(org)} + 2Cl^- \quad (2)$$

$$[IrCl_6]^{3-}{}_{(aq)} + 3(R_3R'N^+Cl^-)_{(org)} \rightleftarrows \left[(R_3R'N^+)_3IrCl_6{}^{3-}\right]_{(org)} + 3Cl^- \quad (3)$$

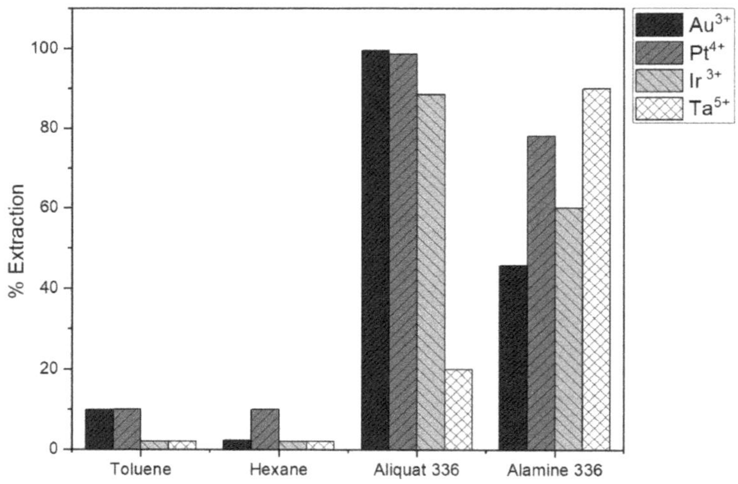

Figure 1. Effect of extractants' solvent/type on the percentage extraction of examined precious/noble metals. Experimental conditions: (i) aqueous phase: single-metal solutions containing 50 mg/L Au^{3+}, Ir^{3+}, Pt^{4+}, or Ta^{5+}; (ii) organic phase: [Toluene] = 9.44 M, [Hexane] = 7.66 M, [Aliquat 336 in toluene] or [Alamine 336 in toluene] = 0.5×10^{-3} M; O/A ratio = 1/1; O/A interaction/contact time = 30 min; temperature 25 ± 1 °C; rotation 45 rpm.

On the other hand, Ta^{5+} was successfully extracted (90%) by Alamine 336 since its active nitrogen atom reacts with inorganic complexes to form amine salts, which can undergo ion exchange reactions with other anions. Reactions 4 and 5 describe the corresponding mechanism, i.e., the protonation of amine followed by the ion exchange extraction procedure [43], making it ideal for the low pH value of the solution that results from the acid digestion of Ta-bearing samples.

$$(R_3N)_{org} + HF \rightarrow \left(R_3NH^+F^-\right)_{org} \tag{4}$$

$$2\left(R_3NH^+F^-\right)_{org} + \left(TaF_7^{2-}\right) \rightarrow [(R_3NH)_2TaF_7]_{org} + 2F^- \tag{5}$$

2.1.2. Optimal Extractant Concentrations

The effect of the optimal extractants' concentration is presented in Figure 2. The highest extraction rate was achieved by applying equal or greater than the respective required stoichiometric concentrations, as determined by Reactions 3, 4, 5, and 7. In total, 99.5% of Au^{3+} was extracted by using 0.27×10^{-3} M Aliquat 336, 100% of Pt^{4+} was extracted by using 0.625×10^{-3} M Aliquat 336, and 95.4% of Ir^{3+} was extracted by using 0.65×10^{-3} M Aliquat 336, while 94.4% of Ta^{5+} was extracted by using 0.54×10^{-3} M Alamine 336. The optimal extractant concentration was stoichiometrically required for most examined metals; however, for the case of Ta, a double concentration of the stoichiometric concentration was needed for the optimal extraction due to the ability of H_2O to strip Ta back to the aqueous phase [24].

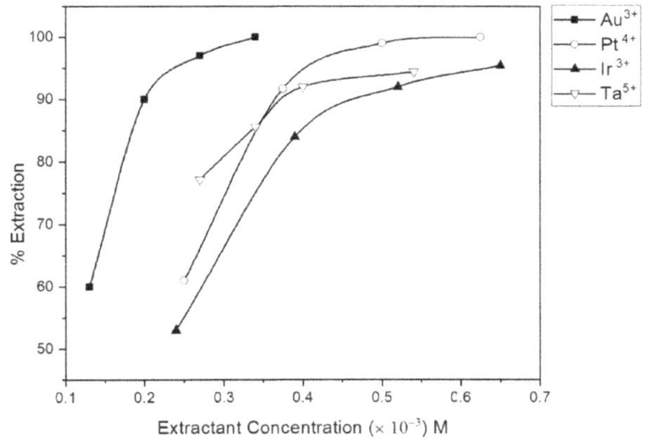

Figure 2. Effect of extractants' concentration on the percentage extraction of studied precious/noble metals. Experimental conditions: (i) aqueous phase: single-metal solutions containing 50 mg/L of Au^{3+}, Ir^{3+}, Pt^{4+}, or Ta^{5+}; (ii) organic phase: [Aliquat 336 in toluene] = 0.13–0.65 × 10^{-3} M for Au^{3+}, Ir^{3+} and Pt^{4+}, [Alamine 336 in toluene] = 0.27–0.54 × 10^{-3} M for Ta^{5+}; O/A = 1/1; O/A interaction/contact time = 30 min; temperature 25 ± 1 °C; rotation 45 rpm.

2.1.3. Optimal Extraction Phase Ratio

The results of the three examined organic/aqueous (O/A) phase volume ratios on the extraction rate of the examined metals are presented in Figure 3. Au^{3+}, Pt^{4+}, and Ta^{5+} exhibited better extraction results (>85%) at O/A = 1/1, and Ir^{3+} at O/A = 1/2. The slight difference in the extraction percentages of Ir^{3+} (i.e., 88% with O/A = 1/1 and 93.5% with O/A = 1/2) is based on the variation in the distribution of the chloroanions during the ion exchanges, which follows the order $[MCl_6]^{2-}$ > $[MCl_4]^{2-}$ > $[MCl_6]^{3-}$ overall lowering the Ir^{3+} extraction percentage in a lower Aliquat 336 concentration [36]. Due to the fact that Ir and Pt coexist in the Pt/Ir marker bands of the examined single-use medical devices, we selected O/A = 1/1 ratio as the optimum to be used for the next steps of the experimental investigation.

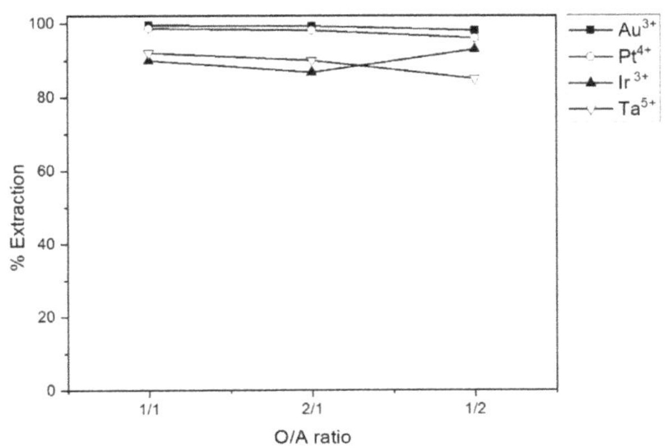

Figure 3. Effect of O/A ratio (volume of respective phases) on the percentage extraction of examined precious metals. Experimental conditions: (i) aqueous phase: single-metal solutions, containing 50 mg/L Au^{3+}, Ir^{3+}, Pt^{4+}, or Ta^{5+}; (ii) organic phase: [Aliquat 336 in toluene] = 0.135–0.54 × 10^{-3} M for Au^{3+}, [Aliquat 336 in toluene] = 0.25–1 × 10^{-3} M for Pt^{4+}, [Aliquat 336 in toluene] = 0.26–1.2 × 10^{-3} M for Ir^{3+}, [Alamine 336 in toluene] = 0.2–0.8 × 10^{-3} M for Ta^{5+}; total volume = 30 mL; O/A interaction/contact time = 30 min; temperature 25 ± 1 °C; rotation 45 rpm.

2.1.4. Optimal Extraction Time

The effect of the contact time variation for the O/A phase interaction is presented in Figure 4. Au^{3+} and Pt^{4+} exhibited almost total quantitative extraction (>98%) after 30 min. Ir^{3+} extraction increased from 88% at 30 min to 98% at 60 min. Once again, since Ir and Pt co-exist in the marker bands of the examined single-use medical devices, the 30 min time period was selected as the optimum extraction time for both metals. The percentage extraction of Ta^{5+} dropped from 90% at 10 min to 77% at 30 min and remained almost constant (78%) at 60 min. The longer the interaction time used, the more Ta was stripped back into the aqueous phase [24]. As a result, 10 min was selected as the optimum extraction/contact time for tantalum.

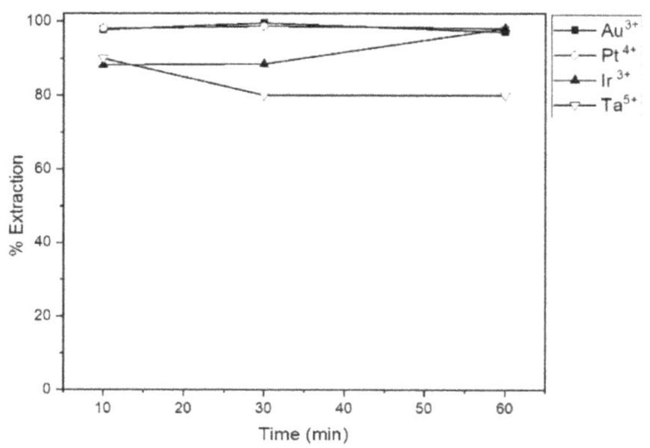

Figure 4. Effect of contact time variation in the O/A phase interaction on the percentage extraction of examined precious metals. Experimental conditions: (i) aqueous phase: single-metal solutions, containing 50 mg/L Au^{3+}, Ir^{3+}, Pt^{4+}, or Ta^{5+}; (ii) organic phase: [Aliquat 336 in toluene] = 0.27–0.52 × 10^{-3} M for Au^{3}, Ir^{3+} or Pt^{4+}, [Alamine 336 in toluene] = 0.27 × 10^{-3} M for Ta^{5+}; O/A ratio = 1/1; temperature 25 ± 1 °C; rotation 45 rpm.

2.1.5. Extraction Selectivity

Aliquat 336 proved a suitable extractant due to the selective extraction of Au^{3+} (99.4%) over the other typical metals of the relevant waste stream, namely, Fe, Cr, Ni, and Ti, as presented in Figure 5a. In the acidic aqueous solution, these potential metallic interferences exist mainly in the respective ionic forms, i.e., Fe^{3+}, Cr^{3+}, Ni^{2+}, and Ti^{4+}. On the other hand, Au occurs in the solution as the cholo-complex form, i.e., $AuCl_4^-$. Thus, the ammonium functional group of Aliquat 336 ($R_3R'N^+$) can bind with gold by electrostatic attraction but not with the other metallic cations, as the repulsive forces are favored in the latter case [26]. The extraction step was then repeated, loading the solution obtained from the processing of the raw medical samples by the acid digestion procedure, as previously presented by Kokkinos et al., 2023 (Table S1). The extraction percentages of the samples of the actual medical devices (Figure 6a) did not differ by more than 1% from the experimental results performed with simulated solutions (Figure 5a).

Aliquat 336 also selectively extracted Pt (88.3%) and Ir (94.2%) without significantly affecting the presence of Fe, Cr, Ni, and Ti in the aqueous phase, as presented in Figure 7a. Similarly to Au, in the aqueous solution, Pt and Ir exist mainly as chloro-complexes ($PtCl_6^{2-}$, $IrCl_6^{3-}$). In contrast, the other metals exist in their cationic forms and, thus, cannot be electrostatically bound to Aliquat 336 (Table S2). The extraction was then repeated using the acid-digested solution from the raw medical samples. Due to the low pH value, adjusting the solution's pH was necessary for the extraction. The extraction percentages of the samples of actual medical devices (Figure 6b) did not differ by more than 1% compared to the experiments with simulated synthetic solutions (Figure 5a).

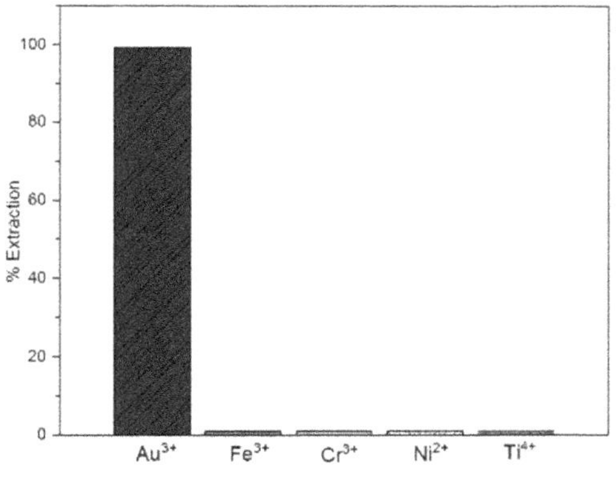

(**a**)

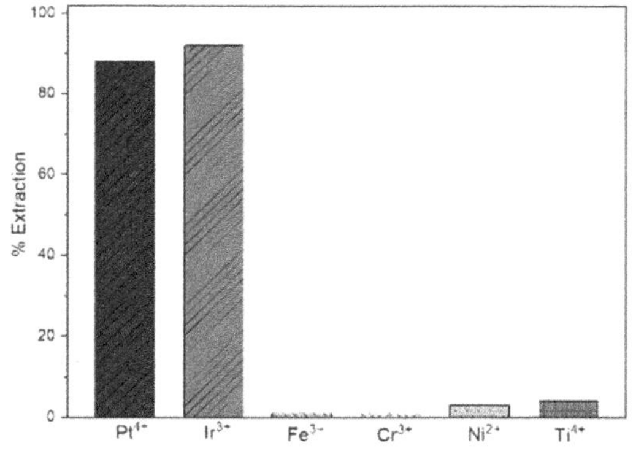

(**b**)

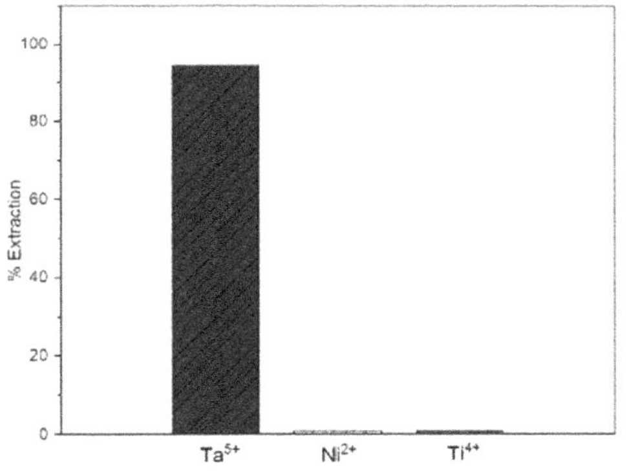

(**c**)

Figure 5. Selective extraction in synthetic solutions. (**a**) Aliquat 336 selective extraction of Au over Fe, Cr, Ni, and Ti. Experimental conditions: (i) aqueous phase: Au^{3+} 42.2 mg/L, Fe^{3+} 220 mg/L, Cr^{3+} 51.8 mg/L, Ni^{2+} 37.6 mg/L, Ti^{4+} 28 mg/L; (ii) organic phase: [Aliquat 336 in toluene] = 0.23×10^{-3} M; O/A = 1/1; O/A interaction time = 30 min; temperature 25 ± 1 °C; rotation 45 rpm. (**b**) Aliquat 336 selective extraction of Pt and Ir over Fe, Cr, Ni, and Ti. Experimental conditions: (i) aqueous phase: Pt^{4+} 1708.3 mg/L, Ir^{3+} 317 mg/L, Fe^{3-} 9.1 mg/L, Cr^{3+} 0.6 mg/L, Ni^{2+} 480 mg/L, Ti^{4+} 405 mg/L; organic phase: [Aliquat 336 in toluene] = 20.8×10^{-3} M; O/A = 1/1; O/A interaction time = 30 min; temperature 25 ± 1 °C; rotation 45 rpm. (**c**) Alamine 336 selective extraction of Ta over Ni and Ti. Experimental conditions: (i) aqueous phase: Ta^{5+} 90 mg/L, Ni^{2+} 90 mg/L, Ti^{4+} 90 mg/L; (ii) organic phase: [Alamine 336 in toluene] = 0.97×10^{-3} M; O/A ratio = 1/1; O/A interaction time = 30 min; temperature 25 ± 1 °C; rotation 45 rpm.

Similarly, Alamine 336 extracted 92% of Ta selectively over the Ni and Ti present in the original aqueous solution (as <1% of Ni and Ti co-extracted) (Figures 5c and 6c).

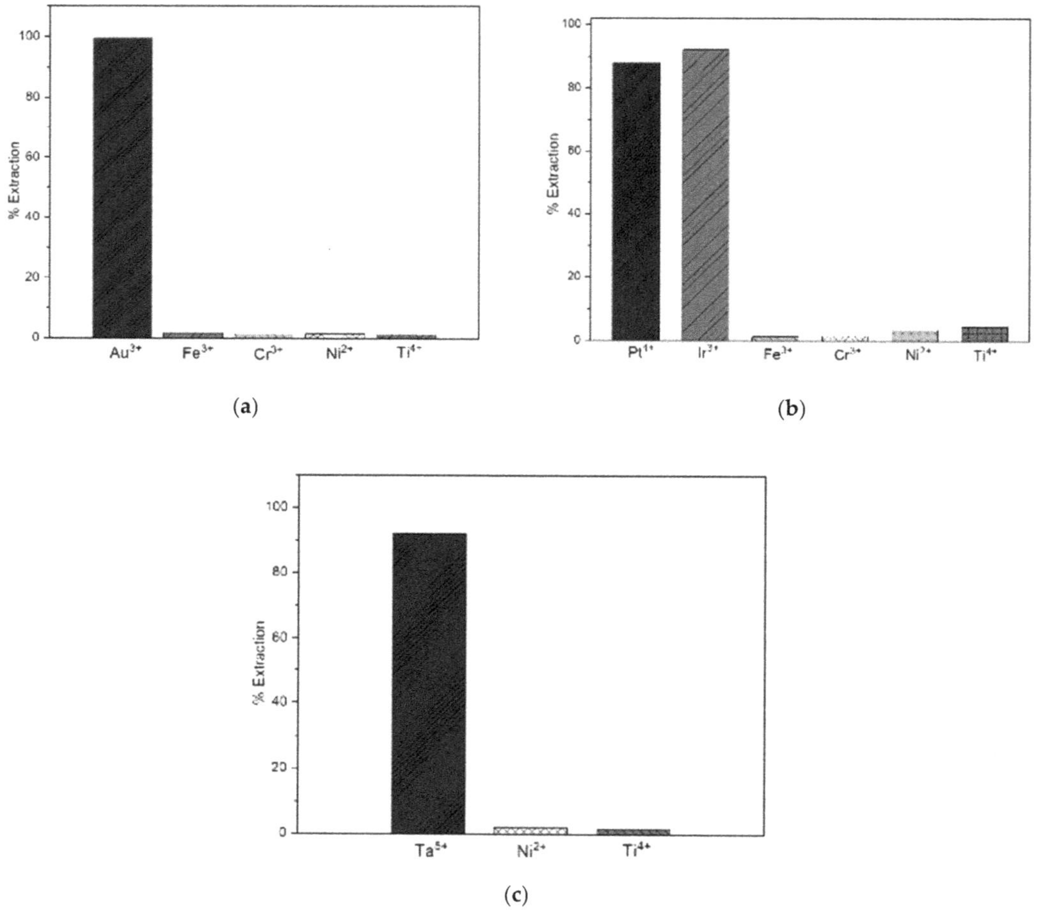

Figure 6. Selective extraction of examined metals in real solutions. (**a**) Aliquat 336 selective extraction of Au over Fe, Cr, Ni, and Ti. Experimental conditions: (i) aqueous phase: Au^{3+} 42.2 mg/L, Fe^{3+} 220 mg/L, Cr^{3+} 51.8 mg/L, Ni^{2+} 37.6 mg/L, Ti^{4+} 28 mg/L; (ii) organic phase: [Aliquat 336 in toluene] = 0.23×10^{-3} M; O/A = 1/1; O/A interaction time = 30 min; 25 $\pm$ 1 °C; 45 rpm. (**b**) Aliquat 336 selective extraction of Pt and Ir over Fe, Cr, Ni, and Ti. Experimental conditions: (i) aqueous phase: Pt^{4+} 1708.3 mg/L, Ir^{3+} 317 mg/L, Fe^{3+} 9.1 mg/L, Cr^{3+} 0.6 mg/L, Ni^{2+} 480 mg/L, Ti^{4+} 405 mg/L; (ii) organic phase: [Aliquat 336 in toluene] = 20.8×10^{-3} M; O/A = 1/1; O/A interaction time = 30 min; 25 $\pm$ 1 °C; 45 rpm. (**c**) Alamine 336 selective extraction of Ta over Ni and Ti. Experimental conditions: (i) aqueous phase: Ta^{5+} 90 mg/L, Ni^{2+} 90 mg/L, Ti^{4+} 90 mg/L; organic phase: [Alamine 336 in toluene] = 0.97×10^{-3} M; O/A = 1/1; O/A interaction time = 30 min; 25 $\pm$ 1 °C; 45 rpm.

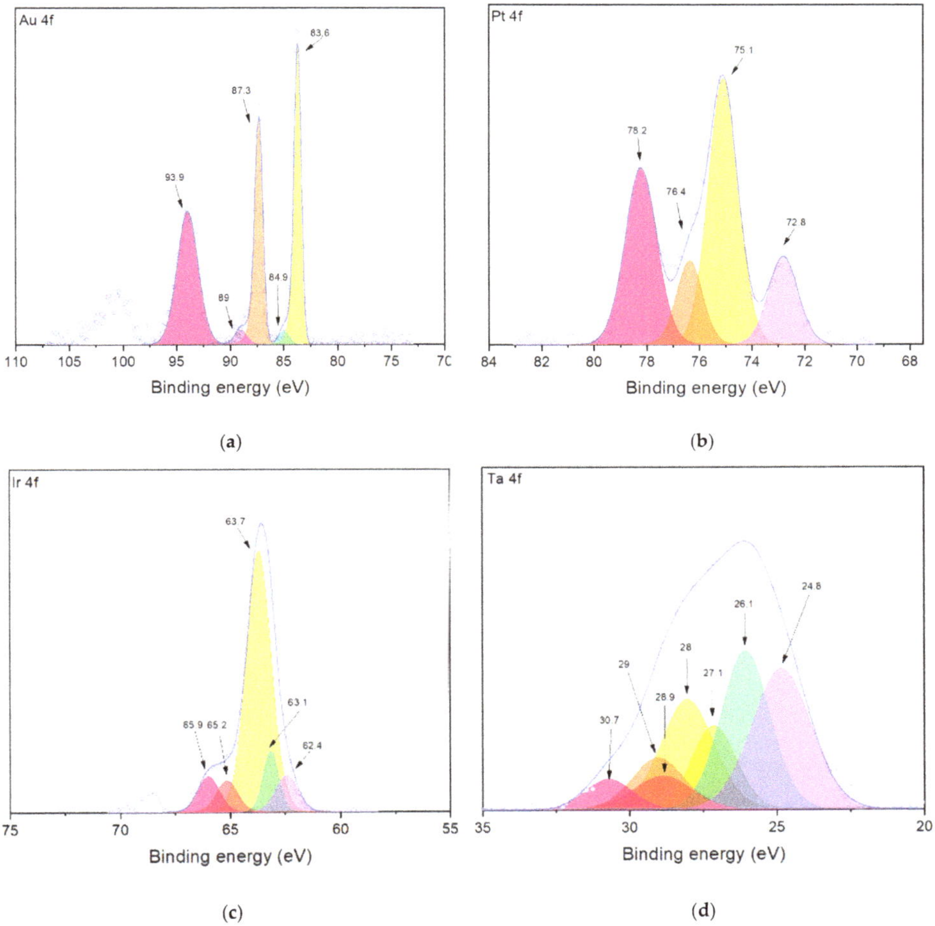

Figure 7. High-resolution XPS spectrum of examined noble metals' precipitates, indicating their speciation (**a**) Au, (**b**) Pt, (**c**) Ir, and (**d**) Ta (each color represents a peak that was obtained in high resolution core-level spectra).

2.2. Stripping Performance

Various parameters were also optimized during the stripping process of noble metals from the organic solvents' solutions obtained during the preceding extraction procedure, which originated from either the simulated aqueous streams or the actual samples. The results are summarized in Table 1. The detailed data of all the conducted stripping experiments are presented in Figures S1–S6 in the Supplementary Material (SM).

Thiourea presented the advantage of stripping Au stoichiometrically, according to reaction 6 [26].

$$\begin{aligned}\left[(R_3R'N^+)AuCl_4{}^-\right]_{(org)} + (NH_2)_2CS_{(aq)} \\ \rightleftarrows AuCl_3(NH_2)_2CS_{(aq)} + \left(R_3R'N^+Cl^-\right)_{(org)}\end{aligned} \tag{6}$$

The complete stripping of Au (>99%) was achieved by the aqueous mixture of thiourea and HCl, as the acidic conditions (pH < 5) favor thiourea to form strong cationic complexes with Au^{3+} [44,45]. Additionally, lower pH values (pH < 3) increase the solubility of thiourea, favoring a potential scale-up of this process [45,46]. Stripping Pt and Ir from the organic phase was significantly more complicated, as they form strong ionic bonds with the quaternary ammonium salts, such as Aliquat 336, through the respective chloro-complexes; these bonds cannot be easily broken [47]. Higher stripping efficiencies of Pt and Ir (>99%) were obtained by using 0.01 M and 0.05 M $HClO_4$, respectively, regarding the synthetic sample (Table 1). The stripping mechanism of Ir and Pt from the loaded Aliquat 336 by the addition of $HClO_4$ attributed to the competitive behavior of ClO_4^- over XCl_6^{2-} (where X = Ir or Pt) for the active sites of Aliquat 336, alongside the lower pH values of $HClO_4$ solutions (pH < 2) that can assist the breaking of strong ionic bonds between the metal chloride ions and Aliquat 336 [48]. This, however, may compromise the reusability of organic Aliquat 336 solution [32].

Table 1. Optimized parameters for stripping metals from the loaded synthetic and actual samples of organic solutions.

Metal Solution (Organic Phase, O)	Stripping Solution (Aqueous Phase, A)	Stripping Conditions	%S	
			Synthetic Sample	Real Sample
$[Au^{3+}] = 0.27 \times 10^{-3}$ M in Aliquat 336	[thiourea] = 0.27×10^{-3} M HCl = 1 M	O/A = 1/1; O/A interaction time = 10 min; 25 ± 1 °C; 45 rpm	>99%	98%
$[Pt^{4+}] = 0.25 \times 10^{-3}$ M in Aliquat 336	$[HClO_4]$ = 0.01 M	O/A = 1/1; O/A interaction time = 30 min; 25 ± 1 °C; 45 rpm	>99%	98.6%
$[Ir^{3+}] = 0.26 \times 10^{-3}$ M in Aliquat 336	$[HClO_4]$ = 0.05 M	O/A = 1/1; O/A interaction time = 10 min; 25 ± 1 °C; 45 rpm	>99%	95%
$[Ta^{5+}] = 0.34 \times 10^{-3}$ M in Alamine 336	[NaCl] = 3 M	O/A = 1/1; O/A interaction time = 10 min; 25 ± 1 °C; 45 rpm	92%	91%

A relatively high concentration of NaCl solution (3 M) was used to strip Ta from the loaded Alamine 336 solution. Although Ta can be stripped using deionized water in multiple cycles, NaCl was promoted because it is particularly efficient for stripping tantalum from amines [43]. Notably, when the interaction time between the two phases was extended, tantalum was not extracted back to the organic solution (Figure S6).

When the organic-to-aqueous phase ratio was examined, the highest stripping efficiencies were achieved for the O/A ratio 1/1 (Figure S5). The optimal phase interaction/contact time was 10 min for all the metals except platinum, which required 30 min to be stripped by the $HClO_4$ solution (Figure S6).

When the same stripping conditions were applied to the actual samples, a comparably high %S was achieved for all the examined metals, as seen in Table 1.

2.3. *Precipitation Performance*

The precipitation mechanisms of the examined noble metals are presented in reactions 9–11. The various parameters of this process are optimized regarding the loaded aquatic phases of the simulated and real samples as they emerge after stripping. The corresponding results are summarized in Table 2. The Supplementary Material presents additional data from all the conducted precipitation experiments in Figure S7 and Tables S3–S5.

Au precipitated through reduction by using ferrous sulfate [49]. Based on reaction 7, the stoichiometric demand of $FeSO_4$ was added to the solution, and using NaOH, the pH was adjusted to 7.3. Au may also be precipitated without ferrous sulfate, but an even higher alkaline environment is needed, i.e., pH 10 (Figure S7). However, this leads to the formation of $Au(OH)_3$ instead of its elemental (metallic) form, resulting in the further processing of this solid product [50].

$$Au^{3+} + 3FeSO_4 \rightarrow Au + Fe_2(SO_4)_3 + Fe^{3+} \tag{7}$$

Regarding the obtained efficiencies, 97% of the gold was precipitated from the synthetic and 90% from the real sample solution; it is also worth noting that the same reaction has been previously applied to precipitate/separate gold from solutions resulting from other metal-containing waste materials, such as printed circuit boards [51].

Table 2. Optimized parameters for the precipitation of noble metals from the aqueous solutions obtained after stripping organic loading (synthetic and real) samples.

Metal Concentration ($\times 10^{-3}$ M)			Precipitation Parameters						%P
			Chemical Agent	Agent Conc. (M)	Time (min)	pH	Temp. (°C)	No of Cycles	
$[Au^{3+}]$	Synthetic sample	0.27	$FeSO_4$	0.77×10^{-3}	30	7.3 *	25 ± 1	1	90%
	Real sample	0.23		0.64×10^{-3}					90%
$[Pt^{4+}]$	Synthetic sample	0.25	$NaBrO_3$	12.5×10^{-3}	30	1.4	25 ± 1	3	98%
	Real sample	8.75		2					98%
$[Ir^{3+}]$	Synthetic sample	0.26	$NaBrO_3$	12.5×10^{-3}	30	8.5 *	70 ± 1	1	90%
	Real sample	1.65		2					94%
$[Ta^{5+}]$	Synthetic sample	0.34	KF	1	30	8.6 *	25 ± 1	1	96%
	Real sample	0.5		2					97%

* pH adjustment by using 4 M NaOH.

$NaBrO_3$ was added to precipitate platinum and iridium, but since they present similar chemical behavior, pH and temperature adjustments to different values were required for their selective precipitation (reaction 8).

$$NaBrO_3 + [MCl_6]^{x-} \rightleftarrows M(OH)_{4y} + Br^-, \text{ where } M = Ir,\ Pt,\ x = 3 \text{ when } M = Ir \text{ and } 2 \text{ when } M = Pt,\ y = 3 \text{ when } M = Ir \text{ and } 4 \text{ when } M = Pt \tag{8}$$

To precipitate the total of 98% platinum, three subsequent additions/cycles of $NaBrO_3$ were applied. After each cycle, the solution was filtered to obtain the precipitated solid metal Pt. To precipitate 94% Ir, the solution was heated to 70 °C for 30 min, and the pH was adjusted to 8.5. This procedure is commonly used in industry to precipitate iridium, as heat and alkaline conditions favor the respective precipitation due to the hydrolysis of metal, which produces insoluble hydroxides [37,52].

Tantalum may be precipitated by using potassium salts or ammonia solution. Ammonia precipitated a lower percentage of tantalum (82%), but it is used commercially because it limits the fluoride concentration in the resulting solid [43]. The KF that was used in this study precipitated 96% of tantalum, according to reaction 9. The obtained solid product of K_2TaF_7 may be further reduced to metallic tantalum, a commercial material [53].

$$H_2TaF_7 + 2KF \rightarrow K_2TaF_7 + 2HF \tag{9}$$

The separation of Pt and Ir was achieved through selective precipitation, as presented in Table 3. Platinum was precipitated by adding $NaBrO_3$ in three consecutive steps. After 87.5% of Pt was precipitated, and $NaBrO_3$ was added again as a fourth step, the solution was heated to 70 °C, and the pH value was adjusted to 8–9.

Table 3. Four consecutive steps of the precipitation process and its selectivity for separating Pt and Ir from the aqueous solution (synthetic sample *).

Precipitation Sequence	Metal Concentration (at the Beginning of Each Cycle, $\times 10^{-3}$ M)		%P
1st	$[Pt^{4+}]$	1.15	70%
	$[Ir^{3+}]$	0.13	<1%
2nd	$[Pt^{4+}]$	0.345	50%
	$[Ir^{3+}]$	0.13	<1%
3rd	$[Pt^{4+}]$	0.17	15%
	$[Ir^{3+}]$	0.13	<1%
4th	$[Pt^{4+}]$	0.144	<1%
	$[Ir^{3+}]$	0.013	90%

* Experimental conditions: 0.002 M $NaBrO_3$, time 30 min per step, pH = 1.4 for the 1st–3rd steps and 8.2 for the 4th, T = 25 ± 1 °C for the 1st–3rd steps and 70 ± 1 °C for the 4th.

2.4. Solids Characterization

According to Figure 7a, the XPS peaks of Au precipitate at 83.6 eV and 87.3 eV can be attributed to the dominant elemental form of Au, and the 84.9 eV and 89 eV peaks to the monovalent form [54]. The existence of Fe (peak 93.9 eV) suggests that further refining of this solid is required. Since the ratio Fe/Au was equal to 3.4, the Fe (62.5% w/w) that was obtained in the Au-rich (18.2% w/w) precipitate was attributed to the added reagent rather than the initial stainless-steel source, which was removed during the stripping step.

The platinum precipitate was rich in oxides and chlorides. According to Figure 7b, the peaks at 75.1 eV and 78.2 eV can be attributed to the dominant tetravalent form Pt(IV), while the peaks at 72.8 eV and 76.4 eV are attributed to the bivalent form Pt(II) [55]. Adding NaOH to adjust the pH of this sample during the precipitation process reduces Pt(IV) in the aqueous solution to Pt(II) because the alkaline solution acts as a reducing agent.

Similarly to platinum, the iridium precipitate is presented as trivalent and tetravalent forms and as oxides or chlorides. According to Figure 7c, the XPS pair of peaks at 62.4 eV/65.2 eV and 63.1 eV/65.9 eV can be attributed to Ir(IV) and Ir(III), respectively [56]. The overlapping peak at 63.7 eV resulted from the Na presence in the solid phase, which precipitated due to the alkaline conditions implemented during the Ir separation step and is considered an impurity.

The ratio Pt/Ir was found to be almost equal to 10, as in the initial sample of the catheter (i.e., 19.5% w/w Pt and 1.9% w/w Ir). Since both metals were in high valence forms, the chloride content was determined at 39.7% w/w. The high chloride content was also favored by Na precipitation (19% w/w). It is worth mentioning that none of the initial impurities (Fe, Cr, Ni, and Ti) was detected in the solid phase.

The resulting solid phase from the Ta precipitation, following the addition of KF and after the pH adjustment to 8.6, consisted mainly of the respective oxides and fluoride complexes (32.9% w/w Ta). The XPS pairs of peaks at 26.1 eV/28 eV and 27.1 eV/29 eV were attributed to the Ta(IV) and Ta(V) oxides, respectively (Figure 7a), while the peaks at 28.9 eV/30.7 eV were attributed to the Ta(V) fluoride [57]. The adjustment of pH during the tantalum precipitation limited the fluoride content in the resulting solid (17.9% w/w), thus reducing the requirement for further processing to be reused [43]. As before, none of the initial impurities (Fe, Cr, Ni and Ti) was detected in the solid phase.

2.5. Suggested Process Flowsheet

Based on the aforementioned optimized processes, a conceptual flow sheet for separating precious metals from the examined single-use medical devices (i.e., guide wire with Au coil, electrophysiology catheter with Ir/Pt markers, and self-expanding stent with Ta markers) is presented in Figure 8. Once the metals of interest are transferred into the aqueous phase, following the process previously described by [1], the metals can be firstly extracted by proper organic solvents, then stripped back to aqueous solutions and finally precipitated in mostly elemental forms, following the optimized conditions, as described in Figure 8.

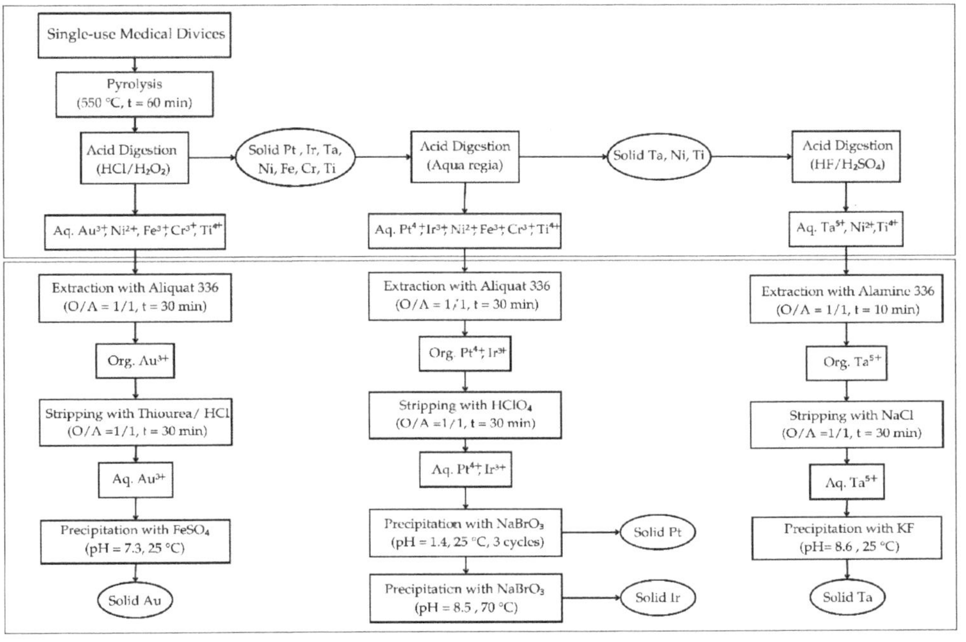

Figure 8. Proposed conceptual flow sheet and optimized conditions for separating the examined noble/precious metals (Au, Ir, Pt, Ta) from the spent single-use medical devices.

3. Materials and Methods

3.1. Reagents

The synthetic solutions were prepared by using 1000 mg/L AAS standard solutions (Thermo Scientific, Waltham, MA, USA) as follows: gold standard in 20% HCl, platinum standard in 20% HCl, iridium standard in 5% HCl, tantalum standard in 2% HF, iron standard in 2% HNO_3, nickel standard in 5% HNO_3, and chromium standard in 5% HCl. The desired concentration of metals and acidity was achieved by adjusting the solution with deionized water and concentrated HCl (37 wt.%, Sigma-Aldrich, Burlington, MA, USA).

The used extractants included n-hexane (95%, Thermo Scientific), toluene (99.5%, VWR Chemicals, Radnor, PA, USA), Aliquat 336 (methyl-tri octyl-ammonium chloride, $R_3R'N^+Cl^-$, [R = octyl/decyl, R′ = CH_3], 88.2–93%, Alfa Aesar, Karlsruhe, Germany), and Alamine 336 (tri-n-octyl/decyl amine, R_3N, 95%, BASF Co., Ludwigshafen, Germany). The various concentrations of commercial extractants Aliquat 336 and Alamine 336 were obtained using toluene as the diluting agent.

For the stripping of metals from the organic phase back into the aqueous phase, different reagents were examined, namely, thiourea (CH_4N_2S 99%, Laborchemie Apolda, Germany), $HClO_4$ (70%, Thermo Scientific), NaCl (99.5%, Sigma-Aldrich), HCl (37 wt.%, Sigma-Aldrich), H_2SO_4 (70%, Thermo Scientific), KOH (99%, Sigma-Aldrich), HNO_3 (70%, Thermo Scientific), and NH_4F (99.9%, Merck, Darmstadt, Germany). Appropriate dilutions were performed by using deionized water.

For the precipitation of respective metals, the following reagents were used: $FeSO_4$ $7H_2O$ (99%, Sigma-Aldrich), $NaBrO_3$ (99.5%, Merck), NH_4Cl (99.5%, Merck), KF (99.9%,

Merck), KCl (99.9%, Merck), NH_3 (32 wt.%, Sigma-Aldrich), and NaOH (99%, Sigma-Aldrich). Proper dilutions with deionized water yielded the desired concentrations.

3.2. Sample Preparation

The metals of interest were obtained by using the following commercially available single-use medical devices: (i) a diagnostic guide wire (ev3 Inc., Minneapolis, MN, USA. Nitrex 0.018″ × 180 cm, N180802) consisting of Au coating, (ii) a diagnostic electrophysiology catheter (Boston Scientific, Marlborough, MA, USA, Viking catheter, M00440000406Fr), consisting of Pt/Ir marker bands, and (iii) a self-expanding stent (MEDTRONIC VASCULAR, Minneapolis, Minnesota, U.S., Protégé^{TM} GPS^{TM} stent, SERP65-14-60-80), consisting of Ta dot markers. According to their manufacturers, the respective metal content of these samples was 79–90% for Au and Ta, 90% for Pt, and 10% for Ir. The non-marker metallic parts of the medical devices consisted of alloys, namely, stainless steel (Fe and Cr) and Nitinol (Ni and Ti). The samples were initially pyrolyzed to remove the polymer coatings. Then, a selective acid digestion process was applied to efficiently separate the metals of interest from the (residual) solid waste. The pyrolysis and digestion processes are the pre-treatment steps for the noble metals recovery from the aforementioned medical devices, optimized in our previously published relevant work [1]. More specifically, three solutions containing the metals of interest were obtained, i.e., (i) a stream rich in Au (42.2 mg/L), obtained by the acid leaching of samples with the use of a HCl/H_2O_2 at room temperature (25 ± 1 °C), (ii) a stream rich in Pt (1708.3 mg/L) and Ir (317 mg/L), obtained by the subsequent leaching of samples with hot (90 °C) aqua regia, and lastly, (iii) after the subsequent leaching of the sample with HF/H_2SO_4 solution, a stream rich in Ta (90 mg/L) was acquired. Due to residual metal alloys, the (i), (ii), and (iii) waste streams also contain the following respective metal impurities, as given in parentheses: Fe (220, 9.1 and 0 mg/L), Cr (51.6, 0.6 and 0 mg/L), Ni (37.6, 480 and 90 mg/L), and Ti (28, 403 and 90 mg/L).

3.3. Solvent Extraction

For cost-saving reasons (higher cost of real samples and lower availability), as well as their high purity, which made the reproduction of synthetic metal solutions of identical composition to real sample solutions possible, preliminary experiments using synthetic metal solutions were carried out to optimize the solvent extraction procedure. Synthetic single-metal solutions containing 50 mg/L of Au^{3+}, Ir^{3+}, Pt^{4+}, or Ta^{4+} were prepared. The concentration of metals in these solutions is based on the solutions' concentrations, resulting from the acid digestion of single-use medical devices, as determined in our previous work [1].

The extraction process investigated how to obtain the optimum conditions by testing the following parameters and extractant agents: toluene, hexane, Aliquat 336, and Alamine 336 concentrations in toluene (three different concentrations in the range of 0.1×10^{-3} to 0.65×10^{-3} M examined); organic/aqueous (O/A) phase volume ratio (O/A = 1/1, 2/1, 1/2); and the time that the O/A phases interacted, being in contact with the metal solutions (10, 30, 60 min). The optimum condition of each parameter is determined by keeping all the other parameters constant. The optimum value was the experimental condition with the highest extractant percentage. The extraction process was performed in 50 mL falcon polypropylene conical centrifuge tubes, rotated vertically at 45 rpm, using an LLG-uniLOOPMIX2 disc rotator. A fixed total volume of 50 mL was used. At the end of the extraction step, the two phases were separated using a separating funnel. All these extraction experiments were performed in a single run at room temperature (25 ± 1 °C).

The optimum conditions obtained from the experiments, utilizing the synthetic metal solutions, were applied subsequently to the metal solutions obtained from the examined commercially available single-use medical devices. The experimental procedure followed was the same as the aforementioned.

Selectivity Tests

The selectivity of the optimal extraction conditions for the case of Au investigated the Fe, Cr, Ti, and Ni metals, which were also present in the loaded solution obtained from the processed raw medical samples by the acid digestion procedure, as previously presented [1]. A simulated aqueous solution of 25 mL was used, containing 42.2 mg/L Au^{3+}, 220 mg/L Fe^{3+}, 51.8 mg/L Cr^{3+}, 37.6 mg/L Ni^{2+}, and 28 mg/L Ti^{4+} extracted with Aliquat 336 (0.23×10^{-3} M, O/A = 1/1, 30 min), following the previous experimental extraction procedure described for the actual medical samples.

Aliquat 336's selectivity for Au over Pt and Ir was also tested. A 25 mL aqueous solution containing 50 mg/L of all three metals was extracted with 1.04×10^{-3} M Aliquat 336, O/A = 1/1, for 30 min.

The selectivity of Alamine 336 for Ta over Ni and Ti was also tested. An aqueous solution of 25 mL containing 90 mg/L of all three metals was extracted with 0.97×10^{-3} M Alamine 336, O/A = 1/1, for 30 min.

The selectivity of Aliquat 336 for Pt over Ir was investigated. We used 25 mL of an aqueous solution, containing 50 mg/L of both Pt^{4+} and Ir^{2+}, extracted with 1.04×10^{-3} M Aliquat 336, O/A = 1/1, time = 30 min (organic solution ROS1).

In addition, the selectivity of Aliquat 336 for Pt and Ir over Fe, Cr, and Ni metals was studied by applying the following experimental conditions: 25 mL solution containing 1708.3 mg/L Pt^{4+}, 317 mg/L Ir^{2+}, 9.1 mg/L Fe^{3+}, 480 mg/L Ni^{2+}, 0.6 mg/L Cr^{6+}, and 405 mg/L Ti^{4+}, whose pH was adjusted to 0.5 with the addition of 4 M NaOH (as the initial pH was below 0.5), extracted with an organic solution of Aliquat 336 20.8×10^{-3} M, O/A = 1/1, for 30 min (organic solution ROS2).

3.4. Stripping

Stripping was employed to produce a liquor containing high concentrations of noble metals from the above-obtained extractant organic phase. The stripping process was investigated to obtain the optimum conditions by testing the following parameters: stripping agents thiourea, $HClO_4$, and NaCl; stripping concentrations: 2.7×10^{-3}–8.1×10^{-3} M of thiourea, 0.001–0.5 M of $HClO_4$, and 0.5–4 M of NaCl; organic/aqueous (O/A) phase volume ratio: O/A = 1/1, 2/1, and 1/2; and time that the O/A phases interacted: 10, 30, and 60 min. The optimum condition of each parameter was determined by keeping all the other parameters constant. The condition at which the highest stripping percentage was obtained is mentioned as the optimum. The optimum conditions were also applied to the metal solutions obtained from the examined commercially available single-use medical devices. All the experiments followed the same procedure, as described in Section 2.3.

Selective Stripping

As the organic solutions resulting from the solvent extraction of Pt and Ir (ROS1 and ROS2) were the only ones containing more than one metal, separating those metals was subsequently examined in the stripping phase.

The resulting organic solution (ROS1) was stripped with 0.002–0.1 M $HClO_4$, O/A = 1:1, and contact time = 10 min to examine the selective separation of Pt and Ir. Subsequently, the ability to separate selectively iridium from platinum was examined by stripping an organic solution of the two metals, applying the following conditions: 50 mg/L of both Pt^{4+} and Ir^{3+}, volume 25 mL, O/A = 1:1, and aqueous solution of 0.002 M $HClO_4$ in 3 cycles, with each extraction cycle lasting for 10 min.

The organic solution (ROS2) was stripped with 25 mL of 2 M $HClO_4$ solution, O/A = 1:1, for 30 min.

3.5. Precipitation

The aqueous solutions obtained from the stripping experiments underwent precipitation experiments. The impact of the precipitation agent, pH, and temperature (only for the case of Ir and Pt) was examined to determine the optimal precipitation conditions for each metal ion. The following agents were studied for the precipitation of Au: $FeSO_4$ (6.4×10^{-3} M), $NaBrO_3$, and NH_4Cl (1.3×10^{-3} M each). The pH gradually increased at various values in the range of 2.3–10 by adding 4 M NaOH stepwise, followed by 3 min conditioning. We used 15 mL of Au^{3+} aqueous solution, and the experiments took place in 100 mL glass beakers, whereas the pH value was monitored using a Mettler Toledo FiveGo portable pH meter. When a significant precipitate was visually observed, the test was terminated. The solution was subjected to settling until the pH was stable, indicating that the precipitation process was completed. Then, the slurry was vacuum filtered using MCE gridded membrane filters (pore size 0.45 μm, diameter 47 mm, Hawach Scientific, Xi'an, China), and the filtrate was collected after measuring the volume. The precipitates were washed with water and dried at 110 °C before weighing.

For the precipitation of Ir and Pt, 25–30 mL of the respective aqueous solution was placed into 100 mL beakers, and $NaBrO_3$ (12.5×10^{-3} M), and NH_4Cl (25×10^{-3} M) were examined as additional reagents; the pH value was adjusted to 8 with the addition of 4 M NaOH solution. Furthermore, the optimum precipitation temperature and pH value were examined by heating the Ir and Pt solution to 70 °C, using an RSLab-2C Heating Plate, after adding $NaBrO_3$. A 20 mL aqueous solution was used to study the precipitation of Ta, and the following three precipitation agents were examined: 1 M KF, 1 M KCl, and 4 M NH_3. The optimal pH value for all three agents was determined by adjusting it to the range of 8.5–9 by using 4 M NaOH. Teflon beakers were used when using KF.

The experimental conditions where the highest precipitation rates were obtained are considered optimum values. These optimum conditions are also applied to the resulting metal solutions from the examined commercially available single-use medical devices.

Selective Precipitation

As an initial sample, 50 mL of an aqueous solution containing 225 mg/L Pt^{4+} and 25 mg/L Ir^{3+} was used. These concentrations aimed to simulate the actual ratio of the respective metals in the marker bands (i.e., 90% Pt and 10% Ir). Moreover, $NaBrO_3$ (12.5×10^{-3} M) was added to precipitate platinum over iridium selectively. After 30 min, the solution was filtered, and the same procedure was repeated two more times until no further precipitation occurred. Afterward, the solution was heated to 70 °C, and the pH was adjusted to 7.5 with the addition of NaOH (4 M) to achieve the precipitation of iridium. Reagent $NaBrO_3$ was also necessary in this case, but no further addition was required due to its excess in the platinum-free solution (unreacted quantity).

3.6. Metals Determination

The metal ion concentrations of Au^{3+}, Ir^{3+}, Pt^{4+}, and Ta^{5+}, as well as of Fe^{3+}, Cr^{6+}, Ti^{4+}, and Ni^{2+} in the aqueous phase before and after extraction, stripping, and precipitation was measured by flame atomic absorption spectrophotometry, by using a Perkin–Elmer AAnalyst 800 instrument (Perkin-Elmer, Waltham, MA, USA). Their respective concentrations in the organic phase were calculated according to the respective mass balances. The calculation of the extraction (%E), stripping (%S), and precipitation (%P) percentages are based on the following equations:

$$\%\text{E or }\%\text{P} = \frac{[A]_{aq}^{i} - [A]_{aq}^{f}}{[A]_{aq}^{i}} \times 100\% \quad (10)$$

$$\%\text{S} = \frac{[A]_{aq}^{e}}{[A]_{org}^{i}} \times 100\% \quad (11)$$

where [A] is the A ion concentration, the subscripts *aq* and *org* refer to the aqueous and organic phases, and the superscripts *i* and *f* refer to the initial and final concentrations after the extraction, stripping, or precipitation procedures are completed. It must be noted, however, that the calculation of extraction (%E), stripping (%S), and precipitation (%P) percentages assumes that the mass balances are accurate and that all species involved are quantified correctly.

3.7. Solids Characterization

The dry solids produced from the selective precipitation tests of the actual samples (i.e., the examined commercially available single-use medical devices) were subjected to X-ray photoelectron spectroscopy (XPS) to detect the elemental composition of the produced solid samples and metals' valence. XPS measurements were performed using the Axis Ultra DLD system (Kratos Analytical Ltd., Wharfside, UK), which was equipped with monochromatic Al Ka radiation as the X-ray source. The passing energy was kept constant at 40 eV and calibrated for charge-induced shifts, considering the C1s peak (originating from the surface contamination of carbon) to be at 284.6 eV.

4. Conclusions

The results presented in this study encourage the application of recycling techniques in waste streams rich in precious/noble metals and set the basis for a complete method for separating and recovering those metals from the spent single-use medical devices, a treasure of high economic value. The proposed complete route goes through the application of initial pyrolysis and acid digestion, followed by solvent extraction, stripping, and precipitation. Pyrolysis was capable of enriching the samples with the targeted metals, though a small change in their oxidation states was observed. Acid digestion was fully able to successfully separate Au using a 50% *v*/*v* aqua regia solution for 30 min at room temperature and the Pt/Ir using concentrated aqua regia for 72 h under heating. Dissolution of Ta required a different leaching solution, i.e., a 50% *v*/*v* HF/H_2SO_4 mixture for 10 h under heating. Aliquat 336 in toluene (organic to aqueous ratio 1/1 for 30 min) was successful in extracting Au, Pt, and Ir, while Alamine 336 in toluene (organic to aqueous ratio 1/1 for 10 min) was successful in extracting Ta from the acid solutions. A thiourea and HCl solution (organic to aqueous ratio 1/1 for 30 min) was able to strip Au, a $HClO_4$ solution (organic to aqueous ratio 1/1 for 30 min) was able to strip Pt and Ir, and a NaCl solution (organic to aqueous ratio 1/1 for 10 min) was able to strip Ta from the organic liquids. Finally, the metals were successfully precipitated: $FeSO_4$ was used to precipitate Au and $NaBrO_3$ to precipitate Pt in three cycles, while the pH adjustment (8.5) of the same aqueous solution achieved the precipitation of Ir. Ta was precipitated with KF and pH adjustment (8.6). The total recovery of 88% for Au, 86% for Pt, 84% for Ir, and 80% for Ta can be obtained, achieving thus a high uptake of precious metals from the examined actual samples.

The complete method offers an alternative to landfill for medical waste, fulfilling the circular economy principles. The recovered precious metals can be recycled and reused. Additionally, further investigation can be performed to determine the capacity of used organic solvent regeneration for several subsequent extraction/stripping cycles, which would contribute to the sustainability of recycling and refining of the examined precious metals. The conceptual proposed flow sheet, as presented in this research, can be easily modified and used for other waste streams, considered valuable resources containing the examined metals, or can be expanded further to include more metals of high technological interest.

Supplementary Materials: The following supporting information can be downloaded at https://www.mdpi.com/article/10.3390/recycling9060118/s1, Table S1: Examination of Aliquat 336 selectivity in the extraction of Au over Fe, Cr, Ni, and Ti. Table S2: Examination of Aliquat 336 selectivity in extraction of Pt and Ir over Fe, Cr, Ni, and Ti. Figure S1: Effect of stripping agents on the percentage stripping of precious metals. (a) stripping of $Au^{#+}$. Experimental conditions: organic phase: $[Au^{3+}] = 0.27 \times 10^{-3}$ M in Aliquat 336, aqueous phases: [Thiourea] = 0.27×10^{-3} M, [HCl] = 0.5 M, [KOH] = 1 M. (b) Stripping of Pt^{4+}. Experimental conditions: organic phase: $[Pt^{4+}] = 0.25 \times 10^{-3}$ M in Aliquat 336, aqueous phases: [Thiourea] = 0.5×10^{-3} M, [HCl] = 0.5 M, $[H_2SO_4]$ = 0.5 M, $[HClO_4]$ = 0.01 M. (c) Stripping of Ir^{3+}. Experimental conditions: organic phase: $[Ir^{3+}] = 0.26 \times 10^{-3}$ M in Aliquat 336, aqueous phases: [Thiourea] = 0.52×10^{-3} M, [HCl] = 0.5 M, $[H_2SO_4]$ = 0.5 M, $[HClO_4]$ = 0.05 M. (d) Stripping of Ta^{5+}. Experimental conditions: organic phase: $[Ta^{5+}] = 0.34 \times 10^{-3}$ M in Alamine 336; aqueous phases: $[HNO_3]$ = 1 M, $[NH_4F]$ = 1 M, [NaCl] = 3 M; O/A = 1/1; O/A interaction/contact time = 30 min; temperature 25 ± 1 °C; rotation 45 rpm. Figure S2: Effect of stripping agent concentrations for the stripping of Au. (a) Effect of thiourea concentration. Experimental conditions: organic phase: $[Au^{3+}] = 0.27 \times 10^{-3}$ M; aqueous phase: [thiourea] = 0.27–0.81 $\times 10^{-3}$ M; O/A = 1/1; O/A interaction time = 30 min; temp. 25 ± 1 °C; 45 rpm. (b) Effect of HCl concentration. Experimental conditions: organic phase: $[Au^{3+}] = 0.27 \times 10^{-3}$ M; aqueous phases: [thiourea] = 0.27×10^{-3} M, [HCl] = 0–1 M, O/A = 1/1; O/A interaction time = 30 min; 25 ± 1 °C; 45 rpm. Figure S3: Effect of $HClO_4$ concentration in the stripping of platinum and iridium.Experimental conditions: organic phases: separate solutions $[Pt^{4+}] = 0.25 \times 10^{-3}$ M, $[Ir^{3+}] = 0.26 \times 10^{-3}$ M in Aliquat 336; aqueous phases: $[HClO_4]$ = 0.001–0.05 M; O/A = 1/1; O/A interaction time = 30 min; 25 ± 1 °C; 45 rpm. Figure S4. Effect of NaCl concentration for the stripping of tantalum. Experimental conditions: organic phase: $[Ta^{5+}] = 0.34 \times 10^{-3}$ M in Alamine 336; aqueous phases: $[HClO_4]$ = 0.5–4 M; O/A = 1/1; O/A interaction time = 30 min; 25 ± 1 °C; 45 rpm. Figure S5: Effect of volume phase ratio O/A on the percentage stripping of examined precious metals. Experimental conditions: organic phase: single-metal solutions, containing 50 mg/L Au^{3+}, Ir^{3+}, Pt^{4+} in Aliquat 336, and Ta^{5+} in Alamine 336; aqueous phase; [Thiourea] = 0.135–0.54 $\times 10^{-3}$ M, [HCl] = 0.5–2 M for Au^{3+}, $[HClO_4]$ = 0.005–0.02 M for Pt^{4+}, $[HClO_4]$ = 0.01–0.04 M for Ir^{3+}, [NaCl] = 3 M for Ta^{5+}, Total volume = 30 mL, O/A interaction time = 30 min. 25 ± 1 °C; 45 rpm. Figure S6: Effect of time on the stripping procedure of examined precious metals. Experimental conditions: organic phase: single-metal solutions containing 50 mg/L Au^{3+}, Ir^{3+}, Pt^{4+} in Aliquat 336, and Ta^{5+} in Alamine 336; aqueous phase: [Thiourea] =0.27×10^{-3} M for Au^3, $[HClO_4]$ = 0.01 M for Pt^{4+}, $[HClO_4]$ = 0.05 M for Ir^{3+}, [NaCl] = 3 M for Ta^{5+}; O/A = 1/1; 25 ± 1 °C; 45 rpm. Figure S7: Effect of pH on the percentage precipitation of Au^{3+} with and without the addition of ferrous sulphate. Experimental conditions: $[Au^{3+}] = 0.27 \times 10^{-3}$ M, $[FeSO_4] = 0.77 \times 10^{-3}$ M, time = 30 min, 25 ± 1 °C. Table S3: Optimization conditions for the precipitation of platinum. Table S4: Optimization conditions for the precipitation of iridium. Table S5: Optimization conditions for the precipitation of tantalum.

Author Contributions: Conceptualization, A.X., E.K. and C.P.; methodology, A.L.; validation, A.Z. investigation, A.L. and T.T.; resources, A.Z.; data curation, A.L., E.K. and C.P.; writing—original draft preparation, A.L. and C.P.; writing—review and editing, E.K., C.P., A.Z. and E.P.; visualization, A.L. and C.P.; supervision, A.Z.; project administration, E.P.; funding acquisition, A.Z. All authors have read and agreed to the published version of the manuscript.

Funding: The project "Collection, processing, and metallurgical recovery of critical raw materials (Au, Pt, Ir, Ta) from discarded medical material" (Project code: KMP6-0084436) was implemented under the framework of the Action "Investment Plans of Innovation" of the Operational Program "Central Macedonia 2014 2020", which is co-funded by the European Regional Development Fund in a partnership agreement with Greece.

Data Availability Statement: Data are contained within the article.

Conflicts of Interest: The authors declare no conflicts of interest.

References

1. Kokkinos, E.; Prochaska, C.; Lampou, A.; Peleka, E.; Simeonidis, K.; Vourlias, G.; Zouboulis, A. Recovery of Noble Metals (Au, Pt, Ir, and Ta) from Spent Single-Use Medical–Technological Products. *Minerals* **2024**, *14*, 90. [CrossRef]
2. Xing, W.D.; Lee, M.S. A Process for the Separation of Noble Metals from HCl Liquor Containing Gold(III), Palladium(II), Platinum(IV), Rhodium(III), and Iridium(IV) by Solvent Extraction. *Processes* **2019**, *7*, 243. [CrossRef]
3. Kokkinos, E.; Merachtsaki, D.; Lampou, A.; Prochaska, C.; Peleka, E.; Simeonidis, K.; Vourlias, G.; Zouboulis, A. Investigating the Recovery of Noble Metals from Single-Use Medical Technology-Specific Waste Streams. *Mater. Proc.* **2023**, *15*, 15027. [CrossRef]
4. Lindagato, P.; Li, Y.; Yang, G. Save the giants: Demand beyond production capacity of tantalum raw materials. *Miner. Econ.* **2023**, *36*, 535–541. [CrossRef]
5. Niinomi, M. *Metals for Biomedical Devices*; Elsevier: Amsterdam, The Netherlands, 2019. [CrossRef]
6. Cowley, A.; Woodward, B. A Healthy Future: Platinum in Medical Applications. *Platin. Met. Rev.* **2011**, *55*, 98–107. [CrossRef]
7. Nzulumike, A.N.; Thormann, E. Fibrin Adsorption on Cardiovascular Biomaterials and Medical Devices. *ACS Appl. Bio. Mater.* **2023**, *6*, 2667–2676. [CrossRef]
8. Park, J.-S.; Yim, K.H.; Jeong, S.; Lee, D.H.; Kim, D.G. A Novel High-Visibility Radiopaque Tantalum Marker for Biliary Self-Expandable Metal Stents. *Gut Liver* **2019**, *13*, 366–372. [CrossRef]
9. Kapoor, A.; Vora, A.; Nataraj, G.; Mishra, S.; Kerkar, P.; Manjunath, C. Guidance on reuse of cardio-vascular catheters and devices in India: A consensus document. *Indian Hear. J.* **2017**, *69*, 357–363. [CrossRef]

10. Gilden, D.J.; Scissors, K.N.; Reuler, J.B. Disposable products in the hospital waste stream. *West J. Med.* **1992**, *156*, 269–272. Available online: http://europepmc.org/abstract/MED/1595242 (accessed on 30 July 2024).
11. Hagelüken, C.; Lee-Shin, J.U.; Carpentier, A.; Heron, C. The EU Circular Economy and Its Relevance to Metal Recycling. *Recycling* **2016**, *1*, 242–253. [CrossRef]
12. Fan, C.; Quan, K.; Han, Z.; Han, F.; Li, Z.; Liu, J.; Liu, X. Recovery and Purification of Iridium from Secondary Resources: A Review. *J. Sustain. Met.* **2023**, *9*, 909–926. [CrossRef]
13. Xing, W.D.; Sohn, S.H.; Lee, M.S. A Review on the Recovery of Noble Metals from Anode Slimes. *Miner. Process. Extr. Met. Rev.* **2020**, *41*, 130–143. [CrossRef]
14. Padamata, S.K.; Yasinskiy, A.S.; Polyakov, P.V.; Pavlov, E.A.; Varyukhin, D.Y. Recovery of Noble Metals from Spent Catalysts: A Review. *Met. Mater. Trans. B* **2020**, *51*, 2413–2435. [CrossRef]
15. Yakoumis, I.; Panou, M.; Moschovi, A.M.; Panias, D. Recovery of platinum group metals from spent automotive catalysts: A review. *Clean. Eng. Technol.* **2021**, *3*, 100112. [CrossRef]
16. Cui, J.; Zhang, L. Metallurgical recovery of metals from electronic waste: A review. *J. Hazard. Mater.* **2008**, *158*, 228–256. [CrossRef]
17. Ding, Y.; Zhang, S.; Liu, B.; Li, B. Integrated process for recycling copper anode slime from electronic waste smelting. *J. Clean. Prod.* **2017**, *165*, 48–56. [CrossRef]
18. Guimarães, R.; Carvalho, J.; Leal, V.; Dias, A.J.G. Characterization, Treatment Proposal and Metal Recovery in Waste of Active Implantable Medical Devices Caracterização, Proposta de Tratamento e Recuperação de Metais dos Resíduos dos Dispositivos Médicos de Implante Ativo. 2014. Available online: http://www.lneg.pt/iedt/unidades/16/paginas/26/30/185 (accessed on 1 August 2024).
19. Boussuge-Roze, J.; Boveda, S.; Mahida, S.; Anic, A.; Conte, G.; Chun, J.K.R.; Marijon, E.; Sacher, F.; Jais, P. Current practices and expectations to reduce environmental impact of electrophysiology catheters: Results from an EHRA/LIRYC European physician survey. *Europace* **2022**, *24*, 1300–1306. [CrossRef]
20. Manda, Y.R.; Baradhi, K.M. Cardiac Catheterization Risks and Complications. StatPearls, Jun. 2023. Available online: https://www.ncbi.nlm.nih.gov/books/NBK531461/ (accessed on 23 July 2024).
21. Chioncel, V.; Andrei, C.L.; Brezeanu, R.; Sinescu, C.; Avram, A.; Tatu, A.L. Some Perspectives on Hypersensitivity to Coronary Stents. *Int. J. Gen. Med.* **2021**, *14*, 4327. [CrossRef]
22. The Global Language of Business. GS1 Healthcare Strategy 2023–2027. 2020. Available online: https://www.gs1.org/docs/healthcare/Strategy/GS1-Healthcare-Strategy-Final.pdf (accessed on 23 July 2024).
23. Dong, H.; Zhao, J.; Chen, J.; Wu, Y.; Li, B. Recovery of platinum group metals from spent catalysts: A review. *Int. J. Miner. Process.* **2015**, *145*, 108–113. [CrossRef]
24. Oeqvist, L.S.; Bourgeois, F. State of the Art on the Recovery of Refractory Metals from Secondary Sources, MSP-REFRAM Project, 2016MSP-REFRAM. pp. 1–104. Available online: https://prometia.eu/wp-content/uploads/2020/12/MSP-REFRAM-D3.2-State-of-the-art-on-the-recovery-of-refractory-metals-from-secondary-resources.pdf (accessed on 2 August 2024).
25. Syed, S. Recovery of gold from secondary sources—A review. *Hydrometallurgy* **2011**, *115–116*, 30–51. [CrossRef]
26. Wei, W.; Cho, C.-W.; Kim, S.; Song, M.-H.; Kwame Bediako, J.; Yun, Y.-S. Selective recovery of Au(III), Pt(IV), and Pd(II) from aqueous solutions by liquid–liquid extraction using ionic liquid Aliquat-336. *J. Mol. Liq.* **2016**, *216*, 18–24. [CrossRef]
27. Geoffroy, N.; Cardarelli, F. A method for leaching or dissolving gold from ores or precious metal scrap. *JOM* **2005**, *57*, 47–50. [CrossRef]
28. Ubaldini, S.; Fornari, P.; Massidda, R.; Abbruzzese, C. An innovative thiourea gold leaching process. *Hydrometallurgy* **1998**, *48*, 113–124. [CrossRef]
29. Cyganowski, P.; Garbera, K.; Leśniewicz, A.; Wolska, J.; Pohl, P.; Jermakowicz-Bartkowiak, D. The recovery of gold from the aqua regia leachate of electronic parts using a core–shell type anion exchange resin. *J. Saudi Chem. Soc.* **2017**, *21*, 741–750. [CrossRef]
30. Witt, K.; Urbaniak, W.; Kaczorowska, M.A.; Bożejewicz, D. Simultaneous Recovery of Precious and Heavy Metal Ions from Waste Electrical and Electronic Equipment (WEEE) Using Polymer Films Containing Cyphos IL 101. *Polymers* **2021**, *13*, 1454. [CrossRef]
31. Schreier, G.; Edtmaier, C. Separation of Ir, Pd and Rh from secondary Pt scrap by precipitation and calcination. *Hydrometallurgy* **2003**, *68*, 69–75. [CrossRef]
32. Sun, P.P.; Lee, M.S. Recovery of Platinum from Chloride Leaching Solution of Spent Catalysts by Solvent Extraction. *Mater. Trans.* **2013**, *54*, 74–80. [CrossRef]
33. Gupta, B.; Singh, I.; Mahandra, H. Extraction and separation studies on Pt(IV), Ir(III) and Rh(III) using sulphur containing extractant. *Sep. Purif. Technol.* **2014**, *132*, 102–109. [CrossRef]
34. Nguyen, T.H.; Lee, M.S. Effect of HCl Concentration on the Oxidation of LIX 63 and the Subsequent Separation of Pd(II), Pt(IV), Ir(IV) and Rh(III) by Solvent Extraction. *Korean J. Met. Mater.* **2016**, *54*, 768–774. [CrossRef]
35. Swain, B.; Jeong, J.; Kim, S.-K.; Lee, J.-C. Separation of platinum and palladium from chloride solution by solvent extraction using Alamine 300. *Hydrometallurgy* **2010**, *104*, 1–7. [CrossRef]
36. Bernardis, F.L.; Grant, R.A.; Sherrington, D.C. A review of methods of separation of the platinum-group metals through their chloro-complexes. *React. Funct. Polym.* **2005**, *65*, 205–217. [CrossRef]
37. Phetla, T.; Muzenda, E.; Belaid, M. A study of the variables in the optimisation of a platinum precipitation process. *World Acad. Sci. Eng. Technol. Int. J. Chem. Mol. Eng.* **2010**, *4*, 573–579.

38. Chen, W.-S.; Ho, H.-J.; Lin, K.-Y. Hydrometallurgical Process for Tantalum Recovery from Epoxy-Coated Solid Electrolyte Tantalum Capacitors. *Materials* **2019**, *12*, 1220. [CrossRef] [PubMed]
39. Nguyen, T.H.; Lee, M.S. A Review on the Separation of Niobium and Tantalum by Solvent Extraction. *Miner. Process. Extr. Met. Rev.* **2019**, *40*, 265–277. [CrossRef]
40. Rodríguez, O.; Alguacil, F.J.; Baquero, E.E.; García-Díaz, I.; Fernández, P.; Sotillo, B.; López, F.A. Recovery of niobium and tantalum by solvent extraction from Sn–Ta–Nb mining tailings. *RSC Adv.* **2020**, *10*, 21406–21412. [CrossRef]
41. Yan, H.; Tang, D. Technical Features and Research Progress of Separating Impurities in Producing Tantalum (Niobium) Oxide by Traditional Technology. *Int. J. Miner. Process. Extr. Met.* **2018**, *3*, 29. [CrossRef]
42. Nguyen, V.T.; Riaño, S.; Binnemans, K. Separation of precious metals by split-anion extraction using water-saturated ionic liquids. *Green Chem.* **2020**, *22*, 8375–8388. [CrossRef]
43. Shikika, A.; Sethurajan, M.; Muvundja, F.; Mugumaoderha, M.; Gaydardzhiev, S. A review on extractive metallurgy of tantalum and niobium. *Hydrometallurgy* **2020**, *198*, 105496. [CrossRef]
44. Gamov, G.A. Complexation of Gold(I) and Gold(III) in solutions. *Coord. Chem. Rev.* **2024**, *520*, 216162. [CrossRef]
45. Kirchnerová, J.; Purdy, W.C. The mechanism of the electrochemical oxidation of thiourea. *Anal. Chim. Acta* **1981**, *123*, 83–95. [CrossRef]
46. Ray, D.A.; Baniasadi, M.; Graves, J.E.; Greenwood, A.; Farnaud, S. Thiourea Leaching: An Update on a Sustainable Approach for Gold Recovery from E-waste. *J. Sustain. Met.* **2022**, *8*, 597–612. [CrossRef]
47. Cleare, M.J.; Charlesworth, P.; Bryson, D.J. Solvent extraction in platinum group metal processing. *J. Chem. Technol. Biotechnol.* **2007**, *29*, 210–224. [CrossRef]
48. Nguyen, T.H.; Sonu, C.H.; Lee, M.S. Separation of Pt(IV), Pd(II), Rh(III) and Ir(IV) from concentrated hydrochloric acid solutions by solvent extraction. *Hydrometallurgy* **2016**, *164*, 71–77. [CrossRef]
49. Djafari, J.; Fernández-Lodeiro, A.; García-Lojo, D.; Fernández-Lodeiro, J.; Rodríguez-González, B.; Pastoriza-Santos, I.; Pérez-Juste, J.; Capelo, J.L.; Lodeiro, C. Iron(II) as a Green Reducing Agent in Gold Nanoparticle Synthesis. *ACS Sustain. Chem. Eng.* **2019**, *7*, 8295–8302. [CrossRef]
50. Mironov, I.V.; Makotchenko, E.V. The Hydrolysis of AuCl 4—And the Stability of Aquachlorohydroxocomplexes of Gold(III) in Aqueous Solution. *J. Solut. Chem.* **2009**, *38*, 725–737. [CrossRef]
51. Sheng, P.P.; Etsell, T.H. Recovery of gold from computer circuit board scrap using aqua regia. *Waste Manag. Res. J. Sustain. Circ. Econ.* **2007**, *25*, 380–383. [CrossRef]
52. Hubert, P.W.; Gerhardus, O.; Ferguson, D.K. Separation and Purification of Iridium. U.S. Patent 3876747A, 8 April 1975.
53. Kolosov, V.N.; Orlov, V.M.; Miroshnichenko, M.N.; Prokhorova, T.Y. Preparation of tantalum powders via the sodium reduction of potassium heptafluorotantalate heat-treated in air. *Inorg. Mater.* **2015**, *51*, 116–121. [CrossRef]
54. Hepperle, P.; Herman, A.; Khanbabaee, B.; Baek, W.Y.; Nettelbeck, H.; Rabus, H. XPS Examination of the Chemical Composition of PEGMUA-Coated Gold Nanoparticles. *Part. Part. Syst. Charact.* **2022**, *39*, 2200070. [CrossRef]
55. Romanchenko, A.; Likhatski, M.; Mikhlin, Y. X-ray Photoelectron Spectroscopy (XPS) Study of the Products Formed on Sulfide Minerals upon the Interaction with Aqueous Platinum (IV) Chloride Complexes. *Minerals* **2018**, *8*, 578. [CrossRef]
56. Pfeifer, V.; Jones, T.E.; Vélez, J.J.V.; Arrigo, R.; Piccinin, S.; Hävecker, M.; Knop-Gericke, A.; Schlögl, R. In situ observation of reactive oxygen species forming on oxygen-evolving iridium surfaces. *Chem. Sci.* **2017**, *8*, 2143–2149. [CrossRef]
57. Simpson, R.; White, R.G.; Watts, J.F.; Baker, M.A. XPS investigation of monatomic and cluster argon ion sputtering of tantalum pentoxide. *Appl. Surf. Sci.* **2017**, *405*, 79–87. [CrossRef]

recycling

MDPI

Article

Technospheric Mining of Cobalt and Nickel from Waste Nickel Furnace Slag by Ascorbic Acid-Assisted Citric Acid Leaching

K. Yamini [1], Laurence Dyer [1], Bona Lim [2] and Richard Diaz Alorro [2,*]

1 Western Australian School of Mines: Minerals, Energy, and Chemical Engineering, Faculty of Science and Engineering, Curtin University, Kalgoorlie, WA 6430, Australia; yamini.kannappan@student.curtin.edu.au (K.Y.); laurence.dyer@curtin.edu.au (L.D.)

2 Western Australian School of Mines: Minerals, Energy, and Chemical Engineering, Faculty of Science and Engineering, Curtin University, Bentley, WA 6102, Australia; bona.lim@curtin.edu.au

* Correspondence: richard.alorro@curtin.edu.au; Tel.: +61-(08)-9266-4527

Abstract: With the growing demand for critical metals, extraction from secondary sources such as slag, tailings, and end-of-life materials has become inevitable. Processing from such secondary sources is described as technospheric mining. Technospheric mining is a broad term for extracting valuable resources from anthropogenic materials that are currently excluded from the material flow. The study utilises technospheric mining to extract cobalt and nickel from nickel furnace slag using organic acids such as ascorbic and citric acid. The experiments were designed using one variable at a time (OVAT) to optimise the different parameters: temperature, time, particle size, and reagent concentration. A maximum recovery of 79% Co and 80% Ni were obtained by leaching the nickel furnace slag using 0.5 mol/L citric acid and 0.5 mol/L ascorbic acid for 6 h at 80 °C. It is proposed that citric acid leaches the surface and ascorbic acid acts as a reducing agent, thereby reducing the slag matrix and leaching the metals trapped in it. The results show that treating nickel furnace slag via a mixture of organic acids is promising, as it is environmentally friendly. Retreating this material would reduce the net waste generated and aid in building a circular economy.

Keywords: technospheric mining; nickel furnace slag; organic acid leaching; cobalt; critical metals

1. Introduction

Clean technologies, such as the electrification of vehicles, are being developed to create a low-carbon society, driving the demand for critical and strategic materials [1]. These vehicles employ certain critical and strategic materials like cobalt (Co) and nickel (Ni) as cathodes in their batteries, further increasing demand for these elements. Besides the battery industry, cobalt is used to make catalysts, dyes, pigments, and magnets and is employed in metallurgical applications such as superalloys for gas turbine engines and cemented carbides [2,3]. Nickel is a major alloying element for stainless steel, shape memory alloys, nickel-plating, and coins, and it is used as a catalyst in the hydrogenation of vegetable oils. With the increasing applications of these elements in diverse fields, there is an expected increase in demand in the upcoming years.

In nature, cobalt exists in the form of sulphide (Linnaeite, $Co^{+2}Co^{+3}{}_2S_4$ and Carrollite, $CuCo_2S_4$), arsenide (Skutterudite, $CoAs_3$), or a combination of both (Cobaltite, CoAsS) due to its chalcophile and siderophile properties [4]. Nickel usually occurs as sulphides, arsenides, silicates, and oxides, of which nearly 60% are found as laterites [5]. Majorly

https://doi.org/10.3390/recycling10020043

exploited nickel-bearing minerals include pentlandite (Fe, Ni)$_9S_8$, millerite NiS, gersdorffite NiAsS, and nickeliferous limonite $FeO(OH){\cdot}nH_2O$ [6].

The primary cobalt supply in mining and processing is geographically concentrated in the Democratic Republic of Congo (DRC) [2]. Mines in the DRC account for more than 70% of the global cobalt production, extracting orders of magnitude more than China [7]. Despite DRC being a major supplier of cobalt to the international market, its state is characterised by high governance risks. About 15 to 20% of cobalt mining in the DRC comes from artisanal mining, which appears problematic for human rights and child labour [8]. Moreover, cobalt is usually mined as a by-product of copper. This, combined with the socio-political instability and heavy geographical concentration [1], can lead to an imbalance in the supply-demand chain of cobalt [9]. This has led to a significant fear of causing a global supply shortage [2,10].

In the case of nickel, vast applications for stainless steel and other nickel-based alloys drive the demand. Albeit laterites constitute around 60% of the land-based nickel reserve, most nickel is extracted from sulphide ores. The global nickel reserves are unevenly concentrated, with Indonesia, Russia and the Philippines representing the bulk of nickel produced from natural minerals [11]. Indonesia produces over 60% of the global nickel supply [12]. The nickel prices have thus been highly dependent on the Indonesian market [12]. Certain aspects that cause price fluctuation include natural disasters, like earthquakes, typhoons and changes in government regulations [12]. One example would be the export ban on nickel ore from Indonesia, which has been an issue since 2014 [12,13]. These geopolitical risks and price fluctuations have posed continued challenges to existing operations and in bringing new operations online [11–13]. With nickel being classified as a critical and strategic mineral by several countries, challenges to meet global demand are well established.

In order to maintain a continuous supply of cobalt and nickel to the global markets and avoid any bottleneck situation, an alternate source for producing cobalt and nickel should be identified. Processing these elements from secondary sources such as slags, tailings, spent batteries, and discarded wastes would be an alternative [14]. Mining or processing valuable resources from a secondary source will provide access to a larger reserve volume than the estimated value from the primary sources [14,15]. The overarching term for extracting metals from secondary sources is termed technospheric mining. The technosphere is proposed as a new component of the environment along with the atmosphere, biosphere, hydrosphere, and lithosphere [16]. The technosphere includes material stocks generated by human activity that are currently excluded from material flow [16]. Technospheric mining describes extracting and recovering valuable resources from the technospheric stocks [16,17].

Johansson et al. identified that recovering metals from these stocks has received greater attention owing to environmental factors [17]. Extracting from technospheric stocks contributes to sustainable development, increases the production base to meet growing demand and diversifies supply chains in times of global environmental problems and a shortage of natural resources [14,18]. Furthermore, processing a waste stream would reduce the volume of net waste generated.

Considering the potential risks in meeting the supply chain and the technospheric streams containing significant proportions of nickel and cobalt, assessing secondary sources for viable production routes is attractive [14,16]. Conventional production of nickel is achieved through beneficiation followed by either hydrometallurgical or combined pyro- and hydrometallurgical extraction routes. Smelting is the primary technology for treating nickel sulphide concentrates, involving an oxidative step to reduce sulphur content and reject iron, resulting in a high-nickel content matte. This is then further processed to recover nickel metal or other high-value components. The primary solid waste streams generated from nickel smelting are slags composed of a solution of iron oxides and silicates along with a significant amount of Ni and Co that are rejected from the process. Even though the desired metal contents in the slag are sometimes low, the volume of waste generated, along with the growing emphasis on the circular economy, highlights the importance of processing mine waste for metal recovery. It is to be noted that an estimated 6–16 tonnes of slag are being discharged in the production of 1 tonne of nickel [19,20].

The main focus of this work is to return cobalt and nickel inventory lost to nickel furnace slag by retreating that material. Moreover, extracting from slag aligns with the principles of circular economy and contributes towards sustainability.

A few studies have been reported for extracting cobalt and nickel from slag, a technospheric stock, using pyrometallurgy, hydrometallurgy, or a combination of both and are reported below.

Yang et al. [21] utilised a matte-slag mixture from the ISAMELT furnace to recover cobalt and copper. They achieved a cobalt recovery of 98% with 2% coke and pyrite at 1260 °C for 2 h [21]. Altundogan and Tumen [22] studied the roasting of copper converter slag with $Fe_2(SO_4)_3$ and then subjected it to water and sulphuric acid leaching. A comparatively higher recovery percentage of 69%, 98%, 57%, and 93% for cobalt, copper, nickel, and zinc was achieved using sulphuric acid, respectively [22]. Altundogan et al. [23] investigated the recovery of Cu, Co, Zn, and Fe from copper converter slag obtained by oxidative leaching with $K_2Cr_2O_7$ and H_2SO_4 lixiviants. The experimental results indicated that the presence of dichromate greatly influences the extraction of metals. The process recovered 81.15% Cu, 12.0% Co, 3.15% Fe, and 10.27% Zn [23]. Anand, Das and Jena's studies showed that using 10% furnace oil as the reducing agent yielded 80% cobalt, 82% copper, and 95% nickel by roasting the copper converting slag at 850 °C and leaching with 1.25 times the stoichiometric amount of ferric chloride for 2 h [24]. Numerous studies reported that the recovery of cobalt and nickel from nickel converter slag using pressure oxidative acid leaching with H_2SO_4 is promising [25–27]. They achieved a cobalt and nickel recovery of greater than 97%, with temperatures ranging from 200–250 °C in 80–120 min [25–27]. Meshram et al. used oxalic and citric acid to recover valuable metals from copper converter slags. They achieved a Co and Ni recovery of 94% and 89% using citric acid and a maximum recovery of 16% cobalt using oxalic acid [5]. Lim et al. (2023) used citric acid and hydrogen peroxide to recover Co and Ni from nickel furnace and converter slag. They achieved a recovery of 82% Ni and 91.7% Co from nickel furnace slag and 65% Ni and 99.1% Co from converter slag using the following experimental conditions: 1M citric acid, 0.5% (v/v) hydrogen peroxide, a pulp density of 2.5%, a particle size f −100 +75 µm, 400 rpm at 60 °C for 6 h [28].

Although relatively high recoveries are achieved using the above methods, it is seen that most of the techniques mentioned above utilise high energy and harsh chemicals. Also, very limited studies have been conducted to extract valuable metals using nickel slag. The study conducted by Meshram et al. (2020) [5] and Lim et al. (2023) [16] has shown promise in recovering these elements from slags using organic acids. Therefore, this study aims to investigate the potential of organic acids to recover cobalt and nickel from nickel furnace slag by studying the effect of concentration, temperature, time, and particle size.

A series of experiments were conducted by employing citric acid and ascorbic acid as leaching reagents. A description of the experimental procedure and the results obtained are discussed below.

2. Results and Discussion

2.1. Sample Characterisation

The particle size distribution indicated that the p_{80} of the slag was 34 µm, which was used for experimentation unless otherwise specified. The elemental composition of the slag was determined using XRF, and is presented in Table 1. For conciseness, only the elements of interest are presented. Major impurities include Si, Al and Mg.

Table 1. Elemental composition of the slag.

Elements	Fe	Ni	Co	Si
Composition (%)	40.7	0.6	0.2	15.9

The ground sample was analysed using XRD to identify the mineralogy of the slag. Figure 1 shows the XRD analysis of the slag, and it is observed that it consists of fayalite (Fe_2SiO_4) and magnetite (Fe_3O_4). It is to be noted that the cobalt, nickel, and any other phases were not detected. It was reported by Jones et al. [29] that Co and Ni could be present either on the surface or entrained in the silicate and oxide matrix as metals or as sulphides [29]. SEM analysis of the sample also shows that silicon, iron, and oxygen are distributed throughout the slag matrix, which comprises the silicate and oxide phases (Figure 2). Previous studies on nickel slag have shown that the nickel was trapped in the matte left behind in the slag [27,28] and that Co exists as $CoFe_2O_4$ and is associated with fayalite [26,28]. TIMA analysis of the nickel furnace slag revealed that it primarily consists of fayalite, with some magnetite present [28]. Additionally, it was reported that a trace amount of godlevskite $(Ni,Fe)_9S_8$ was detected in the furnace slag [28].

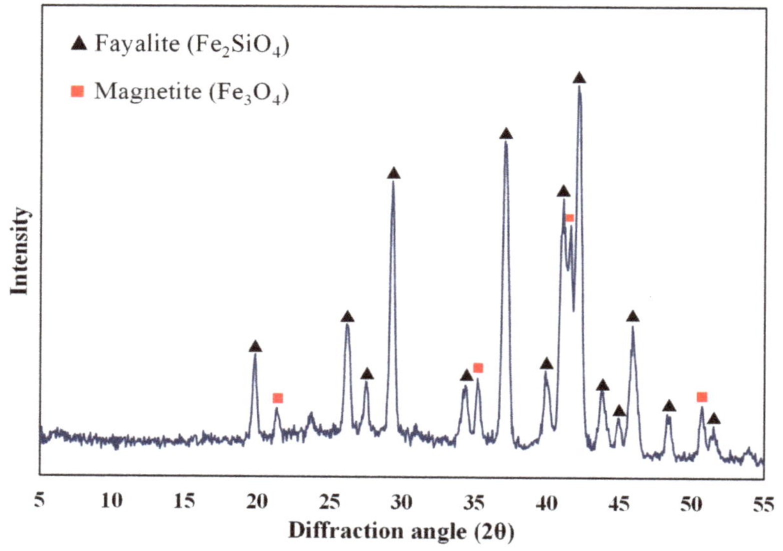

Figure 1. XRD analysis of nickel furnace slag.

2.2. Citric Acid Leaching of Nickel Furnace Slag

Studies carried out by Lim et al. (2023) have shown that 0.5 mol/L CA performed well in extracting Co and Ni from nickel slag [28]. Therefore, experiments were conducted to evaluate the effectiveness of leaching Co and Ni from nickel furnace slag using 0.5 mol/L CA at 50 °C and 80 °C for 6 h. The results are presented in Figure 3.

Figure 3 shows that a 60% Co, 55% Ni, and 66% Fe recovery was achieved by maintaining a temperature of 50 °C. From the data presented above, it can be seen that temperature has a considerable effect on leaching, as a recovery of 69% Co, 67% Ni, and 76% Fe was obtained at 80 °C over 6 h. Also, citric acid tends to be a strong leaching reagent for the chosen sample. Further studies were carried out by adding a strong reducing agent, ascorbic acid.

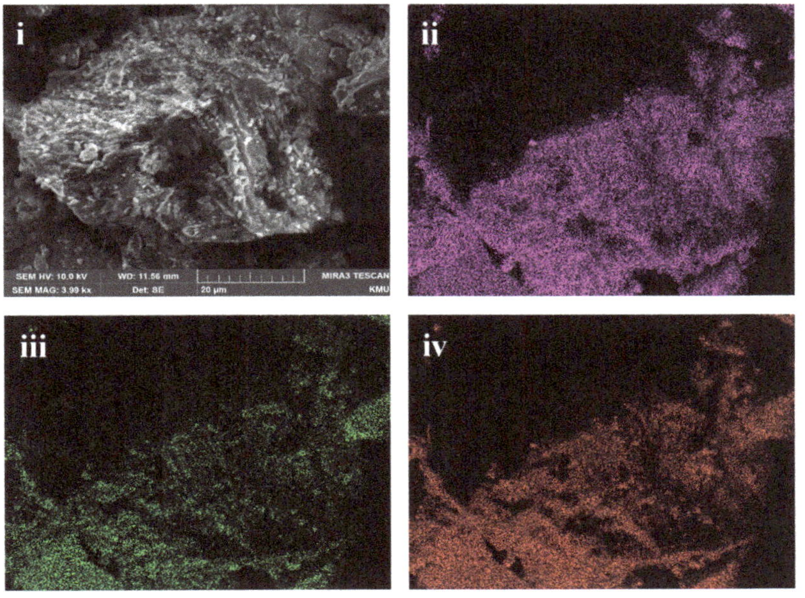

Figure 2. Scanning electron microscope image of nickel furnace slag (**i**) Secondary electron image; (**ii**) Silicon mapping (**iii**) Iron mapping (**iv**) Oxygen mapping using SEM-EDS.

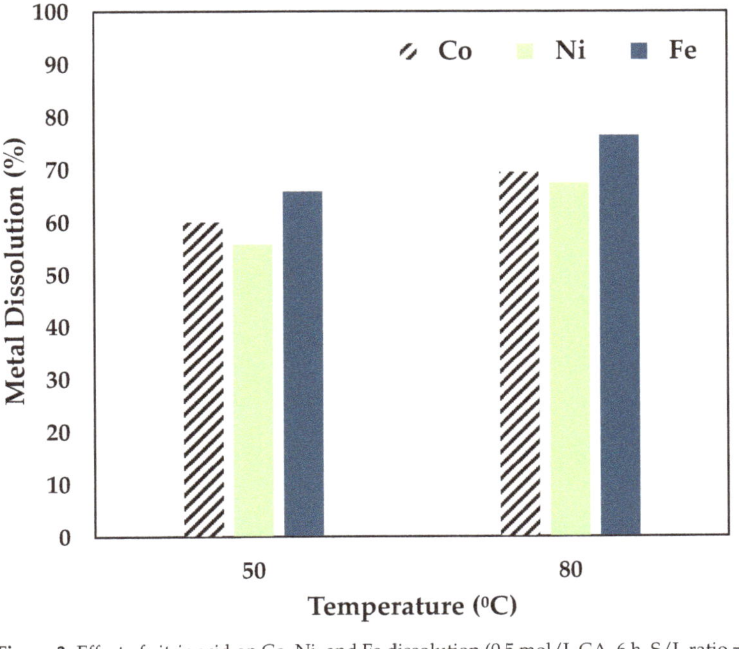

Figure 3. Effect of citric acid on Co, Ni, and Fe dissolution (0.5 mol/L CA, 6 h, S/L ratio = 1:10, 34 μm particles, 250 rpm).

2.3. Ascorbic Acid-Assisted Citric Acid Leaching of Nickel Furnace Slag

2.3.1. Effect of Ascorbic Acid Concentration on Metal Dissolution

The effect of ascorbic acid concentration was investigated by leaching the slag at different temperatures for 6 h using 0.5 mol/L CA with varying concentrations of AA. The results obtained are shown in Figure 4.

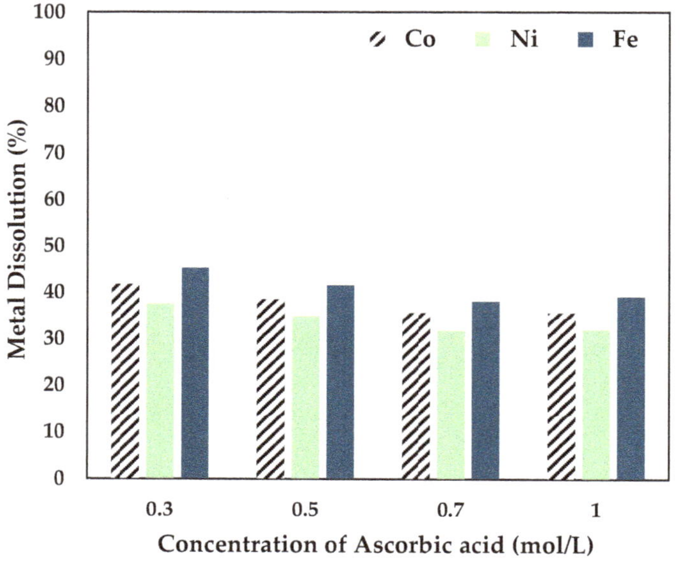

(**a**) 25 °C

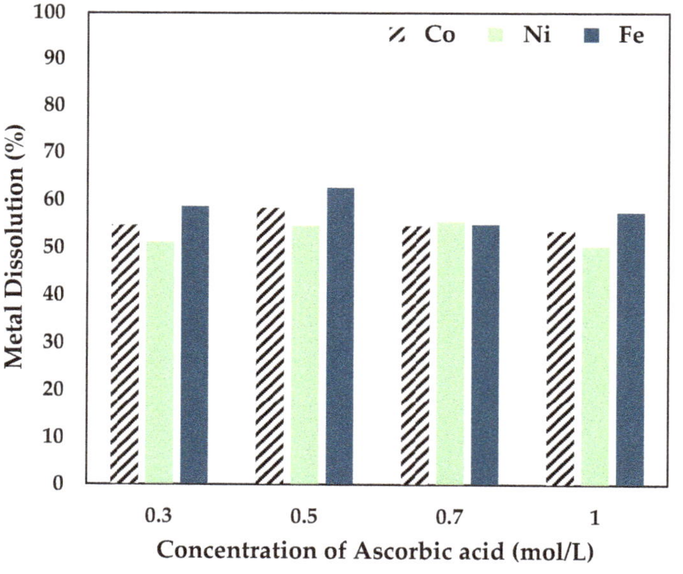

(**b**) 50 °C

Figure 4. *Cont.*

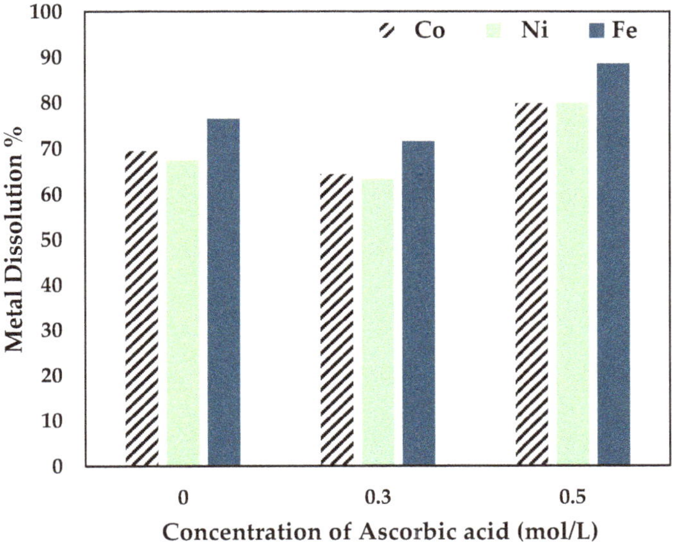

(**c**) 80 °C

Figure 4. Effect of ascorbic acid concentration on Co, Ni, and Fe dissolution at (**a**) 25 °C (**b**) 50 °C (**c**) 80 °C (0.5 mol/L CA, 6 h, S/L ratio = 1:10, particle size = 34 microns, stirring speed = 250 rpm).

At 25 °C, the recovery percentage of Co, Ni, and Fe dropped with the increase in the concentration of AA added. However, at 50 °C, the recovery percentage increased slightly at 0.5 mol/L compared to 0.3 mol/L AA. It is to be noted that the dissolution percentage for 0.7 mol/L and 1 mol/L AA concentrations was lower than the other concentrations at different temperatures. Therefore, at 80 °C, experiments were carried out only up to 0.5 mol/L AA. It was seen that the recovery percentage increased significantly when 0.5 mol/L AA was added to 0.5 mol/L CA. The drop in dissolution percentage at higher concentrations of ascorbic acid is potentially due to the increased ascorbic/citric acid ratio, thereby decreasing the activity and thus the efficacy of the citric acid. Therefore, it was concluded that the optimal combination was 0.5 mol/L AA + 0.5 mol/L CA.

2.3.2. Effect of Temperature and Time on Metal Dissolution

The effect of temperature and time was investigated by leaching the slag at different temperatures and times using 0.5 mol/L CA + 0.5 mol/L AA. The experiments were carried out at 25 °C, 50 °C, and 80 °C for 24 h, and the results obtained are shown in Figure 5.

With the variation of AA concentration, the temperature played a significant role. For 0.5 mol/L AA, the Co and Ni recovery increased for 25 °C and 50 °C till 24 h. However, at 80 °C, Co, Ni, and Fe recovery dropped significantly at 24 h. This shows that a mixture of 0.5 mol/L AA and 0.5 mol/L CA is suitable to operate at 80 °C for 6 h. As seen in Figure 3, with higher concentrations of AA (0.7 mol/L and 1 mol/L), the leaching efficiency of Co and Ni was relatively lesser than 0.3 mol/L and 0.5 mol/L. For 80 °C, the recovery rate increased until 6 h, and the value decreased. This is possibly due to the denaturing tendency of ascorbic acid, which worsens at 76 °C [30]. Also, Jutkus et al. observed that storing ascorbic acid at higher temperatures led to chemical instability over time [31]. Therefore, it is recommended to either run the experiment at a lower temperature or to run the experiment for a smaller duration. Based on the observations, the following operating conditions were chosen for further experimentation: 80 °C, 0.5 mol/L AA + 0.5 mol/L CA, 6 h, and 250 rpm.

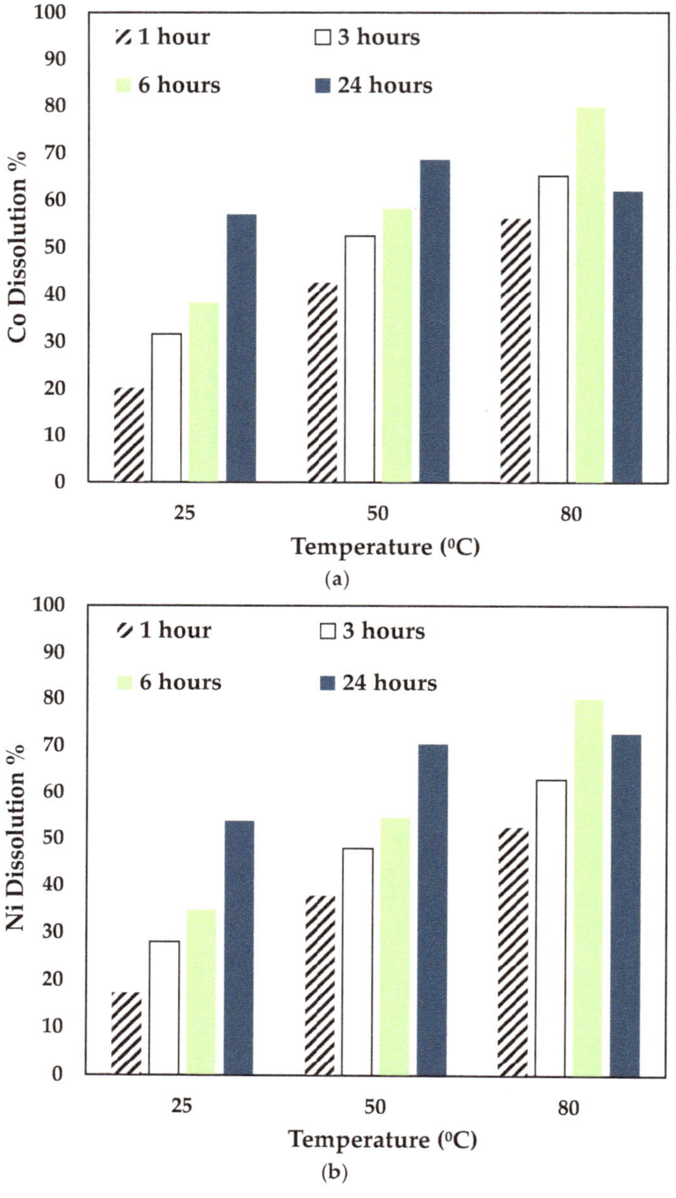

Figure 5. *Cont.*

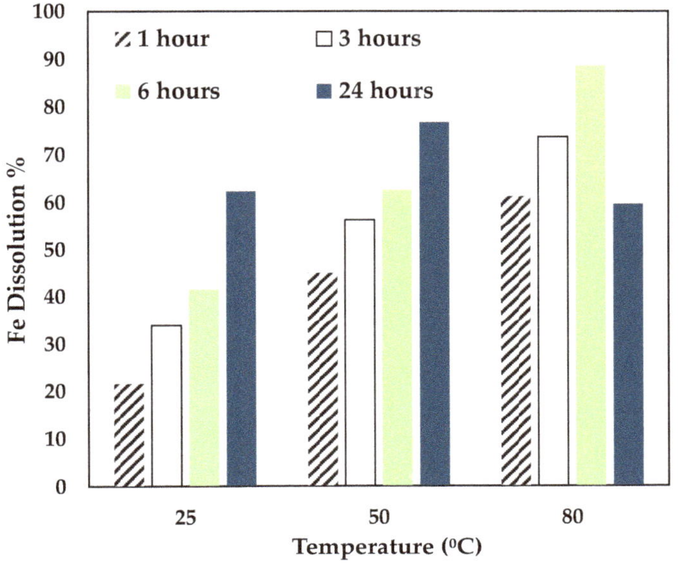

(c)

Figure 5. Effect of temperature on (**a**) Co, (**b**) Ni, and (**c**) Fe dissolution (0.5 mol/L CA + 0.5 mol/L AA, S/L ratio= 1:10, particle size = 34 microns, stirring speed = 250 rpm).

2.3.3. Effect of Particle Size on Metal Recovery

The effect of particle size on the leaching efficiency of Ni and Co was studied by using milled and unmilled particles with a p_{80} of 75 μm and 34 μm, respectively. The experimental conditions were: 80 °C, 0.5 mol/L AA + 0.5 mol/L CA, 250 rpm for 6 h. The results obtained are shown in Figure 6.

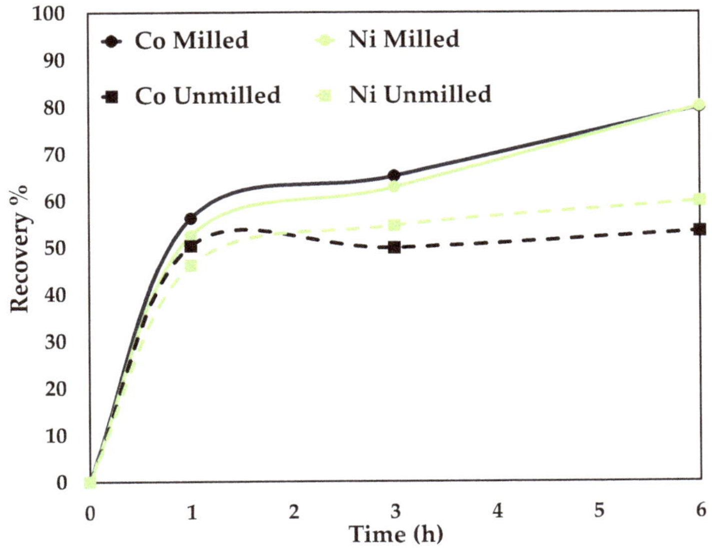

Figure 6. Effect of particle size on Co and Ni dissolution (80 °C, 0.5 mol/L CA + 0.5 mol/L AA, S/L ratio = 1:10, stirring speed = 250 rpm).

A recovery percentage of 53% Co, 59% Ni, and 79.8% Co, 80% Ni was attained for unmilled (p_{80} = 75 microns) and milled (p_{80} = 34 microns), respectively. An optimal particle size screens the unwanted minerals and ensures the liberation of the target elements [32]. Mechanical activation was employed to reduce particle size. The disintegration by high-energy grinding increases the surface area and alters the physicochemical properties, rendering them more reactive in the subsequent processing [33]. The higher recovery percentage is due to the liberation of the target elements [32] and the increased number of particles newly exposed to the leaching conditions [33].

2.4. Leaching Mechanism of the Synergistic System

To understand the effect of the combined reagents, graphs were plotted individually for 0.5 mol/L AA, 0.5 mol/L CA, and 0.5 mol/L AA + 0.5 mol/L CA at 80 °C. The graphs were plotted for Co, Ni, and Fe recoveries under the above-said conditions, shown in Figure 7.

Figure 7 proves that the synergistic system works better than the individual systems, i.e., only citric and ascorbic acid. It is seen from the above graphs that cobalt, nickel, and iron follow the same trend for the three different systems. This shows that it is essential to dissolve Fe by disintegrating the fayalite and magnetite matrix to ensure Co and Ni dissolution [28]. On treating the slag with either citric or ascorbic acid, it is seen that the leaching curve tends to plateau after 3 h. However, with a combination of these reagents, the recovery percentage increases till 6 h. The proposed mechanism for the synergistic system is as follows.

Citric acid provides H^+ ions for protonation, which attacks the fayalite matrix, releasing ferrous ions and silicic acid. The H^+ ions also facilitate the dissolution of magnetite, a mineral containing both Fe^{2+} and Fe^{3+} ions, thereby releasing ferric and ferrous ions into the solution.

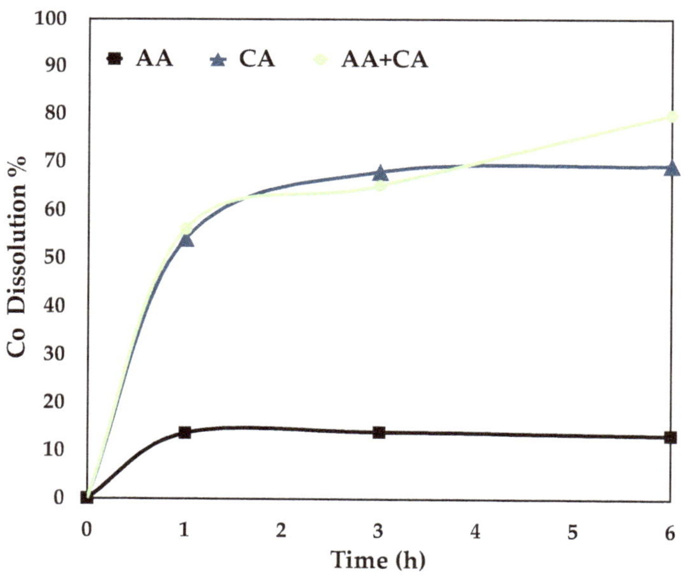

(a)

Figure 7. *Cont.*

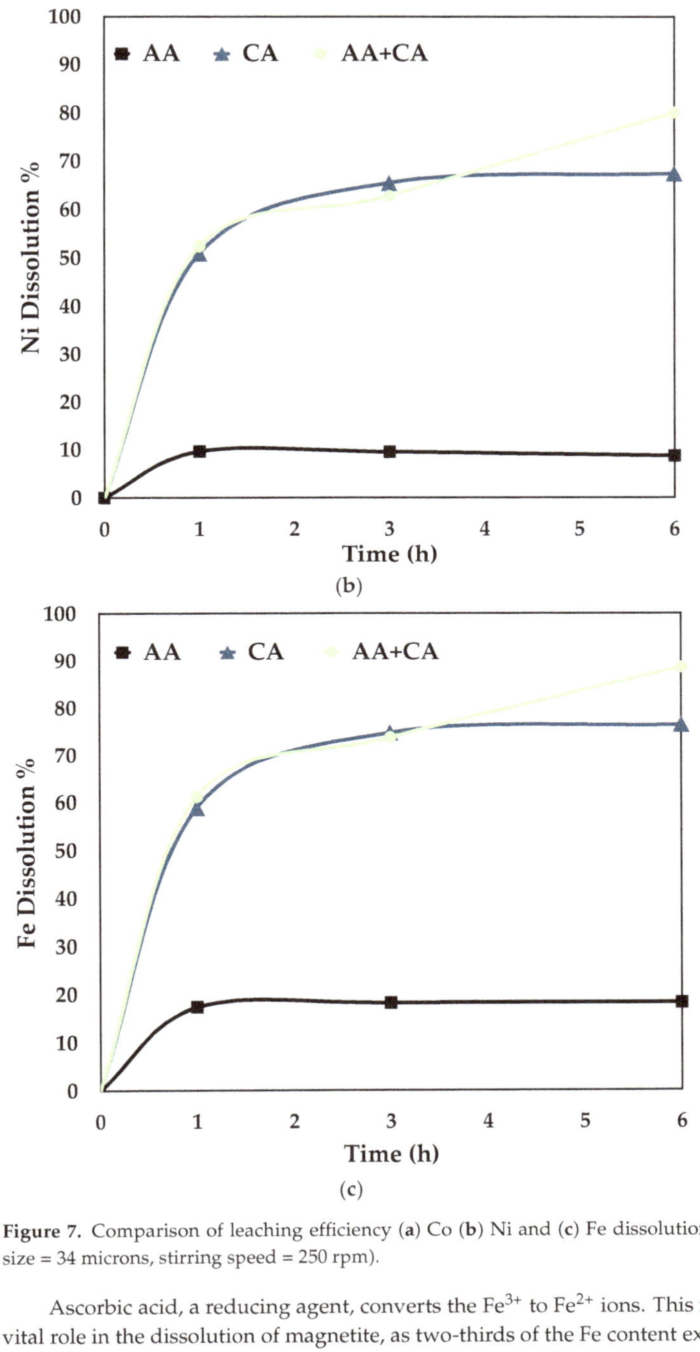

(b)

(c)

Figure 7. Comparison of leaching efficiency (**a**) Co (**b**) Ni and (**c**) Fe dissolution (80 °C, particle size = 34 microns, stirring speed = 250 rpm).

Ascorbic acid, a reducing agent, converts the Fe^{3+} to Fe^{2+} ions. This reaction plays a vital role in the dissolution of magnetite, as two-thirds of the Fe content exists in the ferric state. Furthermore, ascorbic acid forms complexes with silica, promoting the breakdown of the fayalite matrix [34]. In addition, citric acid also acts as a chelating agent and forms stable complexes with both Fe^{2+} and Fe^{3+} ions.

It is evident from Figure 7 that the dissolution of iron is essential for the release of cobalt (Co) and nickel (Ni). Thus, promoting the dissolution of fayalite and magnetite matrices is necessary for efficient leaching of Co and Ni. According to Meshram et al. [5] citric acid forms complexes with both Co and Ni ions. This supports the proposed mechanism's applicability for the extraction of these metals. The reaction equations for the mechanism are summed up in Equations (1)–(6) [5].

$$Fe_2SiO_4 + 4H^+ \rightarrow 2Fe^{2+} + H_4SiO_4 \quad (1)$$

$$Fe_3O_4 + 8H^+ \rightarrow 2Fe^{3+} + Fe^{2+} + 4H_2O \quad (2)$$

$$Fe^{3+} + C_6H_8O_6 \rightarrow Fe^{2+} + C_6H_6O_6 + 2H^+ \quad (3)$$

$$Fe^{3+} + C_6H_8O_7 \rightarrow [Fe(C_6H_5O_7)]^n \quad (4)$$

$$Fe^{2+} + C_6H_8O_7 \rightarrow [Fe(C_6H_5O_7)]^n \quad (5)$$

The metals (M) reacted with citric acid to form soluble metal chelates, and the reaction was expressed as Equation (6) [5].

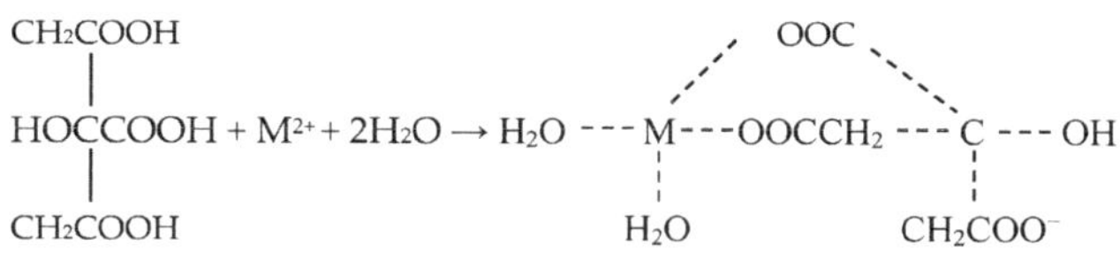

(6)

Overall, it is evident that reducing the slag matrix plays a vital role in efficiently extracting Co and Ni from slag. The combined actions of citric acid and ascorbic acid ensure the breakdown of the mineral structure and the leaching of the target metal ions.

This might be because citric acid, a strong leaching reagent, leaches the cobalt and nickel on the surface and converts them into their corresponding metal salts [5]. Meanwhile, ascorbic acid, a strong reducing agent, causes a disturbance in the slag matrix and reduces it [35]. This helps in leaching the cobalt and nickel trapped in the slag matrix. The increased iron recovery is due to the reduction of the slag matrix. Ascorbic acid also tends to reduce the metal citrate complexes. It is evident from the results that reducing the slag matrix is vital for extracting Co and Ni efficiently from the slag.

However, the decreased recovery after 6 h can be attributed to a number of reasons. The chemical instability of ascorbic acid at higher temperatures plays a major role [30,31]; this denaturing tendency of ascorbic acid causes the reprecipitation of some of the iron into the system. This, in turn, decreases the cobalt and nickel recovery as they tend to follow the same trend as iron (Figure 7). In addition, the system loses ascorbic acid due to its instability, which tends to decrease the reaction rate. It is proposed that iron reprecipitation occurs much faster than forming iron ascorbate, thereby reducing the recovery of iron, cobalt, and nickel. Based on these observations, it is said that the optimal operating conditions for this system are 6 h at 80 °C, which also helps prevent the decomposition of AA. However, further investigations are required to evaluate the extent of AA decomposition, its impact on reagent recyclability, and the overall feasibility of this approach for achieving more sustainable circular processing.

The above process not only treats waste but also provides better utilisation of resources, providing the potential to generate multiple products and thus contributing to a circular economy. The leach liquor obtained can be further processed through solvent extraction to separate Fe, Co, and Ni; however, this approach has not been tested with the specific solution obtained from this study. Given the significant concentration of Fe in the leach liquor, there is an opportunity to create a low-value iron stream that could be used for other applications. By extracting these valuable elements from furnace slag, this method enhances resource efficiency which is beneficial for nickel-producing industries globally. Additionally, it not only aids in reducing waste but also has the potential to generate an additional revenue stream.

3. Materials and Methods

3.1. Sample Preparation and Characterisation

The nickel furnace slag used in this study was obtained from a processing plant in Western Australia. The furnace slag was initially pulverised using a Rocklabs Limited Model C pulveriser. The slag was then mechanically activated using a Restch PM type 100 planetary ball mill, which had a capacity of 50 cm^3. A charge-to-ball ratio of 1:10 was employed. Then, the ground sample was analysed using X-ray diffraction (XRD) (Pan Analytical, Worcestershire, UK) to identify the mineralogy of the slag. Also, Scanning Electron Microscopy–Energy Dispersive Spectroscopy (SEM-EDS) was carried out to understand the elemental distribution of the sample using an MIRA3 Tescan (Brno, Czech Republic) at John de Laeter Centre in Curtin University. The elemental composition of the ground sample was determined using X-ray fluorescence (XRF) and was conducted by Bureau Veritas Laboratory, Perth. The particle size distribution of the slag was carried out using laser sizing (Mastersizer 3000, Malvern, Worcestershire, UK). The concentration of the elements in the dissolved solution was determined using an Inductively Coupled Plasma–Optical Emission Spectroscopy (Agilent 5100 Synchronous Vertical View-SVDV, Boulder, CO, USA).

3.2. Reagent Preparation

The leaching reagents used in this study were 100% pure L-ascorbic acid ($C_6H_8O_6$) (Daintree Quality Herbs, USA) and 100% anhydrous citric acid ($C_6H_8O_7$) (Sigma Group Companies Pty Ltd., Perth, Australia) in a solid crystalline form. Based on the solubility of these reagents in water, the concentrations of the reagents were fixed. The individual solutions were prepared with known concentrations of these reagents in deionised water. A previous study using nickel slag indicated that 0.5 mol/L citric acid as a leaching reagent was effective at elevated temperatures [28]. Therefore, in this study, further experiments were conducted to explore the potential of the synergistic system. Thus, different ascorbic acid concentrations ranging from 0.3–1 mol/L ascorbic acid were added to 0.5 mol/L citric acid. In the upcoming sections, ascorbic and citric acid have been abbreviated as AA and CA, respectively, for ease.

3.3. Leaching Experiments

Batch leaching experiments were conducted using a three-necked round-bottom flask. The flask was fitted with a condenser on one neck and a thermometer on the other to monitor the temperature of the slurry throughout the experiment. An overhead stirrer with a collapsible impeller was fitted to ensure mixing and was set to rotate at a fixed speed of 250 rpm. The effect of variables influencing the leaching process, such as temperature, reagent concentration, particle size, and time was investigated. The reagent was taken in the three-necked round-bottom flask and placed in a heating mantle to achieve and maintain the desired reaction temperature during leaching. Once the desired temperature was attained, an aliquot of slag was weighed and added to the flask. A solid-to-liquid ratio of 1:10 was maintained. The nickel furnace slag was subjected to leaching using a mixture of ascorbic and citric acid. Solution samples were collected at 1, 3, 6, and 24 h. The pH of the solution samples were measured and recorded. The collected samples were centrifuged and vacuum filtered using a Wattman filter paper of 7–11 microns. At the end of the experiment, solid-liquid separation was done using a vacuum filter and the leach residue was washed with distilled water, which yielded a dark red-violet filtrate and a black residue. The resultant dark red solution is due to the formation of ferrous ascorbate.

4. Conclusions

In this study, experiments were carried out to determine if nickel slag is a viable source for recovering cobalt and nickel from nickel furnace slag in an environmentally benign manner using organic acids such as ascorbic and citric acid. The parameters varied were temperature, the concentration of the reagents used, time, and particle size. It was observed that the recovery of Co and Ni generally increased with temperature. Interestingly, the dissolution percentage of Co and Ni decreased at longer residence times at 80 °C and while using 0.5 mol/L of ascorbic acid and citric acid. This is plausibly due to the denaturing tendency of ascorbic acid at 76 °C, as reported by Jutkus et al. (2015) [31]. In addition, altering the physicochemical properties of the slag by mechanically activating it aided in significantly increasing the leaching efficiency of Co and Ni. A combination of the two reagents at specific concentrations attained a much higher recovery than their individual counterparts. This is because the citric acid provides H^+ ions that attack the fayalite and magnetite matrices, releasing metal ions into the solution. Ascorbic acid aids in reducing these ions, while citric acid complexes with the released metal ions to enhance dissolution. Optimal recovery of 79% Co and 80% Ni was obtained by subjecting the nickel furnace slag of p_{80} 34 μm to 0.5 mol/L citric acid and 0.5 mol/L ascorbic acid for 6 h at 80 °C. This result shows that breaking down the Fe matrix to dissolve Co and Ni is essential, as these elements are trapped in the slag matrix. The outcomes of this work highlight the potential of organic acids as effective leaching reagents owing to their capability to attain high recoveries from a low-grade material under atmospheric conditions. Moreover, extracting critical and strategic minerals from slag adds more value to the supply chain of these elements. It also enables us to reprocess the material and reduce the net waste generated, thereby contributing to sustainability and aiding in a circular economy. Additionally, a comprehensive evaluation of the process considering the supply chain of both the waste products and reagents is to be done to enhance the overall circularity of the process.

Author Contributions: Conceptualization, K.Y., R.D.A. and L.D.; Data curation, K.Y.; Formal analysis, K.Y.; Investigation, K.Y.; Methodology, K.Y. and B.L; Project administration, R.D.A.; Supervision, R.D.A. and L.D.; Visualization, K.Y., R.D.A. and L.D.; Writing—original draft, K.Y.; Writing—review & editing, R.D.A., L.D. and B.L. All authors have read and agreed to the published version of the manuscript.

Funding: This research received no external funding.

Data Availability Statement: Data available upon request.

Acknowledgments: The authors thank the Western Australian School of Mines for supporting this research and Mujesira Vukancic and Lahiru Basnayaka for supplying the lab equipment and ICP analysis.

Conflicts of Interest: The authors declare no conflict of interest.

References

1. Watari, T.; Nansai, K.; Nakajima, K. Review of Critical Metal Dynamics to 2050 for 48 Elements. *Resour. Conserv. Recycl.* **2020**, *155*, 104669. [CrossRef]
2. Beatty, D.; Fu, X.; Bustamante, M.; Gaustad, G.; Babbitt, C.; Kirchain, R.; Roth, R.; Olivetti, E. Cobalt Criticality and Availability in the Wake of Increased Electric Vehicle Demand: A Short-Term Scenario Analysis. In *Minerals, Metals and Materials Series*; Springer International Publishing: Cham, Switzerland, 2019; pp. 355–357.

3. Ober, J.A. *Geological Survey Mineral Commodity Summaries 2018*; U.S. Geological Survey: Reston, VA, USA, 2018. [CrossRef]
4. Kovacheva-Ninova, V.K.; Savov, G.M.; Vassileva, V.; Vutova, K.; Petrov, E.; Petrov, D. Trends in the Development of Cobalt Production. *Electrotech. Electron.* **2018**, *53*, 84–94.
5. Meshram, P.; Prakash, U.; Bhagat, L.; Abhilash; Zhao, H.; van Hullebusch, E.D. Processing of Waste Copper Converter Slag Using Organic Acids for Extraction of Copper, Nickel, and Cobalt. *Minerals* **2020**, *10*, 290. [CrossRef]
6. Water Resources Division; U.S. Geological Survey. *Geological Survey Mineral Commodity Summaries 2015*; U.S. Geological Survey: Reston, VA, USA, 2015. [CrossRef]
7. *US. Geological Survey. Geological Survey Mineral Commodity Summaries 2022*; U.S. Geological Survey: Reston, VA, USA, 2022. [CrossRef]
8. van den Brink, S.; Kleijn, R.; Sprecher, B.; Tukker, A. Identifying Supply Risks by Mapping the Cobalt Supply Chain. *Resour. Conserv. Recycl.* **2020**, *156*, 104743. [CrossRef]
9. Mudd, G.M.; Weng, Z.; Jowitt, S.M.; Turnbull, I.D.; Graedel, T.E. Quantifying the Recoverable Resources of By-Product Metals: The Case of Cobalt. *Ore Geol. Rev.* **2013**, *55*, 87–98. [CrossRef]
10. Shedd, K.B.; Mccullough, E.A.; Bleiwas, D.I. Global Trends Affecting the Supply Security of Cobalt. *Min. Eng.* **2017**, *69*, 37–42.
11. Zheng, S.; Zhou, X.; Zhao, P.; Xing, W.; Han, Y.; Hao, H.; Luo, W. Impact of Countries' Role on Trade Prices from a Nickel Chain Perspective: Based on Complex Network and Panel Regression Analysis. *Resour. Policy* **2022**, *78*, 102930. [CrossRef]
12. Lim, B.; Kim, H.S.; Park, J. Implicit Interpretation of Indonesian Export Bans on Lme Nickel Prices: Evidence from the Announcement Effect. *Risks* **2021**, *9*, 93. [CrossRef]
13. Wang, X.Q.; Wu, T.; Zhong, H.; Su, C.W. Bubble Behaviors in Nickel Price: What Roles Do Geopolitical Risk and Speculation Play? *Resour. Policy* **2023**, *83*, 103707. [CrossRef]
14. Yamini, K.; Dyer, L. Extraction of Rare Earth Elements from Low-Grade Ore and Process Waste Stream. In Proceedings of the 26th World Mining Congress, Brisbane, Australia, 26 July 2023; pp. 619–629.
15. Arndt, N.T.; Fontboté, L.; Hedenquist, J.W.; Kesler, S.E.; Thompson, J.F.H.; Wood, D.G. Future Global Mineral Resources. *Geochem. Perspect.* **2017**, *6*, 1–171. [CrossRef]
16. Lim, B.; Alorro, R.D. Technospheric Mining of Mine Wastes: A Review of Applications and Challenges. *Sustain. Chem.* **2021**, *2*, 686–706. [CrossRef]
17. Johansson, N.; Krook, J.; Eklund, M.; Berglund, B. An Integrated Review of Concepts and Initiatives for Mining the Technosphere: Towards a New Taxonomy. *J. Clean. Prod.* **2013**, *55*, 35–44. [CrossRef]
18. Krook, J.; Baas, L. Getting Serious about Mining the Technosphere: A Review of Recent Landfill Mining and Urban Mining Research. *J. Clean. Prod.* **2013**, *55*, 1–9. [CrossRef]
19. Wu, Q.; Wu, Y.; Tong, W.; Ma, H. Utilization of Nickel Slag as Raw Material in the Production of Portland Cement for Road Construction. *Constr. Build. Mater.* **2018**, *193*, 426–434. [CrossRef]
20. Zhang, G.; Wang, N.; Chen, M.; Cheng, Y. Recycling Nickel Slag by Aluminum Dross: Iron-Extraction and Secondary Slag Stabilization. *ISIJ Int.* **2020**, *60*, 602–609. [CrossRef]
21. Yang, X.; Zhang, J.; Zhang, J.; Hu, J.; Li, J.; Zhang, L.; Chen, Y.; Wang, C. Efficient Recovery of Copper and Cobalt from the Matte–Slag Mixture of ISA Furnace by Injection of Coke and Pyrite. *Metall. Mater. Trans. B Process Metall. Mater. Process. Sci.* **2018**, *49*, 3118–3126. [CrossRef]
22. Altundogan, H.S.; Tumen, F. Metal Recovery from Copper Converter Roasting with Ferric Sulphate. *Hydrometallurgy* **1997**, *44*, 261–267. [CrossRef]
23. Altundogan, H.S.; Boyrazli, M.; Tumen, F. A Study on the Sulphuric Acid Leaching of Copper Converter Slag in the Presence of Dichromate. *Miner. Eng.* **2004**, *17*, 465–467. [CrossRef]
24. Anand, S.; Das, R.P.; Jena, P.K. Reduction-Roasting and Ferric Chloride Leaching of Copper Converter Slag for Extracting Copper, Nickel and Cobalt Values. *Hydrometallurgy* **1981**, *7*, 243–252. [CrossRef]
25. Liao, Y.L.; Shi, G.C.; Huang, F.R.; Zhang, Y. Recovery of the Valuable Metals from Complex Converter Slag at Elevated Temperature with Sulfuric Acid Solution. *J. Min. Metall. Sect. B Metall.* **2019**, *55*, 359–370. [CrossRef]
26. Huang, F.; Liao, Y.; Zhou, J.; Wang, Y.; Li, H. Selective Recovery of Valuable Metals from Nickel Converter Slag at Elevated Temperature with Sulfuric Acid Solution. *Sep. Purif. Technol.* **2015**, *156*, 572–581. [CrossRef]
27. Li, Y.; Perederiy, I.; Papangelakis, V.G. Cleaning of Waste Smelter Slags and Recovery of Valuable Metals by Pressure Oxidative Leaching. *J. Hazard. Mater.* **2008**, *152*, 607–615. [CrossRef] [PubMed]
28. Lim, B.; Aylmore, M.; Grimsey, D.; Alorro, R.D. Technospheric Mining of Critical and Strategic Metals from Nickel Slag—Leaching with Citric Acid and Hydrogen Peroxide. *Hydrometallurgy* **2023**, *219*, 106066. [CrossRef]
29. Jones, R.T.; Hayman, D.A.; Denton, G.M. Recovery of Cobalt, Nickel, and Copper from Slags Using DC-ARC Furnace Technology. In Proceedings of the International Symposium on Challenges of Process Intensification, 35th Annual Conference of Metallurgists, Montreal, QC, Canada, 24–28 August 1996.

30. Degradation of Vitamins, Probiotics and Other Active Ingredients Caused by Exposure to Heat, Water and Sunlight. Available online: https://www.nutraceuticalbusinessreview.com/news/article_page/Degradation_of_vitamins_probiotics_and_other_active_ingredients_caused_by_exposure_to_heat_water_and_sunlight/145924#:~:text=Vitamin%20C%20and%20heat,of%20Scientific%20and%20Technology%20Research.&text=The%20negative%20effects%20of%20heat,more%20at%20170%20%C2%B0F (accessed on 4 December 2023).
31. Jutkus, R.A.L.; Li, N.; Taylor, L.S.; Mauer, L.J. Effect of Temperature and Initial Moisture Content on the Chemical Stability and Color Change of Various Forms of Vitamin C. *Int. J. Food Prop.* **2015**, *18*, 862–879. [CrossRef]
32. Ma, N.; Houser, J.B. Recycling of Steelmaking Slag Fines by Weak Magnetic Separation Coupled with Selective Particle Size Screening. *J. Clean. Prod.* **2014**, *82*, 221–231. [CrossRef]
33. Baláž, P. Mechanical Activation in Hydrometallurgy. *Int. J. Miner. Process.* **2003**, *72*, 341–354. [CrossRef]
34. Fenoglio, I.; Martra, G.; Coluccia, S.; Fubini, B. Possible Role of Ascorbic Acid in the Oxidative Damage Induced by Inhaled Crystalline Silica Particles. *Chem. Res. Toxicol.* **2000**, *13*, 971–975. [CrossRef]
35. Li, L.; Lu, J.; Ren, Y.; Zhang, X.X.; Chen, R.J.; Wu, F.; Amine, K. Ascorbic-Acid-Assisted Recovery of Cobalt and Lithium from Spent Li-Ion Batteries. *J. Power Sources* **2012**, *218*, 21–27. [CrossRef]

recycling

MDPI

Article

Effective Recovery of Gold from Chloride Multi-Metal Solutions Through Anion Exchange

Isabel F. F. Neto, Márcia A. D. Silva and Helena M. V. M. Soares *

REQUIMTE/LAQV, Departamento de Engenharia Química, Faculdade de Engenharia, Universidade do Porto, Rua Dr. Roberto Frias, 4200-465 Porto, Portugal; isabel.f.f.neto@gmail.com (I.F.F.N.)
* Correspondence: hsoares@fe.up.pt; Tel.: +351-225-081-650; Fax: +351-225-081-449

Abstract: Leachates from electronic waste, slag dusts generated during the processing of electronic waste, sweeping jewelry, and municipal solid-waste incineration residues contain a myriad of base metals, such as aluminum (Al: 10–2000 mg/L), copper (Cu: 10–1000 mg/L), iron (Fe: 10–500 mg/L), nickel (Ni: 0.1–500 mg/L), lead (Pb: 1–500 mg/L), tin (Sn: 1–100 mg/L), and zinc (Zn: 5–500 mg/L), which are present at much higher quantities than Au (0.01–10 mg/L), which raises several drawbacks to the efficient recycling of Au with high purity using hydrometallurgical strategies. The aim of this work was to study the efficiency and selectivity of two strong basic anion exchange (DOW™ XZ-91419.00 and Purogold™ A194) resins to recover Au from a chloride multi-metal solution containing these metals. For both resins, the adsorption kinetic and equilibrium parameters for Au(III), determined at 1.12 mol/L HCl, Eh = 1.1 V, and 25 °C, proceeded according to a pseudo-second order and a Langmuir isotherm (qmax was 0.94 and 1.70 mmol/g for DOW™ XZ-91419.00 and Purogold™ A194 resins, respectively), respectively. Continuous adsorption experiments of Au (48 µmol/L; 2.0%) from a chloride multi-metal solution evidenced high Au retention capacity and selectivity to Au over Al, Cu, Fe, Ni, and Zn but low selectivity to Au over Ag and Sn for both resins. Concentrated (>3.3 mmol/L) and pure (>94%) Au eluates were obtained for both resins.

Keywords: gold recovery; gold selectivity; gold purification; multi-metal solution; quaternary amine resin; mixed amine resin

1. Introduction

Gold (Au) is a precious metal and it is very important because of its unique physical and chemical properties [1,2]. In the recent years, there has been a tendency for the utilization of precious metals, specifically Au, not only in traditional jewelry materials but also in new emerging applications [3,4], such as manufacturing of electronic parts and devices and corrosion-resistant materials. In electrical and electronic components, Au is used for connectors, switches, relay contacts, connecting wires, and connection strips mainly due to its corrosion resistance and electrical conductivity [1,5].

However, Au resources in the world are scarce and are mainly associated with other metal deposits [3]. In addition to that, the market values of Au are very high and are constantly rising [6]. With the growing demand of Au and reducing amount of its nature resources, it is important to recycle and reuse Au from secondary resources and industrial waste [2,4,7]. The recovery and reuse of Au is of great importance from the economic, environmental, and sustainability perspectives. Thus, the development of procedures to recovery Au from secondary resources has become an attractive and imperative issue [6].

Among the recovery methods of Au, the hydrometallurgical method is one of the most effective to recover this metal from scraps, such as electronic waste, slag dusts generated during the processing of electronic waste, and sweeping jewelry. These hydrometallurgical processes result in high volumes of solution containing low concentrations of Au together with several other metals present at high concentrations. For recovering Au from these multi-metal solutions, various purification methods such as those based on precipitation and/or solvent extraction (SX) procedures have been used. However, these strategies pose several drawbacks. For example, in the case of precipitation methods, the separation of Au is difficult and incomplete; as a consequence, several dissolution and precipitation steps are needed to achieve the purity specifications, which increases the operating costs. In the case of SX, the slow kinetics of Au extraction associated with the high risk of fire due to the use of solvents, which are highly flammable, and the high volumes of effluents generated during the SX purification process constitute serious disadvantages. These procedures definitively do not contribute to a cleaner, more efficient, and more sustainable recovery of Au from multi-metal solutions containing a low grade of Au. Therefore, the breakthrough in the development of more environmentally-friendly and highly selective Au-separation procedures is urgent.

Ion-exchange technology has been recognized as a powerful purification method and provides the possibility of enriching final Au eluate after elution if the resin has a high affinity for Au. This method has proved to be interesting from a technological and economical point of view. The low cost, easy handling, energy efficiency of the elution/regeneration process, and especially its suitability for application to solutions in which Au is present at trace levels are the main advantages [2,8–10]. Since cyanide has been the most widely used reagent for Au extraction from ores and secondary sources [11], the basic anion exchange resins, available on the market, have been developed for the purification of Au from cyanide solutions [1,12,13]. These resins have potentially high loading capacities and loading rates for Au, few are likely to be poisoned by organics, and they are easily regenerated [12]. However, due to the number of incidents involving cyanide, recycling industries are being pressed for substituting cyanide with other alternative strong Au complexing agents. Hydrochloric acid (HCl) offers the best prospects for commercialization and the chloride chemistry of Au is well understood. Under acidic oxidant conditions, the predominant complex ion of Au(III) species is $AuCl_4^-$ and basic anion exchange resins, which contain amine functional groups, are positively charged. Thus, the adsorption of the Au complexes onto the resins could be explained by an ion-exchange interaction according to the following reaction:

$$\vdash N^{+}R_{n}Cl^{-} + [AuCl_{4}]^{-} \Longleftrightarrow \vdash N^{+}[R_{n}AuCl_{4}]^{-} + Cl^{-} \quad (1)$$

where the symbol $\vdash$ denotes the inert matrix (polymer structure) of the resins and n takes values between 1 and 3 according to the resin. When $n < 3$, R is replaced by hydrogen atoms (the number of hydrogen atoms is equal to $3 - n$).

Presently, there is a large gap between the current scientific information available in the literature, mainly focused on the study of the ability of anion exchange resins for purifying Au using batch conditions [14,15] or column continuous experiments using mono and/or bi-metallic solutions [9,16,17], and the reality. The selective separation of Au from leachates, such as e-waste leachates, using anion exchange resins raises a much more complex issue as these leachates are multi-metal solutions containing Au together with other potential interfering metals, which, under acidic chloride media, form negative chlorocomplexes species that may compete for the anion exchange resins. The simultaneous adsorption of a considerable amount of foreign metal anions chlorocomplexes species together with $AuCl_4^-$ species may affect the effective separation and recovery of Au under dynamic systems (packed bed adsorption columns). This is an open and challenging research topic that needs to be investigated in order to offer more environmentally friendly and effective solutions for recovering Au from solid waste containing various metals and constitutes the main novelty of this work. Based on all of these facts, in this work, the ability performance of two strong basic anion exchange resins to recover and purify Au from an Au chloride multi-metal solution was studied. In addition to the study of the kinetic and thermodynamic abilities of the resins, the selectivity of each resin for Au relative to other metals was evaluated under dynamic conditions, which is crucial information at the moment of choosing a resin for purifying Au from low-grade Au multi-metal leachates.

2. Results and Discussion

2.1. Batch Adsorption Kinetics and Isotherms Experiments

Envisaging the future application of the resins for recovering Au from solutions using column experiments, adsorption kinetics and isotherms were studied for both resins in order to obtain information about the properties of the resins for Au sorption.

The evolution of Au retention by both resins was studied through the time using an Au concentration of 30 mmol/L (Figure 1). Since both resins are strong anionic resins, they are completely positively charged up to pH 8–9 at a particular S/L ratio. Therefore, only one S/L ratio was used to study the adsorption ability of both resins. For the Purogold^{TM} A194 resin, the experimental data show that approximately seven hours (420 min) were needed for the Au adsorption equilibrium to be reached (Figure 1B), while for the DOW^{TM} XZ-91419.00 resin forty minutes were enough (Figure 1A). Pseudo-first order and type I pseudo-second order kinetic models were adjusted to the experimental points through the respective linear plots. The calculated parameters are presented in Table 1. Both models had a good correlation with the experimental data, as can be seen from the R^2 values. However, even though high R^2 values were achieved for the pseudo-first order kinetic model, this model did not adjust well to the experimental data, as can be seen in Figure 1. On the other hand, the type I pseudo-second order kinetic model adjusted perfectly to the experimental data (Figure 1), which shows that the kinetics of Au adsorption to the resins were better described by a pseudo-second order kinetics. For adsorption isotherms and for both resins, the variation in q_e as a function of C_e is shown in Figure 2, where the Langmuir and Freundlich isotherms were adjusted to the experimental results. The parameters of each model [q_{max} and affinity constants (K_L and K_F for Langmuir and Freundlich models, respectively)] were calculated by the respective linear plots and the values are presented in Table 2. For both resins, the fitting of both models had a good correlation with the experimental data as it is evidenced by the correlation factors. However, the Langmuir model adjusted better to the experimental data (R^2 values $\geq$ 0.993) than the Freundlich model (R^2 values between 0.881 and 0.970), as evidenced in Figure 2. These results showed that the Langmuir isotherm model was more appropriate to describe the adsorption behavior of Au to both resins and suggests that the surface of the resins can be regarded as a homogeneous surface for Au adsorption. According to the Langmuir model, the maximum adsorption capacity of each resin to Au was 0.94 and 1.70 mmol/g for the DOW^{TM} XZ-91419.00 and Purogold^{TM} A194 resins, respectively. Additionally, the values of the Langmuir constant (k_L) showed a similar trend evidencing that the Purogold^{TM} A194 resin had the highest affinity for Au followed by the DOW^{TM} XZ-91419.00 resin (Table 2).

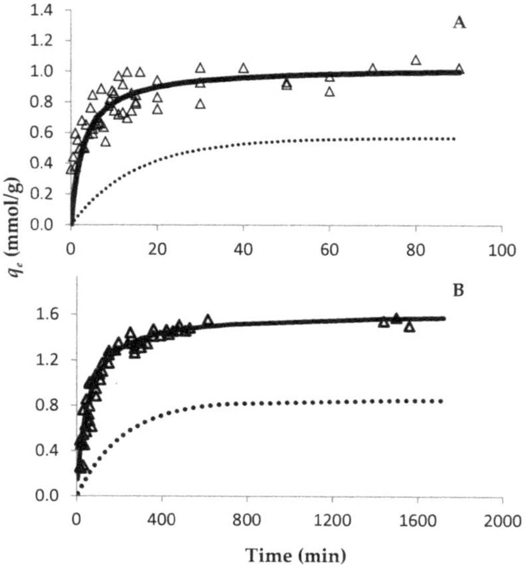

Figure 1. Effect of Au(III) on adsorption time on the DOW^{TM} XZ-91419.00 (**A**) and Purogold^{TM} A194 (**B**) resins. An Au(III) concentration of 30 mmol/L was used. Experimental conditions: 1.12 mol/L HCl at Eh = 1.1 V and 25 °C. Open triangles represent experimental data. Pseudo-first order (dotted line) and type I pseudo-second order (full line) kinetic models are represented. All points collected from two independent experiments were included in the figure.

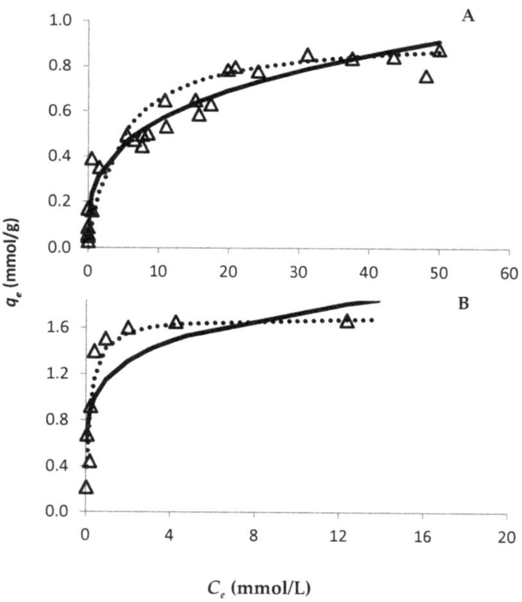

Figure 2. Effect of Au(III) concentration on its adsorption, during 24 h on the DOW^{TM} XZ-91419.00 (**A**) and Purogold^{TM} A194 (**B**) resins. Experimental conditions: 1.12 mol/L HCl at Eh = 1.1 V and 25 °C. Open triangles correspond to experimental data. The lines represent the fittings for Langmuir (dotted line) and Freundlich (full line) adsorption models. The results were obtained with points collected at least in three different assays.

Table 1. Parameters for kinetic models for Au-DOW^{TM} XZ-91419.00, Au-Purogold^{TM} A194, and Au-Purogold^{TM} S992 resins. Experimental conditions: 1.2 mol/L HCl at Eh = 1.1 V and 25 °C.

	Pseudo-First Order		Type I Pseudo-Second Order	
	DOW^{TM} XZ-91419.00	**Purogold^{TM} A194**	**DOW^{TM} XZ-91419.00**	**Purogold^{TM} A194**
Correlation factor	0.919	0.980	0.986	0.998
q_{max} (mmol/g)	0.57	0.85	1.03	1.62
K_1 (min^{-1})	0.668	0.0046	-	-
K_2 (g/mmol.min)	-	-	0.330	0.012

Table 2. Parameters for Langmuir and Freundlich models for Au-DOW^{TM} XZ-91419.00 and Au-Purogold^{TM} A194 resins. Experimental conditions: 1.2 mol/L HCl at Eh = 1.1 V and 25 °C.

	Langmuir Model		Freundlich Model	
	DOW^{TM} XZ-91419.00	**Purogold^{TM} A194**	**DOW^{TM} XZ-91419.00**	**Purogold^{TM} A194**
Correlation factor	0.993	0.998	0.935	0.970
q_{max} (mmol/g)	0.94	1.70	-	-
K_L (L/mmol)	0.22	5.56	-	-
n	-	-	3.29	5.71
K_F (mmol/g)	-	-	0.28	1.16

According to these results, it was possible to conclude that the DOW^{TM} XZ-91419.00 resin has the fastest adsorption kinetics (Table 1) whereas the Purogold^{TM} A194 resin revealed the highest adsorption capacity and affinity for Au (Table 2). In order to understand the differences in the kinetic behavior of both resins, the porosity of both resins was evaluated. Porosity values of 58 and 20% were quantified for the DOW^{TM} XZ-91419.00 and Purogold^{TM} A194 resins, respectively. The higher porosity observed for the DOW^{TM} XZ-91419.00 resin may explain the faster kinetics evidenced by this resin relative to the Purogold^{TM} A194 resin. Additionally, the high percentage of crosslinking agent present in the case of the Purolite A194 resin relative to the DOW^{TM} XZ-91419.00 resin, which results in an increase in the functional group density in the resin but decreases the diffusivity of $AuCl_4^-$ ions through the resin [18], may also explain the lower kinetics of the Purolite A194 resin. Also, the higher bead-size uniformity observed for the DOW^{TM} XZ-91419.00 resin relative to the Purogold^{TM} A194 resin (Figure 3) may contribute to the faster kinetics observed for the DOW^{TM} XZ-91419.00 resin.

The literature reports several studies where different polymer resins were used to recover Au from solution (Table 3). Erim et al., 2013 [4] studied the adsorption capacity of a 1,8-DAN-F blended polymer to Au; a maximum adsorption capacity of 0.60 mmol/g (119 mg/g, as described in the original paper) was achieved. Donia et al. [19] studied the ability of a magnetic adsorbent derived from chemically modified chitosan to recover Au. The maximum adsorption capacity found was 2.8 mmol/g. Pang and Yung [5] proposed the use of multiwalled carbon nanotubes and the maximum adsorption capacity was 0.47 mmol/g (93.5 mg/g, as described in the original paper). Yang et al. [20] studied the ability of cellulose acetyl derivatives to recover Au from acidic chloride solutions and cellulose acetate fibers and an adsorption capacity of 0.56 mmol/g (110 mg/g, as described in the original paper) was calculated. Fan et al. [21] prepared a resin by crosslinking the persimmon residual with formaldehyde and the maximum Au uptake was 2.6 mmol/g. A comparative analysis between the maximum adsorption capacities determined for both resins used in this work shows that they were in the range of values described in the literature for other resins studied for the same application (Table 3).

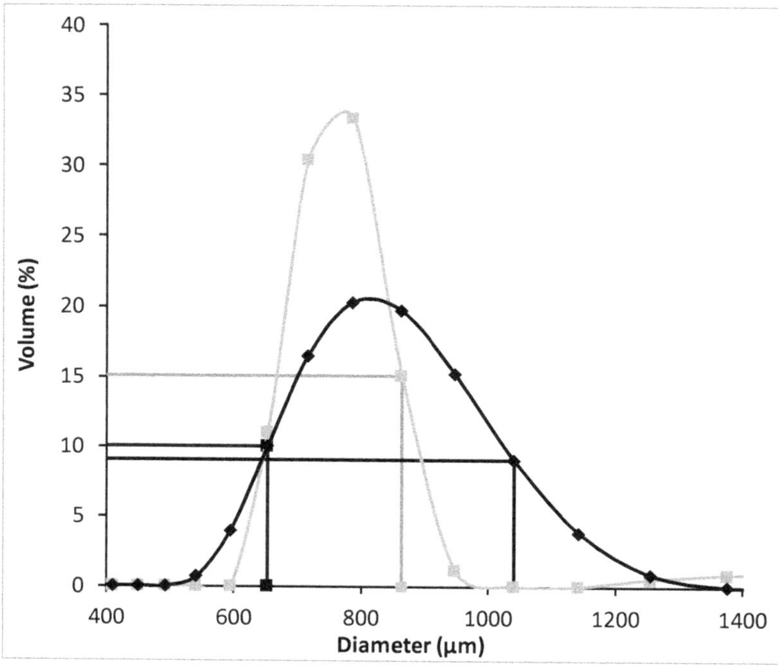

Figure 3. Particle size distribution of the DOW™ XZ-91419.00 (■) and Purogold™ A194 (◆) resins with emphasis on the diameter in which less than 10% and up to 90% of the particles are 653 µm and 863 µm for the DOW™ XZ-91419.00 (■) and 653 µm and 1041 µm for the Purogold™ A194 (◆) resins.

Table 3. Values of adsorption capacity (q_{max}) and affinity constant (K_L) for different polymer resins to Au described in the literature and those used in this work.

Resin	q_{max} (mmol/g)	K (L/mmol)	Reference
1,8-Diaminonaphtalene–formaldehyde/PVC	0.60	3.48	[4]
Magnetic chitosan	2.8	1.32	[19]
Multiwalled carbon nanotubes	0.42	92.8	[5]
Cellulose acetate fibers	0.56	-	[20]
PPF resin	2.57	-	[21]
DOW™ XZ-91419.00	0.94	0.22	This work
Purogold™ A194	1.70	5.56	

2.2. Column Experiments

2.2.1. Optimization of Adsorption and Elution Conditions

In electronic waste, slag dusts generated during the processing of electronic waste, sweeping jewelry, and municipal solid-waste incineration residues, Au is a minor component (usually less than 1%), when compared with the contents of the other metals [22]. Thus, all these facts together will certainly influence the selective recovery of Au from a highly complex multi-element solution. In order to predict the metal affinity series of each resin, column experiments, using a multi-metal synthetic solution (see the composition of the inlet solution in Table 4), were performed with both resins.

Table 4. Composition of the initial multi-metal synthetic solution and the final Au solutions (eluates) after treatment with the DOW^{TM} XZ-I91419.00 and Purogold^{TM} A194 resins and elution with thiourea and separation number (SN) for each metal, relative to Au.

Inlet Solution			Eluates	
			DOW^{TM} XZ-I91419.00 Resin	**Purogold^{TM} A194 Resin**
Metal (%)		**Speciation**	**Metal%**	**Metal%**
Au	2.0	$AuCl_4^-$	94.6	94.1
Ag	1.2	$AgCl_x^{-(x-1)}$	0.84	2.3
Sn	14.7	$SnCl_5^-/SnCl_6^{2-}$	3.9	3.1
Cu	43.0	$Cu^{2+}/CuCl^+$ (80.1%); $CuCl_2$ (19.4%)	0.08	0.02
Fe	24.4	$FeCl_2^+/FeCl^{+2}$ (91.1%) $FeCl_3$ (8.4%)	0.40	0.008
Ni	3.5	$Ni^{2+}/NiCl^+$ (99.6%)	0.04	0.03
Pb	3.1	$PbCl_2$ (53.4%); $PbCl^+$ (10.7%); $PbCl_3^-/PbCl_4^{-2}$ (35.6%)	0.06	0.44
Al	6.6	Al^{3+}	<0.0008	<0.0001
Zn	1.5	$Zn^{2+}/ZnCl^+$ (24.5%); $ZnCl_2$ (31.8%), $ZnCl_3^-/ZnCl_4^{-2}$ (43.7%)	0.08	0.02

Because DOW^{TM} XZ-91419.00 evidenced faster adsorption kinetics than the Purogold^{TM} A194 resin (Table 1), different adsorption kinetic profiles are expected with both resins. After preliminary experiments, continuous adsorption experiments using the following flow rates of 0.83 and 0.42 mL/(min.g) for the DOW^{TM} XZ-91419.00 and Purogold^{TM} A194 resins, respectively, were selected (Figure 4). For both resins, the outlet concentrations of Al, Cu, Fe, and Ni were similar to the inlet concentrations since the beginning of the assays. These results pointed out that no appreciable amount of Al, Cu, Fe, and Ni was adsorbed by either resin, and they were not considered in the figure. Figure 4 evidences that both resins retained Au efficiently for a long period of time. For the DOW^{TM} XZ-91419.00 resin, Au was retained for 3 days (72 h); after this period of time, more than 10% of Au entered the raffinate solution. After 8 h of assay, Zn was not retained by the resin and after 24 h Sn and Ag were also not retained. In the case of the Purogold^{TM} A194 resin, Au was retained for 10 days (223 h). After 24 h, Sn, Ag, and Zn were not retained any more. For both resins, when Au started to leave the resin, the other metals were also not retained for a long time.

The results obtained for both resins suggest that they can be interesting to selectively recover Au from the synthetic chloride multi-metal solution, and 72 and 223 h were the times defined for stopping the adsorption of Au for the DOW^{TM} XZ-91419.00 and Purogold^{TM} A194 resins, respectively.

The adsorption step with the Purogold^{TM} A194 resin was slower than with the DOW^{TM} XZ-91419.00 resin. Moreover, the Purogold^{TM} A194 resin allowed the treatment of 5.8 L of synthetic solution per gram of resin (Figure 4B), while 3.0 L/g was treated with the DOW^{TM} XZ-91419.00 resin (Figure 4A). These results correlate well with the maximum adsorption capacities of both resins: the ratio of the volume of synthetic solutions treated by both resins (5.8/3.0 = 1.9) (Figure 4) is equal to the ratio between the q_{max} values (1.70/0.94 = 1.8) achieved by isotherm studies (Table 2).

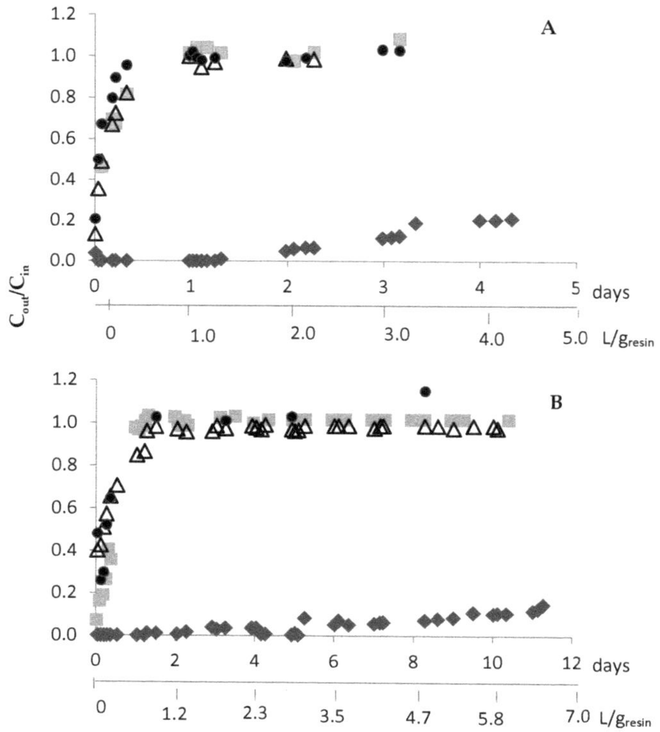

Figure 4. Ratio between the metal concentrations in the outlet, C_{out}, and in the inlet solution (1.12 mol/L HCl at Eh = 1.1 V), C_{in}, vs. time for the DOW^{TM} XZ-91419.00 (**A**) and Purogold^{TM} A194 (**B**) resins: Au (♦), Ag (■), Sn (△) and Zn (•). Flow rates of 0.83 and 0.42 mL/(min.g) were used for the DOW^{TM} XZ-91419.00 and Purogold^{TM} A194 resins, respectively. These are examples of experiments performed two times.

The elution of Au adsorbed from both resins (Figure 5) was promoted using thiourea (TU). During this desorption stage, Au(III) is reduced to Au(I) and then complexed with TU [23], which is released to the bulk solution according to Equation (2):

$$\vdash N^{+}[R_nAuCl_4]^{-} + 2\,TU \Longleftrightarrow \vdash N^{+}R_nCl^{-} + [Au(TU)_2]^{+}Cl^{-} + \mathrm{Cl_2} \quad (2)$$

For this purpose, a 0.1 mol/L thiourea and 0.5 mol/L H_2SO_4 solution was tested using the same flow rates applied in the adsorption step. For the DOW^{TM} XZ-91419.00 resin, the elution process was shown to be fast; after 65 and 85 min, 85 and 95% of the Au adsorbed was eluted, respectively. After 120 min, almost all of the Au was removed from the resin. On the other hand, for the Purogold^{TM} A194 resin, the elution was slower. After 170 min, 88% of the adsorbed Au was eluted, and after 210 min, the Au elution achieved 92%. For total Au elution, more than 420 min was needed. In order to reduce the time of elution, a higher concentration of thiourea (0.25 mol/L) was tested but no significant improvement was observed.

Based on these results, elution times of 65 and 200 min were defined for the DOW^{TM} XZ-91419.00 and Purogold^{TM} A194 resins, respectively, since final solutions with higher concentrations of Au are achieved. To close the cycle, the direct recovery of Au from thiourea, as Au metallic, can be achieved by reduction using sodium borohydride followed by filtration (not tested in this work) and the remaining filtrate containing thiourea can be reused in another elution cycle [24]. By this way, thiourea can be reused in another elution cycle contributing to reduce the operating cost of the process.

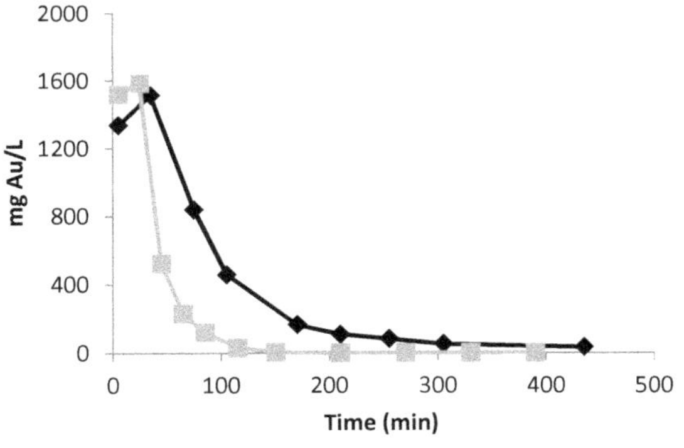

Figure 5. Gold elution profiles for the DOW^{TM} XZ-91419.00 (■) and Purogold^{TM} A194 (◆) resins with 0.5 mol/L H_2SO_4 and 0.1 mol/L thiourea. Flow rates of 0.83 and 0.42 mL/(min.g) were used for the DOW^{TM} XZ-91419.00 and Purogold^{TM} A194 resins, respectively.

2.2.2. Concentration and Selectivity Ability of Resins for Gold

For both resins, additional experiments to evaluate the simultaneous concentration and selectivity (expressed by the separation number, SN) ability of both resins for recovering Au from a chloride multi-metal synthetic solution (Figure 6) were performed using the experimental (adsorption and elution) conditions previously optimized and described in Section 2.2.1. The metal composition of the initial chloride multi-metal synthetic solution (inlet solution), measured by AAS-FA and expressed as a percentage, as well as the chemical metal species predicted by computer chemical simulations are presented in Table 4. Additionally, the metal composition of the eluate solutions, quantified by AAS-FA and expressed as a percentage, is presented in Table 4. Moreover, the SN values for both resins are presented in Figure 6.

The results presented in Table 4 show the high affinity of both resins for Au and are in agreement with the previous adsorption parameters. Final solutions containing Au with high purity (94.6 and 94.1% for DOW^{TM} XZ-91419.00 and Purogold^{TM} A194 resins, respectively) and concentration (3.5 and 3.3 mmol/L, respectively; the amount of Au in the final solutions increased almost 70 times when compared with the concentration in the initial solution, 48 μmol/L) were achieved (Table 4).

For the Au purified solution using the DOW^{TM} XZ-91419.00 resin, Sn was the major (3.9%) contaminant followed by Ag (0.84%) and Fe (0.40%)—whereas for the Au eluate resulting from the Purogold^{TM} A194 resin, Sn was the major contaminant (3.1%), followed by Ag (2.3%) and Pb (0.44%).

Even though Cu (43.0%), Fe (24.4%), and Al (6.6%) were the major metals in the initial chloride multi-metal synthetic solution, together with Ni also present in a significant amount (3.5%), both resins evidenced a high selectivity, SN > 10^3, to Au over these metals (Figure 6). This behavior can be explained by the fact that no anionic chloride species of these metals are formed under the chemical conditions used (Table 4), and thus, no adsorption occurs. As a consequence, less than 0.5% of these metals was present in all final Au eluates (see the metal composition of the eluates in Table 4). On the other hand, both resins evidenced poor selectivity for Ag, SN $\leq$ 63, (Figure 6). As a consequence, even though Ag was a minor (1.2%) constituent in the initial chloride multi-metal solution, no significant purification occurred for the DOW^{TM} XZ-91419.00 XZ resin (the amount of Ag was 0.84% in the eluate vs. 1.2% in the initial solution) and a concentration effect occurred when the Purogold^{TM} A194 resin (the amount of Ag was 2.3% in the eluate vs. 1.2% in the initial solution) was used. Moreover, for Sn (for both resins) and Pb (for Purogold^{TM} A194 resin), SN values lower than 404 were recorded, which denotes the lower selectivity of these resins for these metals relative to Au. This lower selectivity (Figure 6) can be explained by the adsorption of several anionic chloride complexes of these metals (Table 4), which are formed in solution, by a similar mechanism as described for Au but with less attraction to the exchange sites on the resin due to the high charge density that enhances Au's attraction to the exchange sites on the resin, leading to a stronger interaction and preferential uptake.

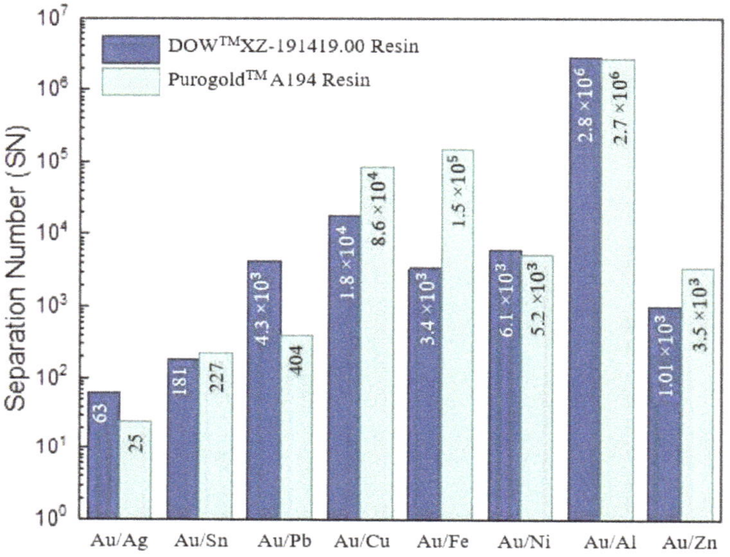

Figure 6. Separation number for the DOW^{TM} XZ-91419.00 and Purogold^{TM} A194 resins with 0.5 mol/L H_2SO_4 and 0.1 mol/L thiourea.

Considering that chlorination is a practicable alternative for Au leaching due to the high dissolution rate [25] and the good selectivity for Au relative to other metals demonstrated by both resins in the present work, it seems plausible to anticipate that the combination of chlorination leaching and purification of Au using these two strong anionic resins presents potentialities for recovering Au with high purity from various sources, such as electronic waste, slag dusts generated during the processing of electronic waste, sweeping jewelry, and municipal solid-waste incineration residues.

3. Materials and Methods

3.1. Reagents and Analytical Techniques

Two strong basic anion exchange resins were used: DOW^{TM} XZ-91419.00 and Purogold^{TM} A194. Table 5 lists some of the main physical and chemical properties of these resins. The DOW^{TM} XZ-91419.00 and Purogold^{TM} A194 resins were kindly given by the DOW (Midland, MI, USA) and Purolite (Philadelphia, PA, USA) companies.

Table 5. Physical and chemical properties of the resins studied in this work.

Physicochemical Properties	DOW^{TM} XZ-91419.00	Purogold^{TM} A194
Matrix	Styrene divinyl benzene gel	Polystyrene structure crosslinked with divinylbenzene
Functional groups	t-butylamine, quaternary amines	Mixed tertiary and quaternary amines
Physical appearance	White spherical beads	Brownish-cream spherical beads
Exchanger type	Strong base	Strong base
Exchange capacity (in chloride form,	0.31	3.0
mol/Kg)	760–1200	800–1300
Particle size (μm)	<768 (5%)	<800 (5%)
Moisture retention (%)	45–55	44–52

All chemicals used in the experiments were of analytical reagent grade. Synthetic mono-metal (containing only Au for kinetics and isotherms adsorption studies) and multi-metal solutions containing Ag, Al, Au, Cu, Fe, Ni, Pb, Sn, and Zn were prepared by solubilizing a suitable amount of the salt of each metal [$HAuCl_4.3H_2O$, $CuSO_4.5H_2O$, $Ni(NO_3)_2{\cdot}6H_2O$, $Fe(NO_3)_3.9H_2O$, $Pb(NO_3)_2$, $Al_2(SO_4)_3{\cdot}16H_2O$, $ZnCl_2$, Ag_2SO_4, and $SnCl_2.2H_2O$] in a 1.12 mol/L HCl solution. Aqueous solutions prepared from analytical grade thiourea, $SC(NH_2)_2$, and sulfuric acid (H_2SO_4) (95%) were used in the elution studies. Sodium hypochlorite (NaClO) was used to raise the redox potential (Eh) of the solutions up to 1.1 V in order to stabilize Au as $[AuCl_4]^-$.

The metal (Ag, Al, Au, Cu, Fe, Ni, Pb, Sn, and Zn) concentration was measured by atomic absorption spectrometry with flame atomization (AAS-FA) in a Perkin Elmer AAnalyst 400 spectrometer (Perkin Elmer, Norwalk, CT, USA) or an Analytik JenaA NovAA350 spectrometer (Konrad-Zuse, Germany). A nitrous oxide flame was used for the determination of Al and Sn and an acetylene flame for the remaining metals.

For all solutions, the redox potential was measured with a Pt indicator electrode using a Metrohm 744 millivoltmeter (Metrohm AG, Herisau, Switzerland).

3.2. Batch Experiments

Kinetic and isotherm adsorption batch experiments were performed in solutions containing Au(III), prepared from the $HAuCl_4{\cdot}3H_2O$ salt, in 1.12 mol/L HCl, at an Eh of 1.1 V; under this Eh and pH conditions, all Au is present as $[AuCl_4]^-$ (Figure 7). Because the DOW^{TM} XZ-91419.00 and Purogold^{TM} A194 resins are strong basic anion exchange resins, they are completely positively charged up to at least pH 8–9 and present a maximum ion-exchange capacity up to this pH [26]. Therefore, only one solid/liquid (S/L) ratio of 10 g/L was used to study the adsorption ability of both resins. For this purpose, 0.5 g of wet resin was accurately weighed into a flask and 50 mL of each solution containing an adequate amount of Au(III), prepared from the $HAuCl_4{\cdot}3H_2O$ salt, in 1.12 mol/L HCl, at an Eh of 1.1 V was used. The flasks were shaken in a bath with temperature control (OLS200, Grant, Royston, United Kingdom) at 25 °C and 150 rotations per minute (rpm).

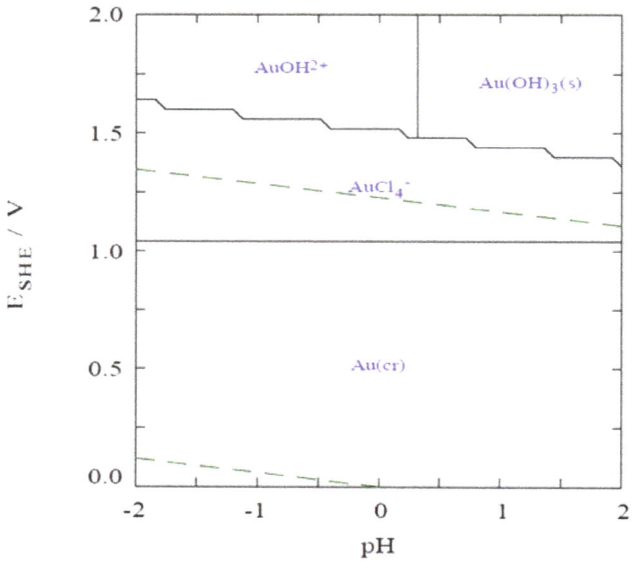

Figure 7. Pourbaix-type diagram for the $Au^{3+}/Cl^{-}/e^{-}$ system, as function of pH. Temperature of 25 °C; $[Au^{3+}]$ = 48 µmol/L; $[Cl^{-}]$= 1.12 mol/L.

3.2.1. Adsorption Kinetics

For kinetic studies, a synthetic 30 mmol/L Au(III) solution was used. At different intervals of time, an amount of 0.4 mL of the suspension (solution plus resin) was taken from the flask under agitation. Then, the resin was separated from the suspension by filtration and not returned back to the flask. The respective concentration of Au was measured in the filtrate by AAS-FA.

Langergren pseudo-first order and type I pseudo-second order models were fitted into the experimental points using linear regression (Equations (3) and (4), respectively):

Langergren pseudo-first order:

$$\log\left(q_e - q_t\right) = \log q_e - \frac{k_1}{2.303}t \tag{3}$$

Type 1 pseudo-second order:

$$\frac{t}{q_t} = \frac{1}{k_2 {q_e}^2} + \frac{t}{q_e} \tag{4}$$

3.2.2. Adsorption Isotherms

For isotherm studies, flasks containing solutions with increasing concentrations of Au(III) were shaken during 24 h. After this period of time, each solution was separated from the resin by filtration. The concentration of Au was analyzed by AAS-FA. The quantity of metal adsorbed per mass of resin (q_e) was calculated according to Equation (5), where C_0 and C_e are the initial and equilibrium metal concentrations (mmol/L), respectively, m is the mass of resin (g), V is the volume of solution (L), and q_e is the equilibrium adsorption capacity (mmol/g).

$$q_e = \frac{(C_0 - C_e)V}{m} \tag{5}$$

Langmuir and Freundlich isotherm models were tested. The Langmuir isotherm assumes adsorption on a homogeneous surface (monolayer) with a negligible interaction between adsorbed molecules and allows the estimation of the maximum metal uptake (q_{max}) and the Langmuir constant (k_L) according to Equation (6).

$$q_e = \frac{q_{max} k_L C_e}{1 + k_L C_e} \quad (6)$$

The Freundlich isotherm assumes adsorption on heterogeneous surfaces. The Freundlich constant (k_F) and n are Freundlich parameters that can be estimated from experimental data according to Equation (7). Values of n greater than 1 indicate favorable conditions for adsorption.

$$q_e = k_F C_e^{\frac{1}{n}} \quad (7)$$

3.3. Column Experimental Conditions

For experiments performed in continuous mode, a chloride multi-metal synthetic solution was prepared in 1.12 mol/L HCl, at an Eh of 1.1 V, with the following metal compositions: 4.8×10^{-5}, 5.4×10^{-5}, 1.2×10^{-3}, 3.2×10^{-3}, 2.1×10^{-3}, 2.9×10^{-4}, 8.7×10^{-5}, 5.9×10^{-4}, and 1.1×10^{-4} mol/L of Au, Ag, Al, Cu, Fe, Ni, Pb, Sn, and Zn, respectively.

For the DOW^{TM} XZ-91419.00 and Purogold^{TM} A194 resins, a precise amount of resin, equivalent to 0.5 bed volumes (1.5 mL), was placed in the chromatographic glass column (di = 6.6 mm, h = 100 mm) and wetted with water in accordance with the manufacturer's specifications [27]. The air pockets were removed from the bed column by attaching the water line at the bottom of the column and simultaneously the resin was washed. For both resins, flow rates of 0.83 and 0.42 mL/(min.g), respectively, were used. The flow rates were controlled using a peristaltic pump MS-Reglo (Ismatec, Glattbrugg, Switzerland). Between adsorption and elution steps, the resins were washed with 30 mL of deionized water. After elution, each resin was washed with 60 mL of deionized water and reused for the next adsorption and elution cycle. All experiments were performed at room temperature and, for each experiment, at least two independent experiments were conducted.

Samples were collected from the outlet solution at different intervals of time to measure the metal concentration by AAS-FA. Elution of the retained metals was performed using two concentrations (0.1 and 0.25 mol/L) of thiourea and a fixed concentration (0.5 mol/L) of H_2SO_4.

For all solutions (inlet, raffinate, and eluates from both resins), the concentration of all metals (Ag, Al, Au, Cu, Fe, Ni, Pb, Sn, and Zn) was measured by AAS-FA.

For gold purification (concentration and selectivity ability of resins for gold) assays, the adsorption and elution steps had a duration of 72 h and 65 min, respectively, for the DOW^{TM} XZ-91419.00 resin, and a duration of 223 h and 200 min, respectively, when the Purogold^{TM} A194 resin was used. For both resins, the elution of Au was performed with 0.1 mol/L thiourea and 0.5 mol/L H_2SO_4.

For each resin, the separation number (SN) for each metal, relative to Au, was calculated according to Equation (8).

$$\text{Separation number} = \frac{[Au]_{eluate}/[Metal]_{eluate}}{[Au]_{feed}/[Metal]_{feed}} \quad (8)$$

3.4. Computer Chemical Simulations

The Eh–pH diagrams of dissolved Au-chloride species at 25 °C were drawn using the SPANA software (Visual Basic, Stockholm, Sweden, version 5.5.43), whereas metal chemical speciation simulations were carried out using MINEQL+ software (version 4.5) [28]. Software generates chemical equilibrium concentrations of the species considered in the model by the program reactions, based on component stability constants [29] and the total molar metal concentrations measured by AAS.

3.5. Porosity Determination

Total porosity (Φ) is defined as the fraction of the total volume occupied by pores and can be expressed as follows:

$$\Phi = \left(1 - \frac{\rho_{apparent}}{\rho_{real}}\right) \times 100\% \quad (9)$$

where Φ is the total porosity (%), ρ_{real} is the density of the material without pores being considered, and ρ_{apparent} is the density including the pores [30].

The determination of ρ_{real} was carried out using a custom-developed gas picnometry method with helium as the working gas. The pycnometer consists of two containers: one designed to hold the sample and another with a known internal volume, serving as a reference. Both containers were sealed and helium was introduced into the system at a pressure of 1 bar. The initial helium volume in the reference container was recorded. Then, the valve of the container holding the sample was opened, allowing the helium to flow into it. The remaining helium volume in the reference container was then measured, and the volume of the sample container was determined by the difference. Finally, the sample was weighed, and its density was calculated by dividing the measured mass by the obtained volume [31]. The ρ_{apparent} was determined using the PoreMaster Mercury-intrusion apparatus [32].

4. Conclusions

In this work, a comparative study of the ability of two strong basic anion exchange resins to adsorb and purify Au from a chloride multi-metal solution was performed.

Firstly, Au(III) adsorption kinetic and equilibrium parameters were determined at a temperature of 25 °C in 1.12 mol/L HCl at Eh 1.1 V and 25 °C using batch experiments. The Au adsorption proceeded according to a pseudo-second order model for both resins. The kinetics of Au adsorption was faster for the DOW^{TM} XZ-91419.00 resin [K_2 = 0.330 g/(mmol.min)] than for the Purogold^{TM} A194 resin [K_2 = 0.012 g/(mmol.min)]. The experimental data fitted well to a Langmuir isotherm model, and maximum adsorption capacities of 0.94 and 1.70 were determined for the DOW^{TM} XZ-91419.00 and Purogold^{TM} A194 resins, respectively.

In a second attempt, the efficiency and selective ability of the resins to retain Au(III) was evaluated by performing continuous column experiments with a synthetic chloride multi-metal solution (1.2 mol/L HCl and Eh = 1.1 V). Final Au solutions with a purity grade of at least 94% were achieved due to the high selectivity revealed by both resins to Au over Al, Cu, Fe (except for the DOW^{TM} XZ-91419.00 resin), Pb (except for the Purogold^{TM} A194 resin), Ni, and Zn. However, even though eluates with similar amounts of Au (about 3.5 mmol/L) were produced for both resins, there were some differences in their composition. For the DOW^{TM} XZ-91419.00 resin, Sn was the major contaminant, followed by Ag and Fe whereas, for the Purogold^{TM} A194 resin, Ag and Sn were the major contaminants.

The results obtained in this work evidence that both resins are interesting options for purifying Au from chloride multi-metal solutions containing a low grade of Au, particularly in the following conditions: (i) DOW^{TM} XZ-91419.00 resin is suitable to purify Au from solutions containing Al, Ag, Cu, Fe, Ni, Pb, and Zn; (ii) Purogold^{TM} A194 resin is suitable to purify Au from solutions containing Al, Cu, Fe, Ni, Pb, and Zn.

Author Contributions: Conceptualization: H.M.V.M.S.; methodology: I.F.F.N. and M.A.D.S.; validation: I.F.F.N. and M.A.D.S.; formal analysis, I.F.F.N. and M.A.D.S.; resources: H.M.V.M.S.; original draft preparation: I.F.F.N. and H.M.V.M.S.; writing—review and editing: H.M.V.M.S.; supervision: H.M.V.M.S.; project administration: H.M.V.M.S. All authors have read and agreed to the published version of the manuscript.

Funding: This work is financially supported by national funds through the FCT/MCTES (PIDDAC), under the project PTDC/CTA-AMB/3489/2021—RECY-SMARTE—Sustainable approaches for recycling discarded mobile phones, with DOI 10.54499/PTDC/CTA-AMB/3489/2021 (https://doi.org/10.54499/PTDC/CTA-AMB/3489/2021).

Data Availability Statement: The data that support the findings of this study are available from the corresponding author upon reasonable request.

Conflicts of Interest: The authors declare no conflicts of interest.

References

1. Ok, Y.S.; Jeon, C. Selective adsorption of the gold–cyanide complex from waste rinse water using Dowex 21K XLT resin. *J. Ind. Eng. Chem.* **2014**, *20*, 1308–1312. [CrossRef]
2. Dong, Z.; Liu, J.; Yuan, W.; Yi, Y.; Zhao, L. Recovery of Au(III) by radiation synthesized aminomethyl pyridine functionalized adsorbents based on cellulose. *Chem. Eng. J.* **2016**, *283*, 504–513. [CrossRef]
3. Ahamed, M.E.H.; Mbianda, X.Y.; Mulaba-Bafubiandi, A.F.; Marjanovic, L. Selective extraction of gold(III) from metal chloride mixtures using ethylenediamine N-(2-(1-imidazolyl)ethyl) chitosan ion-imprinted polymer. *Hydrometallurgy* **2013**, *140*, 1–13. [CrossRef]
4. Erim, Ü.C.; Gülfen, M.; Aydın, A.O. Separation of gold(III) ions by 1,8-diaminonaphthalene-formaldehyde chelating polymer. *Hydrometallurgy* **2013**, *134–135*, 87–95. [CrossRef]
5. Pang, S.-K.; Yung, K.-C. Prerequisites for achieving gold adsorption by multiwalled carbon nanotubes in gold recovery. *Chem. Eng. Sci.* **2014**, *107*, 58–65. [CrossRef]
6. Gurung, M.; Adhikari, B.B.; Gao, X.; Alam, S.; Inoue, K. Sustainability in the Metallurgical Industry: Chemically Modified Cellulose for Selective Biosorption of Gold from Mixtures of Base Metals in Chloride Media. *Ind. Eng. Chem. Res.* **2014**, *53*, 8565–8576. [CrossRef]
7. Xiong, C.; Zhou, S.; Liu, X.; Jia, Q.; Ma, C.; Zheng, X. 2-Aminothiazole Functionalized Polystyrene for Selective Removal of Au(III) in Aqueous Solutions. *Ind. Eng. Chem. Res.* **2014**, *53*, 2441–2448. [CrossRef]
8. Pilśniak-Rabiega, M.; Trochimczuk, A.W. Selective recovery of gold on functionalized resins. *Hydrometallurgy* **2014**, *146*, 111–118. [CrossRef]
9. Wołowicz, A.; Hubicki, Z. Adsorption characteristics of noble metals on the strongly basic anion exchanger Purolite A-400TL. *J. Mater. Sci.* **2014**, *49*, 6191–6202. [CrossRef]
10. Sousa, R.; Regufe, M.J.; Fiúza, A.; Leite, M.M.; Futuro, A. A systematic review of sustainable gold extraction from raw ores using alternative leaching reagents. *Extr. Ind. Soc.* **2022**, *9*, 101018. [CrossRef]
11. Sethurajan, M.; van Hullebusch, E.D.; Fontana, D.; Akcil, A.; Deveci, H.; Batinic, B.; Chmielarz, A. Recent advances on hydrometallurgical recovery of critical and precious elements from end of life electronic wastes—A review. *Crit. Rev. Environ. Sci. Technol.* **2019**, *49*, 212–275. [CrossRef]
12. Green, B.R.; Kotze, M.H.; Wyethe, J.P. Developments in ion exchange: The mintek perspective. *JOM* **2002**, *54*, 37–43. [CrossRef]
13. Yahorava, V.; Kotze, M.H. Novel application of a gold cyanide selective resin in the PGMs industry. In *Hydrometallurgy*; Canadian Institute of Mining, Metallurgy and Petroleum: Montreal, QC, Canada, 2014; Volume 2, pp. 269–279.
14. Dicinoski, G.W.; Gahan, L.R.; Lawson, P.J.; Rideout, J.A. Application of the shrinking core model to the kinetics of extraction of gold(I), silver(I) and nickel(II) cyanide complexes by novel anion exchange resins. *Hydrometallurgy* **2000**, *56*, 323–336. [CrossRef]
15. Azizitorghabeh, A.; Mahandra, H.; Ramsay, J.; Ghahreman, A. Selective gold recovery from pregnant thiocyanate leach solution using ion exchange resins. *Hydrometallurgy* **2023**, *218*, 106055. [CrossRef]
16. Zhang, H.; Dreisinger, D.B. The recovery of gold from ammoniacal thiosulfate solutions containing copper using ion exchange resin columns. *Hydrometallurgy* **2014**, *72*, 225–234. [CrossRef]
17. Negrea, A.; Mihailescu, M.; Mosoarca, G.; Ciopec, M.; Duteanu, N.; Negrea, P.; Minzatu, V.E. Estimation on fixed-bed column parameters of breakthrough behaviors for gold recovery by adsorption onto modified/functionalized amberlite xad7. *Int. J. Environ. Res. Public Health* **2020**, *17*, 6868. [CrossRef]

18. Mohebbi, A.; Abolghasemi Mahani, A.; Izadi, A. Ion exchange resin technology in recovery of precious and noble metals. In *Applications of Ion Exchange Materials in Chemical and Food Industries*; Inamuddin, R.T.A., Asiri, A.M., Eds.; Springer International Publishing: Cham, Switzerland, 2019; pp. 193–258.
19. Donia, A.M.; Yousif, A.M.; Atia, A.A.; Elsamalehy, M.F. Efficient adsorption of Ag(I) and Au(III) on modified magnetic chitosan with amine functionalities. *Desalination Water Treat* **2014**, *52*, 2537–2547. [CrossRef]
20. Yang, J.; Kubota, F.; Baba, Y.; Kamiya, N.; Goto, M. Application of cellulose acetate to the selective adsorption and recovery of Au(III). *Carbohydr. Polym.* **2014**, *111*, 768–774. [CrossRef]
21. Fan, R.; Xie, F.; Guan, X.; Zhang, Q.; Luo, Z. Selective adsorption and recovery of Au(III) from three kinds of acidic systems by persimmon residual based bio-sorbent: A method for gold recycling from e-wastes. *Bioresour. Technol.* **2014**, *163*, 167–171. [CrossRef]
22. Neto, I.F.; Sousa, C.A.; Brito, M.S.; Futuro, A.M.; Soares, H.M. A simple and nearly-closed cycle process for recycling copper with high purity from end life printed circuit boards. *Sep. Purif. Technol.* **2016**, *164*, 19–27. [CrossRef]
23. Alguacil, F.J.; Adeva, P.; Alonso, M. Processing of residual gold (III) solutions via ion exchange. *Gold Bull.* **2005**, *38*, 9–13. [CrossRef]
24. Farouk, T.; Awadalla, H.; Ronald, E.; Molnar, O.; Gordon, M.R. Recovery of Platinum Group Metals (PGM) from Acidic Solutions by Reudction Precpitation with Sodium Borohydride. U.S. Patent 5304233, 19 April 1994.
25. Aylmore, M.G. Alternative Lixiviants to Cyanide for Leaching Gold Ores. In *Gold Ore Processing: Project Development and Operations*; Elsevier: Amsterdam, The Netherlands, 2016; pp. 447–484.
26. Hubicki, Z.; Kolodynska, D. Selective Removal of Heavy Metal Ions from Waters and Waste Waters Using Ion Exchange Methods. In *Ion Exchange Technology*; IntechOpen: Rijeka, Croatia, 2012. [CrossRef]
27. van Deventer, J.; Bazhko, V.; Yahorava, V. Comparison of gold selective ion exchange resins and activated carbon for the recovery of gold from copper-gold leach liquors. In Proceedings of the Alta 2014 Gold Conference, Perth, Australia, 24–31 May 2014.
28. Schecher, W.D.; Mcavoy, D.C. *MINEQL+: A Chemical Equilibrium Modeling System, Version 4.5 for Windows, User's Manual*, 2nd ed.; Environmental Research Software: Hallowell, ME, USA, 2003.
29. Martell, A.E.; Smith, R.M. *NIST SRD 46*; NIST Standard Reference Database 46 Version 8.0. NIST Critically Selected Stability Constants of Metal Complexes Database. US Department of Commerce, National Institute of Standards and Technology: Gaithersburg, MD, USA, 2004.
30. Rouquerol, J.; Rouquerol, F.; Sing, K. Assessment of Mesoporosity. In *Adsorption by Powders and Porous Solids: Principles, Methodology and Applications*, 1st ed.; Elsevier: San Diego, CA, USA, 1998; pp. 191–236.
31. Columbu, S.; Mulas, M.; Mundula, F.; Cioni, R. Strategies for helium pycnometry density measurements of welded ignimbritic rocks. *Measurement* **2021**, *173*, 108640. [CrossRef]
32. Mukaida, K.-I. Density Measurement of Small Porous Particles by Mercury Porosimetry. *Powder Technol.* **1981**, *29*, 99–107.

recycling

MDPI

Article

Effect of Solvent Pre-Treatment on the Leaching of Copper During Printed Circuit Board Recycling

Ahmed Tarek Ismail Mohamed [1], Giuliana Schimperna [2], Gianluca Cantoni [2], Francesca Demichelis [3], Debora Fino [3], Sara Perucchini [2], Francesca Rubertelli [2] and Francesco Laviano [3,4,*]

1 Department of Energy, Politecnico di Torino, 10129 Turin, Italy; ahmed.mohamed@polito.it
2 Eni SpA—Renewable, New Energies and Material Science Research Center, 28100 Novara, Italy; giuliana.schimperna@eni.com (G.S.); gianluca.cantoni@eni.com (G.C.); sara.perucchini@eni.com (S.P.); francesca.rubertelli@eni.com (F.R.)
3 Department of Applied Science and Technology, Politecnico di Torino, 10129 Turin, Italy; francesca.demichelis@polito.it (F.D.); debora.fino@polito.it (D.F.)
4 INFN—Sez. Torino, 10125 Torino, Italy
* Correspondence: francesco.laviano@polito.it

Abstract: Printed circuit boards (PCBs) are fundamental components of electronic devices, acting as an important source of various valuable metals such as copper, gold, and silver. Efficient recycling methods that offer high recovery rates are essential to the full reutilization of these materials. Hydrometallurgical leaching is a prominent technique for metal recovery, but its efficiency can be significantly enhanced through solvent pre-treatment. In this study, an experimental analysis of the material composition of different categories of PCBs is presented. In addition, the study evaluates the influence of particle size on the subsequent copper leaching process and the efficiency of copper recovery. These investigations aim to better understand the material composition of PCBs and propose an optimized material recovery technique. The study finds that there are significant variances among the different categories of PCBs investigated, allowing a more informed handling process of WEEE. This research suggests that solvent pretreatment using DMSO for PCB particle sizes between 5.6 mm and 2 mm would be a good optimization technique, mitigating the drawbacks of treating fine particles while maintaining appealing recovery efficiency.

Keywords: WEEE; waste management; PCB; solvent pretreatment; circular economy

1. Introduction

Printed circuit boards (PCBs) lie at the heart of modern electronic devices, working as connecting paths between electronic components and as a skeletal support. Estimates of the economic value of materials found in Waste Electrical and Electronic Equipment (WEEE) are almost €48 billion, with PCBs containing 40% of these materials [1]. Found in every electronic device, a PCB is a multilayer substrate with copper layers in addition to solder masks and silkscreens. Fiberglass-reinforced epoxy resin (FR4) is typically used for the fabrication of the substrate, offering mechanical strength and insulation. Moreover, phenolic resin and polyimide are also used depending on the application's requirements [2, 3]. The traces, which are the conductive pathways, are made of copper due to its high conductivity.

A high-efficiency material recovery process is crucial from both an economic perspective and a legislative one, especially in the EU, where many of the materials found within WEEE are on the list of critical raw materials, which recently added copper and nickel among them [4–6].

The complex composition of PCBs, combining metals, resins, and other materials, makes the recycling process complicated. To mitigate the effect of this complex composition, PCBs are usually size-reduced (typically by shredding or milling) to enable higher-efficiency metal extraction [7]. The efficiency of the subsequent treatment processes is highly dependent on the particle size after size reduction [4–7].

One of the most common treatment methods for waste PCBs is hydrometallurgical leaching [8], a process that dissolves metals from solid matrices using aqueous solutions to recover them. This method allows high selectivity to extract valuable metals from complex mixtures, making it highly suitable for recycling PCBs with their complex structures. To perform the leaching process, first, the waste PCBs are crushed and ground to reduce their size, increasing the exposed surface area available for chemical reactions during leaching [2]. Acids, alkalis, and complexing agents are then used for the chemical attack. Common leaching agents used for copper recovery processes are sulfuric acid, nitric acid, and ammonia [9–11]. The efficiency of the extraction process depends, among other factors, on the temperature, pH, and concentration of the leaching agent, as well as the agitation [12–14].

A recent approach to improving the efficiency of hydrometallurgical treatment is adding a chemical pretreatment phase using a solvent after the mechanical reduction of the size of the waste PCBs [15]. The process aims to swell or decompose the resin matrix in PCBs, depending on the organic solvent used, exposing the embedded metal layers, which increases the surface area of metals, consequently improving the efficiency of the leaching process [16]. Moreover, solvent pretreatment was reported as a method for the debromination of plastics found in waste PCBs, as well as other WEEE types [17–21].

Numerous studies have investigated the use of different organic solvents for PCB pretreatment, including but not limited to the use of N-methyl-2-pyrrolidone (NMP) [22], Dimethylformamide (DMF) [23], dimethylacetamide (DMA) [24], and dimethyl sulfoxide (DMSO) [20]. These studies evaluate the factors affecting the efficiency of the pretreatment process, such as the operating temperature, duration of exposure, solvent concentration, and treated particle size. The different pretreatment methods, as well as the optimized conditions for each method, are summarized in Table 1.

DMSO has demonstrated the effective delamination of PCB fragments under specific conditions, such as controlled temperatures (60–170 °C) and solid-to-liquid ratios, with optimal results achieved for smaller particle sizes [22,25,26]. However, in all attempts, the experiments were conducted on relatively large particle sizes rather than powdered PCBs. Furthermore, DMSO presents significant health risks due to its ability to penetrate the skin and the toxicity of its vapors [23].

DMA and DMF have shown promise in treating larger PCB fragments, avoiding the fine crushing process that often leads to metal loss in smaller fractions [24,27]. However, another challenge faced when handling such large pieces of PCBs is the presence of through holes and pins on the PCB, which hinder the swelling process. DMF stands out for its low health risks, ease of regeneration, and compatibility with brominated resins, making it a safer alternative to some solvents [23].

NMP is a strong solvent for PCB treatment, achieving effective resin removal in comparative studies; however, its carcinogenic and reproductive health risks pose significant drawbacks [22,23].

Table 1. Summary of the solvent pretreatment methods for waste PCBs.

Method Used	Type of WEEE	Particle Size	Temperature	Duration	Solid-To-Liquid Ratio	Reference
DMSO	PC-Motherboard	16 mm^2	145 °C	60 min	1:7	Zhu et al. [20]
DMSO	PC-Motherboard	1–1.5 cm^2	90 °C	60 min	1:2	Zhu et al. [25]
DMSO	PC-Motherboard	1–1.5 cm^2	135 °C	10 min	1:2	Zhu et al. [25]
DMSO	PC-Motherboard	2–3 cm^2	90 °C	90 min	1:2	Zhu et al. [25]
DMSO	PC-Motherboard	2–3 cm^2	135 °C	20 min	1:2	Zhu et al. [25]
DMSO	PC-Motherboard	15–20 mm^2	170 °C	30 min	1:2	Zhu et al. [26]
DMSO	PC motherboard	6 mm	90 °C	90 min	1:2	Wath et al. [22]
DMA	PC-Motherboard	1 cm^2	160 °C	75 min	1:10	Dean Kang et al. [27]
DMA	PC-Motherboard	1 cm^2	160 °C	150 min	3:10	Verma et al. [24]
DMA	PC-Motherboard	16 cm^2	160 °C	420 min	3:10	Verma et al. [24]
DMF	Mix PCBs	1 cm^2	135 °C	240 min	300 g/L	Verma et al. [23]
NMP	PC motherboard	8 mm sieve	100 °C	90 min	1:5	Wath et al. [22]
Supercritical ethanol	RAM PCB	10–20 mm	300 °C	60 min	1:20	Preetam et al. [28]

Other techniques also include the use of supercritical fluids, which are considered more eco-friendly and have been effective in removing the organic materials from crushed RAM PCBs at around 300 °C [28–31]. However, the stability of brominated epoxy resins above 250 °C remains a concern due to potential pollutant emissions [23].

In addition to environmental and efficiency considerations, several patents have proposed innovative approaches, such as multistage treatment processes involving combinations of oxidizing agents, leaching agents, and organic solvents [32–35].

The advantages of using different pretreatment techniques are the optimization of recovery rates and the mitigation of the challenges posed by varying PCB compositions. In addition, solvent pretreatment reduces the energy used during the treatment process by eliminating the need for extensive mechanical pretreatment, as well as favoring the environmental aspect of the PCB waste handling process.

A more profound inspection of the reported studies shows the lack of information on the effect of solvent pretreatment on the leaching process at fine particle sizes in the micron grade. Such understanding of the impact of solvent pretreatment on small particle sizes is important since most studies on the hydrometallurgical treatment of PCBs had favorable results at particle sizes in micron grade [8,36–39]. However, PCB treatment using fine particles comes with dust-generation hazards and particle loss [27].

Another gap was also highlighted by [40], which reported that studies focus mainly on the analysis of the dissolved brominated epoxy resin and the regeneration techniques and seldom investigate the effects of swelling on metal leaching and the change in the properties of the treated PCB.

Moreover, an in-depth understanding of PCBs and their material composition is essential for an informed approach to PCB treatment. One recent study investigated the

PCB extracted from computers, laptops, and TVs and compared their material content [41]. Furthermore, they compared the different particle size groups and the prevalence of the materials within these groups based on their liberation after grinding, allowing for targeted material recovery with higher efficiency. As a result, they reported that general metals such as copper, zinc, and aluminum are found in coarse fractions between 0.18 and 0.25 mm, while precious metals and rare earth metals (REM) are present in finer sizes. The results complement those from another study, which also studied the metal content of PCBs from computers, laptops, mobile phones, and TVs [42]. It also investigated the best-performing digestion methods for sample preparation and analysis, concluding that the use of strong acids such as hydrogen fluoride (HF) specified in the USEPA 3052 method ensures the effective dissolution of metals embedded within a silica matrix, as in the case of PCBs. These studies, although insightful, leave a gap regarding our knowledge of the material variation among the PCBs extracted from the same WEEE type, such as personal computers.

Based on the identified gaps in the literature, this study aims to investigate the effect of solvent pretreatment on the efficiency of the subsequent leaching process, evaluating how particle size influences recovery rates to optimize the process conditions. Additionally, the study seeks to address the critical question of material variation across different categories and models of PCBs, offering new insights into a largely underexplored area. The research follows an experimental approach to investigate the research questions and evaluate the optimized particle size for solvent pretreatment on a lab scale.

The novelty of this research lies in exploring the swelling of intermediate particle sizes during the solvent pretreatment of motherboard PCBs using DMSO to optimize the conditions, thus mitigating the risks of handling finer particles as well as increasing the copper recovery efficiency to exceed that reached when treating larger particle sizes. Another contribution of this research is that the information obtained from the analysis of the material composition of the different categories of PCBs would allow a more informed decision on whether the separation of PCBs when treating them is justifiable or whether the material variation among the different categories and models is insignificant. The data can also be used to promote more precise calculations in the case of being inserted in calculation models. These areas of novelty fulfill the study objectives to develop an in-depth understanding of PCBs of different categories and find the optimized treatment conditions using the solvent pretreatment method.

The subsequent section reports the results obtained from the analysis of the samples and outlines important relations and comparisons among the different samples analyzed and treated. In Section 3, the paper discusses the materials analyzed and the preparation of the samples for both the analytical analysis of the component material, as well as the investigation of the effect of solvent pretreatment on the selected particle sizes. Finally, Section 4 concludes the key outcomes of this study.

2. Results and Discussion

2.1. *Characterization of the Various Categories of PCBs*

The data obtained from the investigation of the three different types of PCBs bridge the gap in our knowledge about the differences in the composition between the various types of PCBs. To ensure the data is valuable, it is essential to compare the average content of materials that are economically important or present in large quantities. The average material composition of the three PCB types, based on the XRF analysis data, shows that all three types primarily contain the same materials: copper, silicon, tin, calcium, and aluminum, as shown in Figure 1a.

For these dominant materials, ANOVA tests were carried out. The tests reveal that both copper and silicon are statistically representative and that there is no difference among the three PCB categories, i.e., $F < F_{critical}$. As for tin, aluminum, and calcium, the ANOVA tests have opposite results, meaning that they are not statistically representative, with $F > F_{critical}$. This is especially interesting since the pretreatment and leaching section of this study is focused on copper, thus confirming the representativeness of the results obtained hereafter.

Regarding the presence of copper, which is the most prevalent material in all three analyzed types, CPUs and chipsets are the richest in copper with an average percentage of 20.8%, followed by RAMs at 17.8%, with motherboards showing the lowest amount at only 14.5%, as shown in Figure 1b.

Silicon is most abundant in motherboards, accounting for 13%, although it is present in similar quantities in both RAMs and CPUs, as shown in Figure 1c. In contrast, motherboards are a significant source of aluminum and tin. As shown in Figure 1d, the average aluminum content in motherboards is nearly double that of CPUs and 1.5 times greater than in RAMs. For tin, as depicted in Figure 1e, motherboards are particularly rich, followed by CPUs, which contain approximately half the tin content of motherboards, and RAM, which has a relatively low tin content.

A detailed analysis of the average individual material compositions reveals that CPUs and chipsets are particularly enriched in copper, gold, and silver, while also exhibiting elevated levels of lead. As shown in Figure 1f, the gold content in CPUs and chipsets is approximately twice the average amount found in RAMs, and the silver content is nearly three times higher than that in motherboards, with an even greater disparity when compared to RAMs. The presence of titanium in RAMs and motherboards is worth noting given the importance of titanium as a critical raw material within the EU, as shown in Figure 1g, especially in the case of motherboards, which although contain less than that of the RAMs in terms of quantity, are actually higher since motherboards are usually heavier in weight than RAMs.

Finally, the average lead content in CPUs and chipsets is eightfold that in RAMs and sixfold the average found in motherboards, as shown in Figure 1h. This underlines the importance of correctly treating CPUs and chipsets and employing extra protective procedures to safeguard both the workers and the environment, especially since such elevated levels also surpass those allowed by European regulations. A detailed analysis of each category of the PCBs studied is provided in the supplementary material [43,44].

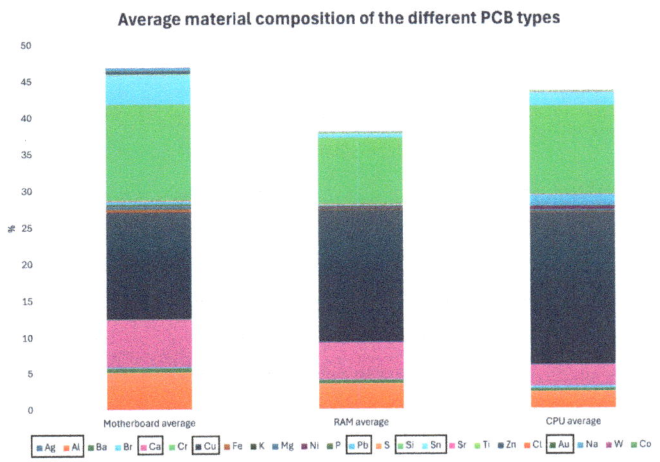

(a)

Figure 1. *Cont.*

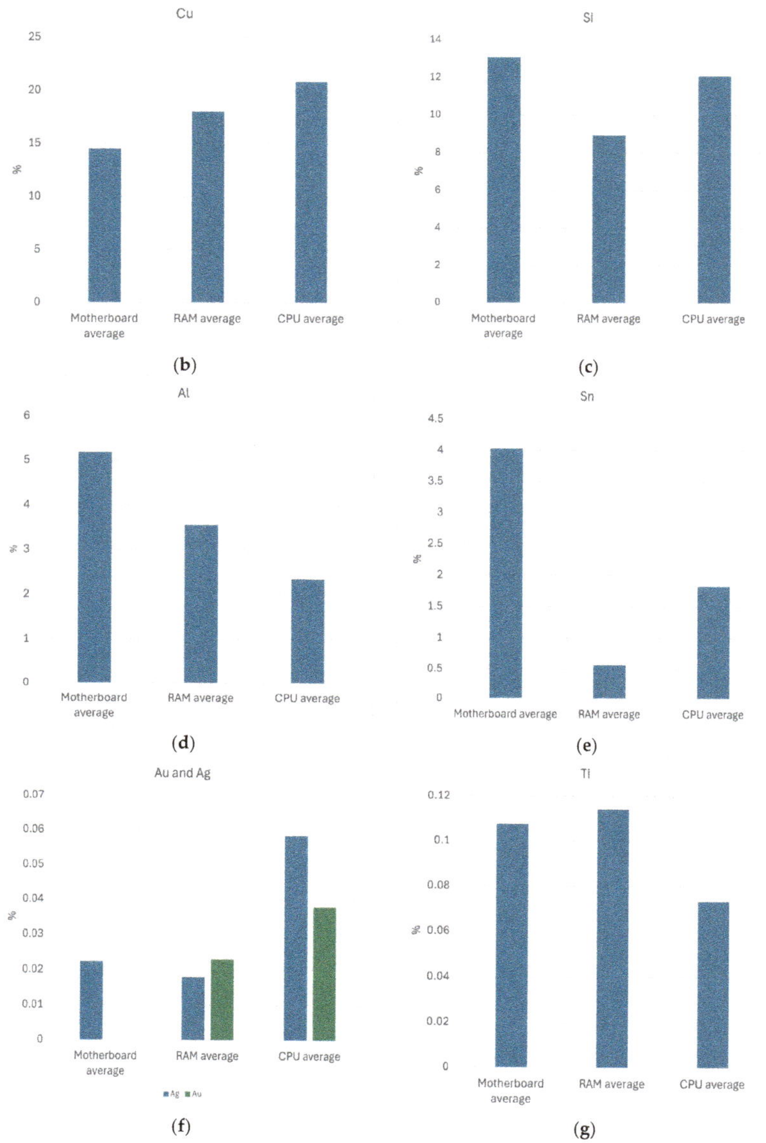

Figure 1. *Cont.*

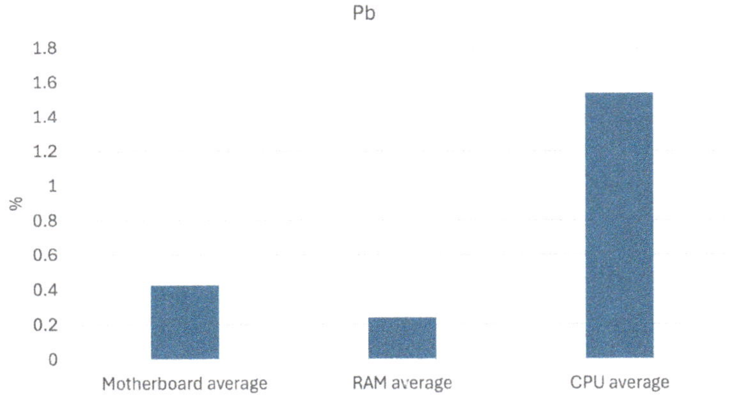

(h)

Figure 1. Comparison of the average content of some of the principal materials in different PCB types (Motherboards, CPUs, and RAMs): (**a**) the overall average composition of three types of PCBs; (**b**) the average content of copper in different PCB types; (**c**) the average content of silicon in different PCB types; (**d**) the average content of aluminum in different PCB types; (**e**) the average content of tin in different PCB types; (**f**) the average content of silver and gold in different PCB types; (**g**) the average content of titanium in different PCB types; and (**h**) the average content of lead in different PCB types.

2.2. *Effect of Solvent Pretreatment of Waste PCBs Using DMSO*

Waste PCBS were first analyzed to determine their material composition and allow for a better understanding of the effect of the subsequent pretreatment and leaching process. The composition of the waste motherboard PCB mixture goes hand in hand with the discussed results in the previous section. The detailed characterization is shown in Table 2.

Table 2. The detailed material composition of the waste motherboard PCB mixture used for the pretreatment and leaching tests.

Material	Al	Ba	Br	Ca	Cu	Fe	Mg	Na	Pb	S	Si	Sn	Sr	Ti	Zn	Zr
Weight %	4.6	0.6	0.1	9.5	12.7	0.1	0.2	0.2	0.5	0.2	13.3	3.2	0.1	0.2	0.03	0.01

2.2.1. Effect of Pretreatment on the Organic Element

The inspection of the pretreated motherboard PCBs using DMSO showed that the organic part, made of resin and glass fibers, is separated from the metals. This swelling effect is seen in both Group 1 with particle sizes between 8 mm and 5.6 mm and Group 2 with particle sizes between 5.6 mm and 2 mm. However, for Group 3 with particle sizes of less than 400 µm, the swelling effect is not visually detectable by eye inspection. Images of the pretreated samples from Group 2 before and after are shown in Figure 2a and b, respectively, and those of Group 3 before and after are shown in Figure 2c and d, respectively.

Further analysis of the treated sample groups was carried out to assess the effect of DMSO on the organic. The calcination process is performed under the conditions mentioned in Section 3.3 as part of the sample preparation process for the chemical analysis.

Figure 2. Images of the solvent-pretreated PCB samples: (**a**) Group 2 with particle size between 5.6 mm and 2 mm before solvent pretreatment; (**b**) Group 2 with particle size between 5.6 mm and 2 mm after solvent pretreatment; (**c**) Group 3 with particle size less than 400 μm before solvent pretreatment; and (**d**) Group 3 with particle size less than 400 μm after solvent pretreatment.

The analysis of all three category groups showed limited to no change in the organic content loss before and after the solvent treatment, as expected from the solvent pretreatment phase using DMSO. For instance, Group 1 had a weight loss percentage representing the organic content equal to 26.7% and 28% before and after treatment, respectively. Similar behavior was exhibited by both Groups 2 and 3, where Group 2 had an organic content of 26.8% and 27% before and after solvent treatment, respectively, while Group 3 had an organic content of 21.8%, which remained constant during the solvent treatment process. This confirmed the results seen visually in Figure 2, which shows that the resin is swollen, liberating the metal particles; however, the resin is not removed. The swelling of the material during the solvent pretreatment phase, along with the release of metal particles, improves the leaching efficiency, as discussed in the following section.

Similarly, the analysis of the inorganic components in the samples, before and after solvent pretreatment, exhibits negligible changes. This stability can be attributed to the heterogeneous nature of the samples.

2.2.2. Effect of Pretreatment on Copper Extraction Efficiency During Leaching

During the preparation of the experiment, different agitation speeds were tested. Agitation between 400 and 510 rpm was empirically found to be best and was subsequently used to carry out leaching experiments. This optimized agitation speed was based on the speed that allowed constant and smooth spinning of the magnetic stirring anchor without any disturbances to the flow. Regarding the copper recovery efficiency for the three sample groups, a significant improvement was noted in the case of solvent-treated samples. For instance, in the case of Group 1, the extraction efficiency rises from 45.5% without solvent treatment to 72.4% after being subjected to solvent treatment. In the case of Group 2, the copper recovery efficiency improved from 47.3% to 87.65% after the sample was treated with DMSO before the leaching process. Finally, for Group 3, the improvement in copper recovery efficiency during the leaching process is less staggering, yet it rises from 93.24% to a 100% copper recovery.

The results obtained go hand in hand with the literature regarding the hypothesis that pretreatment using DMSO as a solvent improves the copper recovery efficiency during leaching. Moreover, they provide a deeper understanding of the relationship between particle size and solvent treatment. In that case, the effect of solvent treatment is more visible on larger particle sizes than on fine grains, which can be seen from the almost 30% increase in efficiency in the case of larger particles (Group 1 and Group 2) compared to the 7% increase in case of fine grains (Group 3), in which case the additional use of resources might not be fully justifiable.

The ICP analysis also showed an increase in the efficiency of recovery of other metals present during the leaching process using nitric acid. The increase also followed the same trend as copper, where the effect of solvent treatment was more evident in larger-sized particles (Group 1 and Group 2) than in finer-sized particles (Group 3). The results are shown in Figure 3. In Figure 3a, the recovery efficiency is almost doubled in the case of magnesium, aluminum, calcium, and barium. For tin, the efficiency increases 40-fold. For Group 2, the efficiency doubles for zinc and triples for magnesium, aluminum, calcium, and barium, with tin experiencing a dramatic 40-fold increase, as shown in Figure 3b. Finally, for Group 3, with its finer particle size, as shown in Figure 3c, the trend is less drastic, with efficiencies improving slightly, showing the limited impact of solvent treatment on the leaching process.

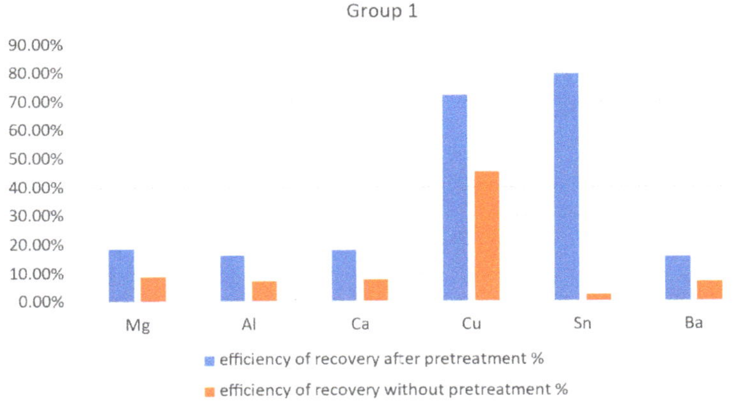

(a)

Figure 3. *Cont.*

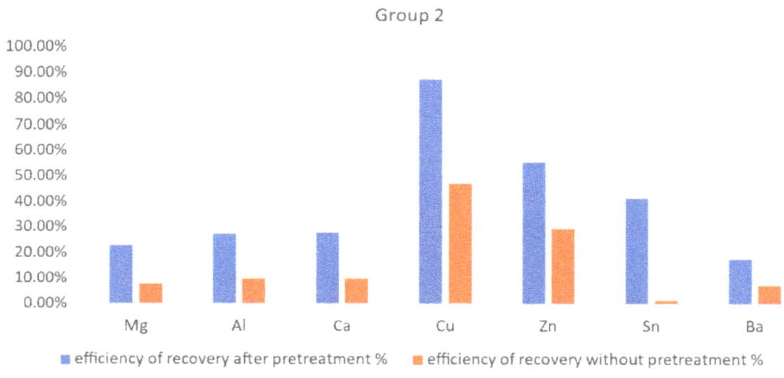

(**b**)

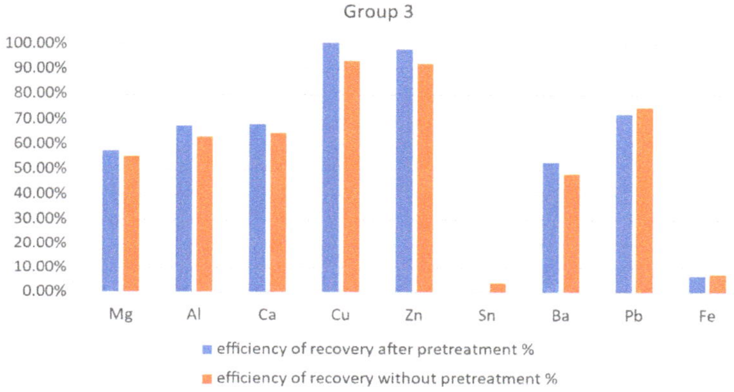

(**c**)

Figure 3. The effect of solvent treatment on the metal recovery efficiency for different particle sizes: (**a**) Group 1 with particle size between 8 mm and 5.6 mm; (**b**) Group 2 with particle size between 5.6 mm and 2 mm; and (**c**) Group 3 with particle size less than 400 μm.

3. Materials and Methods

3.1. Materials Collection and Preparation

Waste PCBs were obtained from personal computers used at the Politecnico di Torino, Italy. The retrieved PCBs were 9 motherboards, 10 Random Access Memories (RAMs), and 11 Central Processing Units (CPUs) and chipsets. The PCBs were mostly produced between 2001 and 2011.

The electronic components mounted on the PCBs were desoldered and removed manually via the indirect heating of the PCBs to fuse the soldering and ease the dismounting of the components, leaving the stripped PCB base, which is a substrate sandwich of glass fibers, metals, and resin. The CPU and RAM were cut manually into 1 cm wide strips and then reduced in size using a universal mill in preparation for their analysis. As for the motherboards, a counter-rotating blade mill was first used to reduce the size of the particles to 5.6 mm. Then, further pulverization of the motherboard PCBs was carried out following the same procedure for the CPU and RAM. At each stage, the particle size was checked using sieves to ensure a homogeneous particle size among the samples. The sieves used were 8 mm, 5.6 mm, and 2 mm. In Figure 4, the samples of the different analyzed PCBs are shown. Each sampled PCB was categorized into one of three categories in the scope of this study. Then, the weight of the stripped PCB was measured and the year of production was researched in the cases where it could be identified. To evaluate the effect of pretreatment on copper leaching, the sourced motherboards were all mixed during the shredding process to guarantee a good blend together and were then used for pretreatment and leaching. The material composition was characterized both before and after the treatment processes, identifying the materials contained in each sample and at each change, which allowed us to compare the effect of the pretreatment on leaching.

(a) (b) (c)

Figure 4. Samples of the different types of PCBs analyzed: (**a**) Motherboard PCB; (**b**) CPU; (**c**) RAM.

3.2. Pre-Treatment and Leaching Method

The pretreatment phase was applied with the aim of swelling the resin and liberating the metals for the subsequent leaching phase. The optimized pretreatment conditions were those reported by [25] for a particle size of 1–1.5 cm^2, which uses DMSO, 1:2 (w/v) at 90 °C for 60 min under constant N_2 flow. Those conditions were chosen as they do not require a long heating time at elevated temperatures. The agitation for the process was established at 400–460 rpm using a heating magnetic stirrer. After the pretreatment, the sample was left to dry overnight in a vacuum furnace at 35 °C. Then, part of the sample was taken for analysis following the previously elaborated process for the ICP and XRF and the rest of the sample was used for leaching.

The leaching process, as established by [36], uses nitric acid (HNO_3) with a concentration of 3 M and 75 g/L pulp density at 75 °C for 120 min. The chemicals used for the solvent pretreatment and the leaching process were all of lab grade.

The experimental setup for both the solvent pretreatment and the leaching process can be seen in Figure 5a, and Figure 5b, respectively. For the solvent pretreatment process, a double-necked round bottom flask was connected to the top of the cooling system using running water and the N_2 flow in an open system setup. The flask was immersed in a hot oil bath with temperature control. Regarding the leaching process, a triple-necked round bottom flask of 250 mL was connected to the top of the cooling system using running water in an open system setup. One of the side openings was fitted with a thermometer for temperature control and the other was used to add the sample. The flask was also immersed in a hot oil bath, like the setup of the solvent pretreatment.

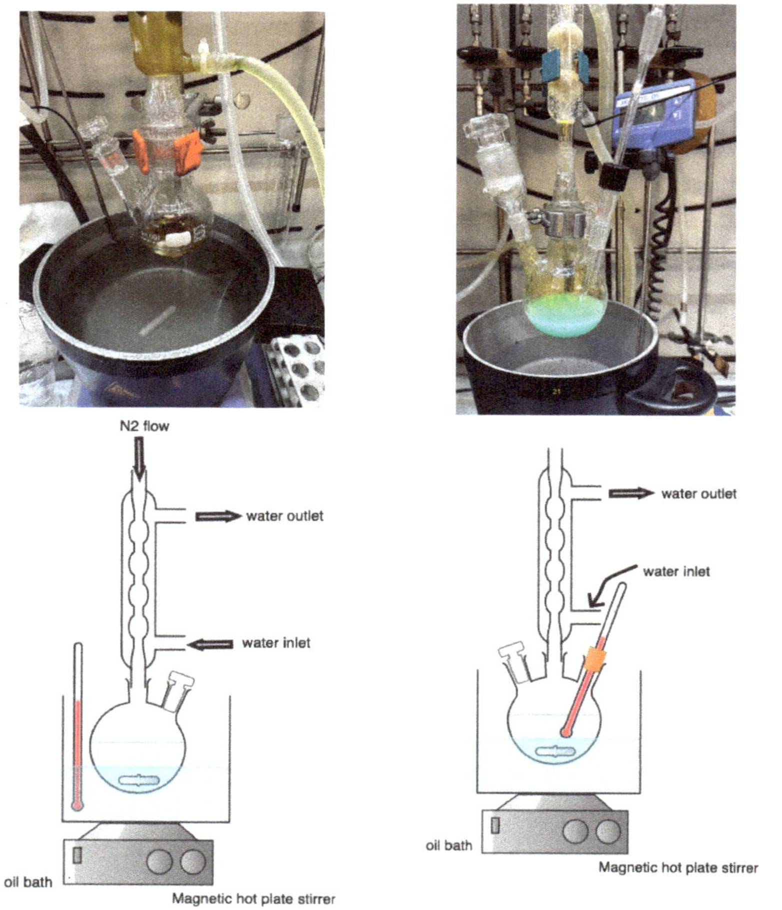

(a) (b)

Figure 5. Experimental setups for the (**a**) solvent pretreatment process and (**b**) leaching process.

The investigation of the effect of solvent pretreatment on the efficiency of leaching was carried out on 3 different particle size ranges; the first group (hereafter referred to as Group 1) had a particle size between 8 mm and 5.6 mm, the second group (hereafter referred to as Group 2) had a particle size between 5.6 mm and 2 mm, and the final group (hereafter referred to as Group 3) contained fine particles with sizes of less than 400 μm. These size groups were chosen based on the gap identified in the literature where the finest particle size of PCBs to be studied for pretreatment was 6 mm [22]. Thus, the first group was large and covered what has been performed in the literature. Group 2 was chosen as it is an intermediate between the fine particle size and the large size reported in the literature. Group 3 was chosen to compare the difference between a powder state and a coarser particle size, which does not suffer from the hazards of handling fine powders. For each group, a sample was prepared and then a part was analyzed to obtain the material content of the untreated sample. Next, the part was pretreated with a subsequent analysis of part of the sample to study the effect of solvent pretreatment. Finally, the pretreated sample underwent leaching and was then analyzed. To study the effect of solvent pretreatment, a part of the originally prepared untreated sample for each group was taken and directly treated by the established leaching process, allowing a comparison of the results with those that underwent pretreatment. A visual representation of the experiments can be seen in Figure 6.

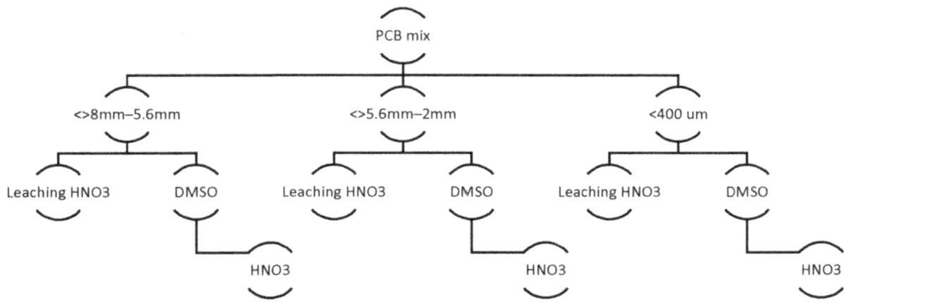

Figure 6. A diagram of the experiments carried out to assess the impact of the solvent pretreatment process.

3.3. Analytical Methods

The sampled PCBs were analyzed using Inductively Coupled Plasma Optical Emission Spectrometry (ICP-OES), Inductively Coupled Plasma Mass Spectrometry (ICP-MS), and X-ray fluorescence spectroscopy (XRF- Axion, Panalytical). For the analysis, the samples were prepared using the previously mentioned process until a fine powder was obtained.

In addition, statistical analysis of the different PCB categories was carried out using the data analysis extension of Excel V.16.89.1, and ANOVA (Analysis-Of-Variance) tests were performed to evaluate the linear correlations among the materials found in the different categories of PCBs, considering only the ones with values F greater than $F_{critical}$ as significant.

4. Conclusions

The objectives established for this study were to investigate the effect of solvent pretreatment on the efficiency of the subsequent leaching process, optimizing the process conditions. This study offers a better understanding of the role that particle size plays in the efficiency of material recovery using DMSO as a solvent pretreatment, bridging the gap in the literature. Although fine particle sizes such as in Group 3 have higher copper recovery efficiency even without the pretreatment step, this method continues to demonstrate favorable results, especially with larger particle sizes, which may help eliminate the need for pulverization and lower the amount of material lost during the pulverizing and transportation process. However, there remains a set of challenges that require further improvements, such as solvent recovery and reuse and the scaling of the current solvent treatment method from the lab scale to an industrial scale. Finally, although DMSO is considered a green solvent, it is not completely eco-friendly. Ongoing research is investigating the use of other green solvents, such as ionic liquids, as alternatives.

This study also investigates the different categories of PCBs and the extent of variation in the material composition among the different categories and models, which helps to better target the extraction of those materials and improve the material recovery rates. The results obtained also show through the ANOVA tests carried out that the different categories are statistically representative when it comes to copper and silicon; however, this is not valid for aluminum, tin, and calcium. The study is in line with the goals of a green circular economy, encouraging urban mining and supporting the recent EU Critical Raw Materials act. The results show that silicon, copper, calcium, and aluminum have an overall dominance in the composition of the different categories and models, despite being present in various percentages.

For future work, we aim to deepen the understanding of the changes undergone by raw materials during the pretreatment and leaching phases by studying the microstructure of the PCBs at each phase. Further scaling up of the study from the lab scale to bench and pilot scales is also planned so as to continue assessing the feasibility of this method on an industrial scale.

Supplementary Materials: The following supporting information can be downloaded at: https://www.mdpi.com/article/10.3390/recycling10030080/s1, Figure S1. The relation between the motherboard weight in grams and the present unit market price in Euros of the motherboards was analyzed. The code of each PCB includes its year of production; Figure S2. The relation between copper content in % and the present unit market price in Euros of the motherboards was analyzed. The code of each PCB includes its year of production; Figure S3. The relation between silver content in % and the unit price of the motherboards in Euros; Figure S4. The material composition of the 9 analyzed motherboard PCBs. The production years of motherboards are indicated between brackets and the thickness of size of the bars represent the amount of material percentage in the sample; Figure S5. Material composition of the 10 analyzed RAM and the average material content. The year of production of each analyzed RAM is shown between parenthesis and the size of the bar represents the material presence within the sample; Figure S6. Visualization of the impact of gold and silver content on the current unit price of the RAM in Euros; Figure S7. Visualization of the relation between the tin content in the analyzed RAMs and their current unit price in Euros; Figure S8. Visualization of the trends in material content of the analyzed RAM showing the fluctuations and unsteady trends. In figure (a) the material content is shown on a log scale while in (b) the graph represents the content percentage of each material; Figure S9. Graphical representation of the material content of the four most present materials in the analyzed RAM samples. The graph is shown in log scale of the content percentage of each material; Figure S10. The material composition of the 11 analyzed CPUs and chipsets; Figure S11. Graphical representation of the gold and silver content in the different analyzed CPUs and chipsets; Table S1. The material composition of the 9 analyzed motherboard PCBs using XRF; Table S2. Material composition of the 10 analyzed RAM PCBs using XRF; Table S3. The material composition of the 11 analyzed CPUs and chipsets using XRF.

Author Contributions: Conceptualization, A.T.I.M., G.S. and F.R.; methodology, A.T.I.M., G.S. and F.R.; software, A.T.I.M.; validation, A.T.I.M., G.S. and F.R.; formal analysis, A.T.I.M., G.S. and F.D.; investigation, S.P. and G.C.; resources, A.T.I.M., G.S., S.P. and G.C.; data curation, A.T.I.M., S.P. and G.C.; writing—original draft preparation, A.T.I.M. and F.D.; writing—review and editing, F.D. and F.L.; visualization, A.T.I.M. and F.D.; supervision, F.L., D.F. and F.R.; project administration, F.L., D.F. and F.R.; funding acquisition, F.L. All authors have read and agreed to the published version of the manuscript.

Funding: This research was funded by Eni S.p.A. and the NODES project, which received funding from the MUR–M4C2 1.5 of PNRR funded by the European Union—Next Generation EU (grant agreement no. ECS00000036).

Data Availability Statement: The original contributions presented in this study are included in the article/supplementary material. Further inquiries can be directed to the corresponding author.

Acknowledgments: A.T.I.M. and F.L. acknowledge financial support by Eni S.p.A.

Conflicts of Interest: Author Giuliana Schimperna, Gianluca Cantoni, Sara Perucchini and Francesca Rubertelli were employed by the company Eni SpA. The remaining authors declare that the research was conducted in the absence of any commercial or financial relationships that could be construed as a potential conflict of interest.

Abbreviations

CPU	Central Processing Unit
DMA	Dimethylacetamide
DMF	Dimethylformamide
DMSO	Dimethyl sulfoxide
ICP-MS	Inductively Coupled Plasma Mass Spectrometry
ICP-OES	Inductively Coupled Plasma Optical Emission Spectrometry
NMP	N-methyl-2-pyrrolidone
PBDE	Polybromodiphenylethers
PC	Personal computer
PCB	Printed circuit Boards
RAM	Random Access Memory
REM	Rare-earth metals
TBBPA	Tetrabromobisphenol A
WEEE	Waste Electrical and Electronic Equipment
XRF	X-ray fluorescence spectroscopy

References

1. Kumar, A.; Holuszko, M.; Crocce, D.; Espinosa, R. E-waste: An overview on generation, collection, legislation and recycling practices. *Resour. Conserv. Recycl.* **2017**, *122*, 32–42. [CrossRef]
2. Cui, J.; Zhang, L. Metallurgical recovery of metals from electronic waste: A review. *J. Hazard. Mater.* **2008**, *158*, 228–256. [CrossRef] [PubMed]
3. Hall, W.J.; Williams, P.T. Separation and recovery of materials from scrap printed circuit boards. *Resour. Conserv. Recycl.* **2007**, *51*, 691–709. [CrossRef]
4. European Commission. *A New Circular Economy Action Plan for a Cleaner and More Competitive Europe*; European Commission: Brussels, Belgium, 2020. Available online: https://eur-lex.europa.eu/legal-content/EN/TXT/?uri=CELEX:52020DC0098 (accessed on 5 November 2024).
5. European Commission. Critical Raw Materials. Available online: https://single-market-economy.ec.europa.eu/sectors/raw-materials/areas-specific-interest/critical-raw-materials_en (accessed on 29 October 2024).
6. European Parliament. *Directive 2012/19/EU on Waste Electrical and Electronic Equipment (WEEE)*; European Parliament: Brussels, Belgium, 2012; pp. 38–71. Available online: https://eur-lex.europa.eu/legal-content/EN/TXT/?uri=CELEX:32012L0019 (accessed on 2 July 2024).
7. Priya, A. *Management of Electronic Waste: Resource Recovery, Technology and Regulation*; Wiley: Hoboken, NJ, USA, 2023; Available online: https://www.wiley.com/en-us/Management+of+Electronic+Waste:+Resource+Recovery,+Technology+and+Regulation-p-9781119894339 (accessed on 16 January 2024).
8. Oke, E.A.; Potgieter, H. Recent chemical methods for metals recovery from printed circuit boards: A review. *J. Mater. Cycles Waste Manag.* **2024**, *26*, 1349–1368. [CrossRef]
9. Trinh, H.B.; Kim, S.; Lee, J. Selective Copper Recovery by Acid Leaching from Printed Circuit Board Waste Sludge. *Metals* **2020**, *10*, 293. [CrossRef]
10. Liu, J.C.; Kao, T.-H. Extraction of Cu and Pb from printed circuit board sludge using ammonia solutions. *Water Sci. Technol.* **2003**, *47*, 167–172. [CrossRef]
11. Kumar, M.; Lee, J.-C.; Kim, M.-S.; Jeong, J.; Yoo, K. Leaching of metals from waste printed circuit boards (WPCBs) using sulfuric and nitric acids. *Environ. Eng. Manag. J. (EEMJ)* **2014**, *13*, 2601–2607. [CrossRef]
12. Ilyas, S.; Anwar, M.A.; Niazi, S.B.; Afzal Ghauri, M. Bioleaching of metals from electronic scrap by moderately thermophilic acidophilic bacteria. *Hydrometallurgy* **2007**, *88*, 180–188. [CrossRef]
13. Birloaga, I.; De Michelis, I.; Ferella, F.; Buzatu, M.; Vegliò, F. Study on the influence of various factors in the hydrometallurgical processing of waste printed circuit boards for copper and gold recovery. *Waste Manag.* **2013**, *33*, 935–941. [CrossRef]
14. Tuncuk, A.; Stazi, V.; Akcil, A.; Yazici, E.Y.; Deveci, H. Aqueous metal recovery techniques from e-scrap: Hydrometallurgy in recycling. *Miner. Eng.* **2012**, *25*, 28–37. [CrossRef]
15. Oke, E.A.; Potgieter, H. Discarded e-waste/printed circuit boards: A review of their recent methods of disassembly, sorting and environmental implications. *J. Mater. Cycles Waste Manag.* **2024**, *26*, 1277–1293. [CrossRef]
16. Cui, J.; Forssberg, E. Mechanical recycling of waste electric and electronic equipment: A review. *J. Hazard. Mater.* **2003**, *99*, 243–263. [CrossRef]

17. Charitopoulou, M.A.; Papadopoulou, L.; Achilias, D.S. Removal of Bromine from Polymer Blends with a Composition Simulating That Found in Waste Electric and Electronic Equipment through a Facile and Environmentally Friendly Method. *Polymers* **2023**, *15*, 709. [CrossRef]
18. Guo, X.; Peng, S.; Jiang, L.; Mo, X.; Zhu, Y.; Liu, Y.; Cai, K.; Song, Q. Removal of polybrominated diphenyl ethers in high impact polystyrene (HIPS) from waste TV sets. *Environ. Sci. Pollut. Res.* **2022**, *29*, 59317–59327. [CrossRef]
19. Schlummer, M.; Popp, L.; Trautmann, F.; Zimmermann, D.; Maurer, A. Recovery of bromine and antimony from WEEE plastics. In Proceedings of the 2016 Electronics Goes Green 2016+ (EGG), Berlin, Germany, 6–9 September 2016; IEEE: Piscataway Township, NJ, USA; pp. 1–5. [CrossRef]
20. Zhu, P.; Chen, Y.; Wang, L.Y.; Zhou, M.; Zhou, J. The separation of waste printed circuit board by dissolving bromine epoxy resin using organic solvent. *Waste Manag.* **2012**, *33*, 484–488. [CrossRef] [PubMed]
21. Council, B.; Haarman, A.; Fedato, S.; Holt, A. Brominated Flame Retardants and the Circular Economy of WEEE Plastics State of Play. 2023. Available online: https://www.bsef.com/wp-content/uploads/2023/09/Brominated-Flame-Retardants-and-the-Circular-Economy-of-WEEE-Plastics.pdf (accessed on 19 January 2024).
22. Wath, S.B.; Katariya, M.N.; Singh, S.K.; Kanade, G.S.; Vaidya, A.N. Separation of WPCBs by dissolution of brominated epoxy resins using DMSO and NMP: A comparative study. *Chem. Eng. J.* **2015**, *280*, 391–398. [CrossRef]
23. Verma, H.R.; Singh, K.K.; Mankhand, T.R. Dissolution and separation of brominated epoxy resin of waste printed circuit boards by using di-methyl formamide. *J. Clean. Prod.* **2016**, *139*, 586–596. [CrossRef]
24. Verma, H.R.; Singh, K.K.; Mankhand, T.R. Delamination mechanism study of large size waste printed circuit boards by using dimethylacetamide. *Waste Manag.* **2017**, *65*, 139–146. [CrossRef]
25. Zhu, P.; Chen, Y.; Wang, L.; Qian, G.; Zhang, W.; Zhou, M.; Zhou, J. Dissolution of Brominated Epoxy Resins by Dimethyl Sulfoxide To Separate Waste Printed Circuit Boards. *Environ. Sci. Technol.* **2013**, *47*, 2654–2660. [CrossRef]
26. Zhu, P.; Chen, Y.; Wang, L.Y.; Qian, G.R.; Zhou, M.; Zhou, J. A novel approach to separation of waste printed circuit boards using dimethyl sulfoxide. *Int. J. Environ. Sci. Technol.* **2013**, *10*, 175–180. [CrossRef]
27. Kang, K.D.; Ilankoon, I.; Chong, M.N.; Wu, T.Y. Exfoliation of coarse printed circuit boards using dimethylacetamide: Production of copper concentrates. *Miner. Eng.* **2023**, *191*, 107963. [CrossRef]
28. Preetam, A.; Modak, A.; Jadhao, R.; Naik, S.N.; Pant, K.K.; Kumar, V. A comprehensive study on the extraction of transition metals from waste random access memory using acetic acid as a chelating solvent. *J. Environ. Chem. Eng.* **2022**, *10*, 2213–3437. [CrossRef]
29. Calgaro, C.O.; Schlemmer, D.F.; Bassaco, M.M.; Dotto, G.L.; Tanabe, E.H.; Bertuol, D.A. Supercritical extraction of polymers from printed circuit boards using CO2 and ethanol. *J. CO_2 Util.* **2017**, *22*, 307–316. [CrossRef]
30. Sanyal, S.; Ke, Q.; Zhang, Y.; Ngo, T.; Carrell, J.; Zhang, H.; Dai, L.L. Understanding and optimizing delamination/recycling of printed circuit boards using a supercritical carbon dioxide process. *J. Clean. Prod.* **2013**, *41*, 174–178. [CrossRef]
31. de Souza, G.B.M.; Pereira, M.B.; dos Santos, L.F.; Alonso, C.G.; Jegatheesan, V.; Cardozo-Filho, L. Management of waste printed circuit boards via supercritical water technology. *J. Clean. Prod.* **2022**, *368*, 133198. [CrossRef]
32. Suslavich, B.T.; Das, A.; Young, C.A. Chemical Liberation of Waste Printed Circuit Boards. WO2022178235A1, 25 August 2022. Available online: https://patents.google.com/patent/WO2022178235A1/en (accessed on 23 September 2024).
33. Kaifua, su. Method for Non-Incineration and Non-Cyanation Treatment of Waste Printed Circuit Boards. JP5766305B2, 19 August 2015. Available online: https://patents.google.com/patent/JP5766305B2/en (accessed on 23 September 2024).
34. Torres Marques, A.; Da Silva Monteiro Bastos, M.M.; Marques Martelo, L.; Soares De Sousa, P.M.; Relvas Vieira Dias, A.C.; Guerreiro Alves Arroja, L.M. Eco-Friendly Method for Recycling Electronic Waste. WO2022123438A1, 16 June 2022. Available online: https://patents.google.com/patent/WO2022123438A1/en (accessed on 23 September 2024).
35. Ma, L.; Shi, Y.; Wang, Q.; Liang, X.; Li, Z.; Liu, L. Method for Harmless Treatment and Resource Comprehensive Recovery of Circuit Board. CN103084369A, 8 May 2013. Available online: https://patents.google.com/patent/CN103084369A/en (accessed on 23 September 2024).
36. Dutta, D.; Panda, R.; Kumari, A.; Goel, S.; Jha, M.K. Sustainable recycling process for metals recovery from used printed circuit boards (PCBs). *Sustain. Mater. Technol.* **2018**, *17*, e00066. [CrossRef]
37. Vijayaram, R.; Nesakumar, D.; Chandramohan, K. Copper Extraction from the Discarded Printed Circuit Board by Leaching. *Res. J. Eng. Sci.* **2013**, *2*, 11–14. Available online: https://www.isca.in (accessed on 29 November 2023).
38. Sethurajan, M.; van Hullebusch, E.D. Leaching and Selective Recovery of Cu from Printed Circuit Boards. *Metals* **2019**, *9*, 1034. [CrossRef]
39. Silvas, F.P.C.; Jiménez Correa, M.M.; Caldas, M.P.K.; de Moraes, V.T.; Espinosa, D.C.R.; Tenório, J.A.S. Printed circuit board recycling: Physical processing and copper extraction by selective leaching. *Waste Manag.* **2015**, *46*, 503–510. [CrossRef]
40. Kang, K.D.; Ilankoon, I.M.S.K.; Dushyantha, N.; Chong, M.N. Assessment of Pre-Treatment Techniques for Coarse Printed Circuit Boards (PCBs) Recycling. *Minerals* **2021**, *11*, 1134. [CrossRef]

41. Priya, A.; Hait, S. Characterization of particle size-based deportment of metals in various waste printed circuit boards towards metal recovery. *Clean. Mater.* **2021**, *1*, 100013. [CrossRef]
42. Priya, A.; Hait, S. Qualitative and quantitative metals liberation assessment for characterization of various waste printed circuit boards for recycling. *Environ. Sci. Pollut. Res.* **2017**, *24*, 27445–27456. [CrossRef] [PubMed]
43. Baldé, C.P.; Kuehr, R.; Yamamoto, T.; McDonald, R.; D'Angelo, E.; Althaf, S.; Bel, G.; Deubzer, O.; Fernandez-Cubillo, E.; Forti, V.; et al. "The Global E-waste Monitor 2024", Geneva/Bonn. 2024. Available online: https://www.itu.int/itu-d/sites/environment (accessed on 16 October 2024).
44. The European Parliament, Directive 2011/65/EU of the European Parliament and of the Council of 8 June 2011 on the Restriction of the Use of Certain Hazardous Substances in Electrical and Electronic Equipment Text with EEA Relevance. 2011. Available online: https://eur-lex.europa.eu/legal-content/EN/TXT/?uri=CELEX:32011L0065 (accessed on 8 September 2018).

recycling

MDPI

Article

Electrolytic Recovery of Indium from Copper Indium Gallium Selenide Photovoltaic Panels: Preliminary Investigation of Process Parameters

Monika Gajec *, Anna Król *, Jadwiga Holewa-Rataj, Ewa Kukulska-Zając and Tomasz Kuchta

Oil and Gas Institute–National Research Institute, ul. Lubicz 25a, 31-503 Kraków, Poland; holewa-rataj@inig.pl (J.H.-R.); kukulska-zajac@inig.pl (E.K.-Z.); kuchta@inig.pl (T.K.)
* Correspondence: gajec@inig.pl (M.G.); krola@inig.pl (A.K.)

Abstract: The European Green Deal emphasizes the development of renewable energy sources to combat climate change. However, as photovoltaic expansion accelerates, so does the potential for increased waste, necessitating effective material recycling strategies. Indium, a scarce and valuable element crucial to the production of photovoltaic panels, underscores the necessity for efficient recycling practices to reduce reliance on virgin resources. In a recent laboratory analysis, a CIGS photovoltaic panel underwent a series of processes including crushing, grinding, and homogenization. The concentration of indium, vital for recycling, was meticulously analyzed using ICP-MS and validated through microscopic and composition analyses. Subsequent extraction utilizing 3 M HCl and H_2O_2, followed by electrolysis, yielded a remarkable up to 52% indium recovery within a 48-h timeframe. Importantly, the study encompassed both averaged panel samples and samples from the absorbing layer, emphasizing the comprehensive approach required for efficient recycling. This underscores the critical importance of optimizing recycling processes to mitigate the environmental impact associated with the disposal of photovoltaic panels. By maximizing indium recovery, not only are environmental impacts reduced, but the long-term sustainability of renewable energy technologies is also ensured. This highlights the interconnectedness of recycling practices with the broader goals of achieving a circular economy and securing the viability of renewable energy systems in the fight against climate change.

Keywords: indium recovery; photovoltaic panel; CIGS; electrolysis

1. Introduction

The sustainable economic development of every country is closely tied to resource availability. Ensuring stable resource supplies is crucial for many industrial sectors that produce essential goods to meet diverse societal needs. Population growth, dynamic industrialization, the transition towards ecological transportation and energy systems, constantly increasing demand, and new technological innovations are the main factors driving increased demand for critical resources.

The development of technologies based on renewable energy sources is also associated with a continuous demand for critical resources. Resource recovery is therefore crucial to ensuring supply. Hence, research is necessary to develop resource recovery technologies. As the popularity of renewable energy sources such as solar energy grows, it is important for all associated infrastructure to be sustainable. Recycling CIGS panels aligns with efforts towards sustainable development by addressing potential issues related to electronic waste accumulation and resource utilization. The production process of photovoltaic panels, especially in terms of resource extraction and processing, generates a certain ecological footprint. Recycling CIGS panels helps minimize this footprint because recovered resources can be reused, reducing the amount of electronic waste going to landfills. It also contributes to efficient energy use since reusing recovered resources eliminates the need for mining and processing new ones, which often involves high energy consumption and provides materials for the production of new photovoltaic panels. This closed-loop approach to material circulation is crucial for the sustainable development of the photovoltaic industry [1–4].

Although crystalline silicon dominates the photovoltaic (PV) market, thin-film technologies such as CIGS (copper indium gallium selenide) panels have gained attention due to their high absorption coefficients, flexibility, and potential for lower production costs. As of recent market data, CIGS panels account for approximately 2–3% of global PV production, with annual production capacities estimated at around 1 GW in the past few years. While relatively small in market share, CIGS panels are expected to generate increasing volumes of waste in the coming decades, particularly as earlier generations of these panels reach their end of life [5,6].

Forecasts suggest that by 2030, thousands of tons of CIGS-based PV waste will enter the recycling stream annually, creating a growing need for recovery strategies targeting their critical raw materials, especially indium and gallium [7].

A typical CIGS photovoltaic module contains approximately [8–10]:

- 65–75% glass (by weight);
- 10–15% polymer layers (e.g., EVA, back contact);
- 5–10% metals such as aluminum, molybdenum, and copper;
- <2% semiconducting layer, which includes copper, indium, gallium, and selenium;
- Indium content typically ranging from 100 to 300 ppm, depending on thickness and manufacturer specifications.

Given the limited global reserves and rising demand for indium in high-tech applications, developing efficient recovery technologies from end-of-life CIGS panels is both environmentally and economically justified.

CIGS panels are a type of second-generation thin-film photovoltaic technology, renowned for their high efficiency, lightweight design, and flexibility. These panels utilize a thin layer of copper indium gallium selenide (CIGS) deposited onto a substrate, enabling the production of versatile and cost-effective solar modules. With their favorable performance in low-light conditions and potential for integration into various surfaces, CIGS panels present a promising alternative to traditional silicon-based solar technologies, particularly in applications where weight and adaptability are crucial factors. However, despite these advantages, a major challenge associated with the widespread deployment of CIGS panels lies in the limited availability of critical raw materials used in their production. Indium, in particular, is a key component of the CIGS semiconductor and is primarily obtained as a by-product of zinc extraction. Both indium and gallium are listed among the European Union's critical raw materials due to their economic importance and supply risks. As a result, the recovery of these elements from end-of-life CIGS panels is essential for promoting resource sustainability, reducing dependence on primary raw materials, and supporting the long-term scalability of CIGS technology in the renewable energy sector [11,12]. Effective waste management and recycling of CIGS solar panels are critical due to the toxic nature of the material, which can pose serious environmental risks and potential health hazards, including lung toxicity. While indium is a critical component in CIGS photovoltaic cells due to its role in enabling high-efficiency energy conversion, its presence in end-of-life modules raises significant environmental and occupational health concerns. Although not as widely recognized as other toxic metals, indium has gained attention in recent years for its potential harmful effects, particularly in its oxidized forms. Exposure to indium—especially in the form of indium oxide, a common by-product in the processing and recycling of CIGS panels—has been linked to serious respiratory issues. Research indicates that inhalation of indium-containing dust or fumes can lead to chronic pulmonary conditions, including interstitial lung disease and alveolar proteinosis. Unlike many metals that pose greater risks in their ionic or soluble forms, indium oxide presents dangers even in its particulate state, further complicating safe recovery operations. As the use of indium continues to expand across the electronics and renewable energy sectors, addressing its toxicological profile becomes crucial. Future recycling strategies for CIGS modules must incorporate not only efficient metal recovery techniques but also protective measures to minimize human exposure and environmental release, ensuring that the transition to clean energy does not come at the cost of health and safety [13–15].

The focus of this article and the associated research on CIGS photovoltaic panels can be attributed to several key factors that make CIGS technology particularly relevant for current and future solar energy applications. Firstly, CIGS solar cells have gained considerable attention in the photovoltaic industry due to their high efficiency and potential for low-cost production. CIGS modules have demonstrated superior energy conversion efficiency compared to many other thin-film technologies, making them a prominent choice for sustainable energy generation. Additionally, CIGS technology is known for its flexibility and lightweight characteristics, which opens up possibilities for a wider range of applications compared to traditional crystalline silicon-based solar cells. These unique features make CIGS modules increasingly popular in both industrial-scale and emerging energy solutions. Secondly, CIGS modules contain valuable and critical materials such as Ga, In, and Ag, which are in high demand for various high-tech applications [7]. As the global transition to renewable energy accelerates, the need for efficient recycling technologies becomes paramount to ensure the sustainability of this growing industry. CIGS panels, due to their material composition, are of particular interest when it comes to recovering these scarce and high-value elements. This has sparked considerable research into improving recycling methods and minimizing environmental impact. Lastly, CIGS technology is widely considered to be at the forefront of thin-film solar cell development, with significant ongoing research aimed at improving the efficiency and lifespan of these panels. Focusing research on CIGS provides an opportunity to not only advance the recycling methods for a widely used technology but also to address the broader challenges of material recovery and waste management in the solar panel industry [11,16].

CIGS panels are known for their lightweight and flexible design, which provides significant advantages over traditional crystalline silicon solar cells. These advantages include enhanced applicability in various installation scenarios, from residential rooftops to more unconventional surfaces. The structure of a CIGS panel is multi-layered, with each layer serving a specific purpose to maximize the efficiency of light absorption and energy conversion. Below, a schematic diagram is presented, illustrating the layers of a CIGS solar panel (Figure 1).

Below is a short description of the individual layers of a CIGS panel.

The **front contact layer** serves a protective function. The front layer in CIGS panels is typically made of materials that are transparent to sunlight, allowing it to penetrate freely into the absorber layer. The most commonly used materials are thin-film layers of silicon dioxide (SiO_2) or zinc oxide (ZnO). In ITO (indium tin oxide) panels, zinc oxide can be replaced by aluminum or gallium.

The **protective layer** is a crucial element ensuring the longevity of CIGS panels while also needing to be sufficiently transparent to allow light to pass through to the absorber layer. The composition of this layer may vary depending on the specific panel design and manufacturer preference.

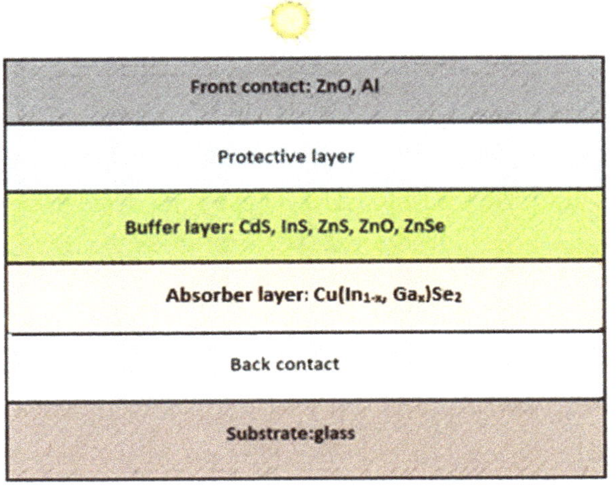

Figure 1. A diagram depicting the layers of a CIGS panel [17–19].

The **buffer layer** plays a role in ensuring the efficiency and stability of panels. In the case of CIGS panels and other thin-film photovoltaic panels, the buffer layer is often used between the absorber layer and the current-collecting layer. The main goals of the buffer layer are:

- Minimizing carrier loss: The buffer layer helps minimize carrier recombination, which can occur at the interface between the absorber layer and the current-collecting layer. Recombination is the process where charge carriers (electrons and holes) combine and lose their energy, which is undesirable as it affects the efficiency of solar energy conversion.
- Improving electrostatic balance: The buffer layer helps maintain electrostatic balance between the absorber layer and the current-collecting layer, which is crucial for effective charge carrier transport.
- Protection of the absorber layer: The buffer layer also serves a protective role, shielding the absorber layer from damage and the effects of atmospheric conditions.
- Improving panel structure coherence: Adding a buffer layer can improve the coherence of panel structures, which is crucial for achieving uniform and efficient solar energy conversion.

In the case of CIGS panels, a semiconductor-based buffer layer, such as CdS, is typically used. However, it is worth noting that the development of photovoltaic technologies is dynamic, and different panels may use different materials in the buffer layer depending on the technology and production process.

The **CIGS absorber layer** plays a crucial role in converting light energy into electrical energy. It is the layer where photons of sunlight are absorbed, leading to the generation of electrical current. The main components of the absorber layer are copper, indium, gallium, and selenium. The absorber layer in CIGS panels is typically very thin, on the order of a few micrometers, which is one of the characteristic features of thin-film photovoltaic panel technology. This structure allows for flexibility, reduced production costs, and application on various types of surfaces.

The **back contact layer** acts as the current-collecting electrode on the reverse side of the panels. It is a crucial layer for efficiently collecting charge carriers generated in the absorber layer as a result of sunlight absorption. Typically, the back contact layer in CIGS panels consists of materials with conducting properties. This layer can be relatively thin, allowing for the free flow of electrical current. Materials such as molybdenum are commonly used in this layer [17–20].

CIGS photovoltaic panels based on copper, indium, gallium, and selenium compounds typically have a long lifespan, often exceeding 25 years. The durability of these panels depends on various factors, such as production quality, operating conditions, the quality of materials used, and applied technologies. Recycling CIGS photovoltaic panels is crucial, especially in the context of growing interest in renewable energy sources. Copper and gallium, which are key components of CIGS panels, are critical resources. Their resources are limited, and recycling allows for the efficient recovery of these valuable materials while also minimizing the need for new, limited resources [21,22]. The significant recovery of indium underscores its crucial role in the production of photovoltaic panels and highlights the necessity for efficient recycling practices to alleviate reliance on virgin resources. Indium, a scarce and valuable element, is a vital component in the production of high-efficiency solar cells, making its recovery essential for the sustainability of renewable energy technologies and the circular economy. Thus, optimizing indium recovery processes not only mitigates environmental impact but also ensures the long-term viability of renewable energy systems. One of the methods used for metal recovery from spent photovoltaic panels is electrolysis. Indium can undergo electrodeposition from a variety of electrolyte solutions, such as cyanide, sulfate, and chloride [21–23].

Gu et al. [24] recovered valuable components from spent CIGS modules by combining electrochemical techniques with dehydration and distillation processes. Initially, the leaching solution of CIGS undergoes a two-step electrodeposition process to recover selenium and copper. Subsequently, the remaining solution is distilled to recycle hydrochloric acid (HCl) and crystallize indium and gallium chlorides. The obtained hydrates are then dehydrated through refluxing with thionyl chloride ($SOCl_2$). Following this step, anhydrous $InCl_3$ is recovered and separated from $GaCl_3$ via a straightforward filtration process. $GaCl_3$ is further separated from $SOCl_2$ through distillation. The recovery rates for indium and gallium are high, at approximately 99.99% and 99.98%, respectively, with corresponding purities of around 99.99%. As part of this study, an attempt was made to simplify this process.

Liu et al. [25] proposed a recycling process for Cu, In, and Ga from CIGS-based thin-film solar panels using a multi-step approach. Initially, the panels are separated layer by layer based on the different thermal strains of materials. Subsequently, annealing is conducted to remove Se, followed by leaching with nitric acid to facilitate individual metal extraction. In the extraction step, In is extracted into the organic phase using di-(2-ethylhexyl) phosphoric acid, leaving Cu and Ga in the aqueous phase. Ga is then extracted using the same agent under different conditions, while nearly pure Cu remains in the residual solution. Metal hydroxide precipitates are formed by adding ammonium hydroxide to the solutions. Under optimized conditions, a recovery rate of over 90% for In, Ga, and Cu can be achieved. Finally, the hydroxides are recycled and converted into metal oxides with a purity exceeding 99% through calcination. This approach offers an effective pathway for recycling and recovering these valuable metals from waste CIGS thin-film solar panels.

One study [26] investigated the potential of benign leaching conditions for recovering primarily silver and indium from production waste of flexible CIGS solar cells while also assessing contamination levels from other industrial elements in the leachate. Selective leaching of contaminants was also explored to obtain purer streams of valuable metals for reuse in new products. The results showed that increasing acid concentration and surface to liquid ratio led to higher leaching yields of Ag and In but also increased contamination. Complete Ag recovery and 85% In recovery were achieved with 2 M HNO_3 and A:L of 1:3 cm^2/mL after 24 h of leaching at room temperature. Leaching with 0.5 M HNO_3 extracted 85% Ag and 30% In under the same conditions, with reduced contamination levels. Additionally, leaching with 0.1 M HNO_3 showed promise for achieving higher Ag purity through an initial step of Zn-selective leaching for 1 h.

The method proposed by Hu et al. [27] outlines an oxidative roasting-leaching-extraction process for recovering valuable metals from spent copper indium gallium selenide materials. Initially, the spent CIGS materials undergo roasting to convert copper, indium, and gallium from selenides to oxides while volatilizing selenium. Subsequently, hydrochloric acid leaching is employed to dissolve the oxides under optimized conditions: 4 mol/L acid concentration, 80 °C temperature, 3 h leaching time, and a liquid–solid ratio of 10 mL/g, resulting in high leaching rates of 99.98% for copper, 93.40% for indium, and 96.86% for gallium. In a subsequent step, indium and gallium are sequentially extracted from the solution using P204 solvent extraction, achieving extraction rates exceeding 99.92% for indium and 99.34% for gallium, while copper remains minimally extracted. Utilizing hydrochloric acid as the stripping agent leads to stripping rates of 99.90% for indium and 99.93% for gallium. These findings demonstrate the efficacy of this method for the efficient recovery and separation of copper, indium, and gallium from spent CIGS materials.

Various methods have been developed for recovering valuable metals from spent CIGS panels, each with unique environmental implications. One approach involves electrochemical techniques combined with dehydration and distillation to separate selenium, copper, indium, and gallium, achieving high purity levels. Another method focuses on layer-by-layer separation based on thermal strains, followed by annealing, acid leaching, and selective extraction using organic solvents, resulting in efficient metal recovery. Additionally, controlled leaching conditions have been explored to recover silver and indium while minimizing contamination. Oxidative roasting to convert metal selenides into oxides, followed by hydrochloric acid leaching and solvent extraction, has also shown high recovery rates. Recycling of CIGS panels, however, requires careful consideration of the environmental impacts of each method. While mechanical and innovative techniques offer potential advantages, they require further optimization to minimize their ecological footprint. Advanced methods, such as electrodeposition and selective recovery, show considerable promise in reducing the impacts of global warming and toxicity. These methods contribute to more efficient material recovery and minimize environmental risks associated with the disposal and recycling of end-of-life solar panels, ultimately supporting resource sustainability and waste reduction [19,28–30].

2. Materials and Methods

2.1. Preparation of the Photovoltaic Panel for Laboratory Studies

For laboratory studies, a CIGS photovoltaic panel from NICE Solar Energy GmbH was used, measuring 27 cm × 39.5 cm and with a thickness of 6 mm. The total weight of the panel was 1826 g (Figure 2).

The panel was broken into smaller pieces using a hammer and then ground in a planetary mill. The resulting mixture was mixed and homogenized, and laboratory samples were taken for analysis (referred to as P1–P6 in the conducted research) (Figure 3).

Considering that the CIGS photovoltaic panel is constructed from various materials including glass and the part that absorbs solar energy, the panel was also examined for the possibility of fragmentation and separation of elements richer in extracted resources.

Therefore, to confirm the presence of indium in the panel, a powdered average sample of the panel was subjected to microscopic analysis as well as composition analysis using a high-resolution scanning electron microscope. The same analysis was also performed on a sample isolated from the dark (energy-absorbing) part of the panel, whose macroscopic image is presented in Figure 4.

Figure 2. The CIGS photovoltaic panel from NICE Solar Energy GmbH, which was utilized for laboratory studies.

Figure 3. The ground and homogenized CIGS photovoltaic panel prepared for laboratory studies.

Figure 4. The absorber layer of the CIGS panel.

2.2. Inductively Coupled Plasma Mass Spectrometry (ICP-MS)

To assess the efficiency of the electrolysis process, samples were collected from the electrolyzed solutions at various predetermined time intervals, with each sample being 100 µL. Subsequently, 1900 µL of 1 M HNO_3 was added to each sample, and they were analyzed using ICP-MS. Additionally, investigations were conducted on the initial sample, comprising ground and homogenized CIGS photovoltaic panels. To achieve this, the sample underwent mineralization using a microwave oven (MARS6, CEM) with 0.5 g of the sample, 6 mL of HNO_3, 2 mL of HCl, and 2 mL of H_2SO_4. The mineralized sample was then diluted and subsequently analyzed using ICP-MS.

Nitric acid (V) 67–69%, hydrochloric acid 36%, and sulfuric acid (VI) 93–98%, used for leaching and mineralization, were produced by ROMIL (high-purity acid dedicated to trace metal analysis). Hydrogen peroxide (35%) used for the experiments was sourced from Merck KGaA. For instrument calibration, the Tuning Solution from LGC Standards containing Ce, Co, Li, Mg, Tl, and Y at a concentration of 1.00 µg/L in 2% HNO_3 was employed.

In the samples, In (m/z 115) was determined using an Agilent 7900 mass spectrometer. Certified reference materials from CPAChem for In at a concentration of 100 mg/L in 2% HNO_3 and a multi-element solution (Ag, Al, B, Ba, Bi, Ca, Cd, Co, Cr, Cu, Fe, Ga, In, K, Li, Mg, Mn, Na, Ni, Pb, Sr, Tl, Zn) at a concentration of 1000 mg/L in 6.5% HNO_3 from Supelco were used for the determination of In contents using ICP-MS. The apparatus was calibrated using the Tuning Solution from LGC Standards containing Ce, Co, Li, Mg, Tl, and Y at a concentration of 1.00 µg/L in 2% HNO_3.

2.3. Scanning Electron Microscopy

Microscopic examinations were conducted using a high-resolution scanning electron microscope JEOL JSM-7500F (JEOL Ltd., Tokyo, Japan) equipped with an attachment for chemical composition analysis in micro areas (AZtecLiveLite Xplore 30, Oxford Instruments NanoAnalysis, High Wycombe, UK). The microscope features the following detectors:

- Secondary Electrons (SE) detector used for imaging the surface morphology of materials.
- Backscattered Electrons (BSE) detector utilized for imaging with chemical composition contrast.
- Transmission Electron Detector (TED) employed for observing "thin" samples in transmission mode.

2.4. Electrolysis Operating Conditions

The setup used for electrolysis is depicted in Figure 5.

A beaker containing the electrolyte solution was placed on a magnetic stirrer. Throughout the process, the solution was stirred at 200 rpm using a magnetic stirrer. The electrolysis was conducted using a potentiostat-galvanostat ATLAS 0931. (Atlas-Sollich Zakład Systemów Elektronicznych, Banino, Poland) Three electrodes were immersed in the solution:

- A chlorosilver electrode type RL-100 was used as a reference electrode. It contains a Ag/AgCl half-panel immersed in a non-exchangeable, potassium chloride solution saturated with silver chloride. The reference half-panel is shielded by an outer chamber filled with an intermediate (protective) solution, which is in contact with the sample under investigation via an external electrolytic connector. A 4.0 M potassium chloride solution is used as the intermediate solution.
- A platinum electrode type EPt-01 served as the counter electrode. Its potential depends on the redox equilibrium in the solution. It cannot be used independently but rather in conjunction with a reference electrode with a potential largely independent of the composition of the solution under study. Both electrodes, when connected to a

pH/millivoltmeter, form a redox potential measurement panel. The indicator element of the electrode is a platinum ring permanently attached to the end of the glass body.

- A carbon electrode, specifically the Glassy Carbon Voltammetry Electrode (BASMF2012-1EA, Bioanalytical Systems, Inc. (BASi®), West Lafayette, IN, USA), was used as the working electrode. Its main advantages include temperature resistance, high hardness and durability, a wide potential window, and chemical stability.

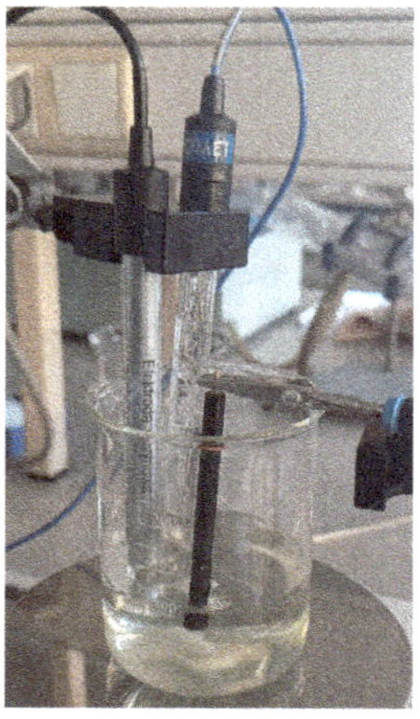

Figure 5. The setup used for electrolysis.

The reagents used for preparing samples of the photovoltaic panel for electrolysis were as follows:

- H_2O_2 (30% concentration, Merck);
- Concentrated HCl (high purity, 35–38%, ROMIL);
- HNO_3 (high purity, 67–69%, ROMIL);
- 1 M HCl—prepared in the laboratory by dissolving concentrated HCl in deionized water;
- 3 M HCl—prepared in the laboratory by dissolving concentrated HCl in deionized water;
- 1 M HNO_3—prepared in the laboratory by dissolving concentrated HNO_3 in deionized water;
- NH_4Cl (analytical grade, POCH);
- Indium beads, diam. 2–5 mm, 99.999% trace metals basis, Sigma Aldrich.

The laboratory equipment used for preparing samples of the photovoltaic panel for electrolysis is listed below:

- Ultrasonic bath with heating function up to 80 °C, SONOREX Bandelin;
- Muffle furnace with temperature gradient setting capability in the range of 30–1800 °C;
- Magnetic stirrer MS11 WIGO;
- Analytical balance Mettler Toledo XP 205.

The conditions for conducting the copper and indium electrolysis process are presented in Table 1.

Table 1. A compilation of conditions for conducting indium electrolysis from solutions after metal extraction from CIGS panel.

Potential [V]	Stirring Speed	Experiment Duration
−0.902	500 rpm	12–72 h, dependent on the concentration of In in the solution

3. Results and Discussion

3.1. Indium Content Analysis in Averaged Panel Samples Versus Samples from the Absorbing Layer

The first stage of laboratory research involved the characterization of the CIGS photovoltaic panel. Preparing the photovoltaic panel for analysis was very challenging due to its composition and diverse fractions, which could not be entirely homogenized in a planetary mill. The aim of grinding and disaggregation of the panel was not to obtain a perfectly homogeneous sample. The panel was prepared in the described manner to facilitate the approximate determination of the analyzed elements using ICP-MS. Additionally, efforts were made to simplify the panel preparation procedure and electrolysis to minimize costs, labor, and time.

Laboratory samples were obtained from the ground and averaged CIGS panel (as described in Section 2.1). These samples underwent mineralization, and after dilution, the content of In was analyzed using ICP-MS.

The obtained research results were compiled in Table 2. They are also presented graphically in Figure 6.

Table 2. Summary of indium content results in averaged panel samples (samples P1-P6). Analysis conducted using ICP-MS on acid-mineralized samples.

The Sample Designation	Indium [mg/g CIGS Panel]
P1	0.1233
P2	0.1525
P3	0.1339
P4	0.0877
P5	0.1461
P6	0.1480
The minimum value	0.0877
The maximum value	0.1525
The median	0.1400
The arithmetic mean	0.1319

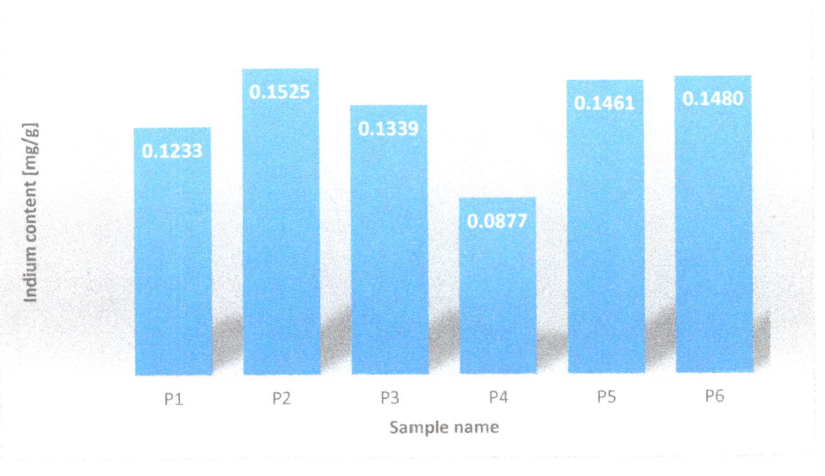

Figure 6. Summary of In content results in averaged panel samples (samples P1–P6). Analysis conducted using ICP-MS on acid-mineralized samples.

The research results indicate that the maximum content of indium in averaged panel samples is 0.152 mg/g. Taking into account the mass of a single panel, which is 1826 g, this allowed us to calculate that from one panel, potentially 277 mg of In can be recovered. The results indicate that the sample was not fully homogenized. To minimize costs and energy consumption associated with sample preparation, the disassembly and grinding of the CIGS panel were deliberately simplified. No thermal or aggressive chemical pre-treatment was applied. Furthermore, in the analysis of In content, no outlier data points were removed. Instead, the results are presented as an approximate concentration range observed in the processed samples from the panel.

CIGS panels at the end of their life cycle are gaining increasing attention as an unconventional yet valuable source of critical raw materials. According to the literature, they contain indium in concentrations of approximately 100–300 ppm—levels that exceed those typically found in natural mineral deposits [8,9]. The concentration of indium obtained in our study (approximately 150 ppm) is consistent with the values reported in the literature, confirming the potential of CIGS panels as a significant secondary source of this element. Figures 7–9 depict the images and graphs obtained for the characterized ground and averaged panel sample, the macroscopic images of which are shown in Figures 2 and 3.

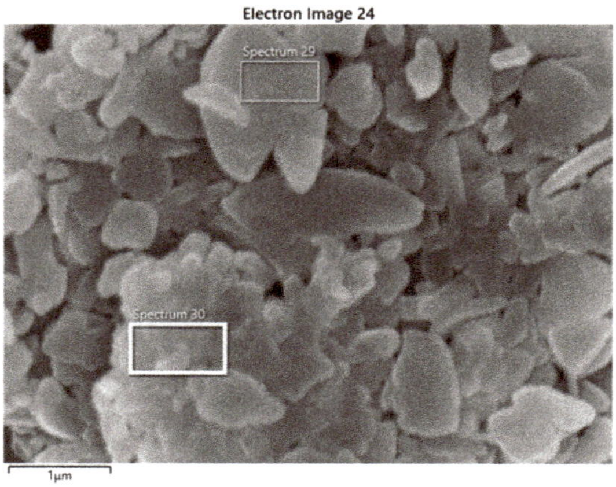

Figure 7. The image from the electron microscope of the averaged CIGS panel sample with marked points at Spectrum 29 and Spectrum 30, where elemental analysis was conducted.

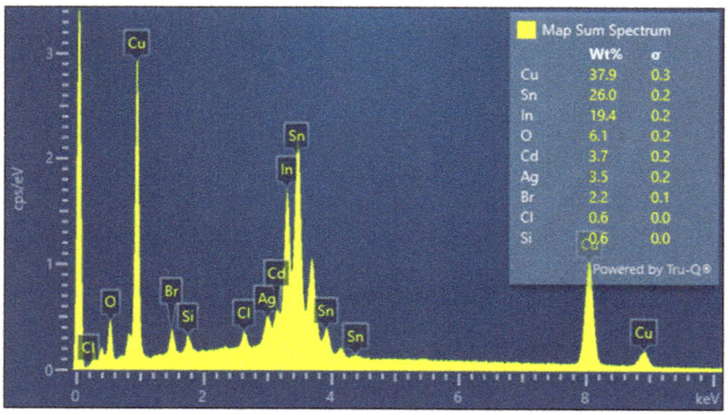

Figure 8. The result of the composition analysis at Spectrum point 29 for the averaged CIGS panel sample.

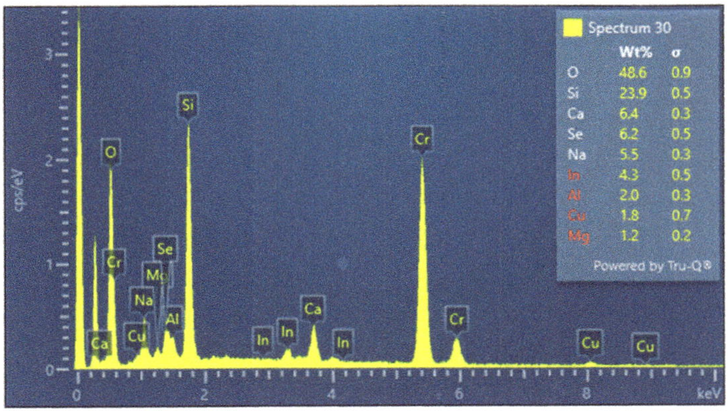

Figure 9. The result of the composition analysis at Spectrum point 30 for the averaged CIGS panel sample.

The results obtained from the analysis of SEM micrographs and compositional analysis for the averaged sample of the CIGS panel (Figures 7–9) confirmed the presence of metals such as Cu, In, and Se on the surface of the sample; however, their distribution was non-uniform. In Figure 9, large contents of Si were also observed, which resulted from the high content of glass comprising the casing of the investigated CIGS panel.

In addition to the characterization of averaged panel samples, investigations were also conducted for samples prepared from the absorbing layer. The absorption layer was not ground; instead, a small fragment of it was analyzed using electron microscopy.

Results obtained via electron microscope for the absorbing layer of the CIGS panel (depicted in Figure 4) are shown below (electron microscope image—Figure 10, and composition analysis results at selected points of the panel marked as Spectrum 7 and Spectrum 8—Figures 11 and 12).

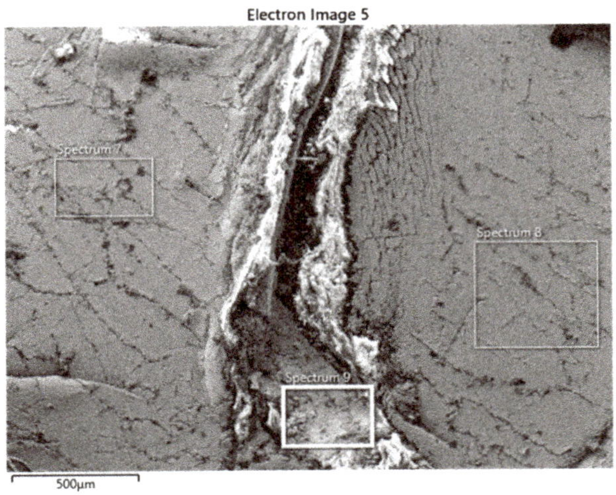

Figure 10. The electron microscope image of the averaged sample of the absorbing layer of the CIGS panel with marked points at Spectrum 7, Spectrum 8, and Spectrum 9, where elemental analysis was conducted.

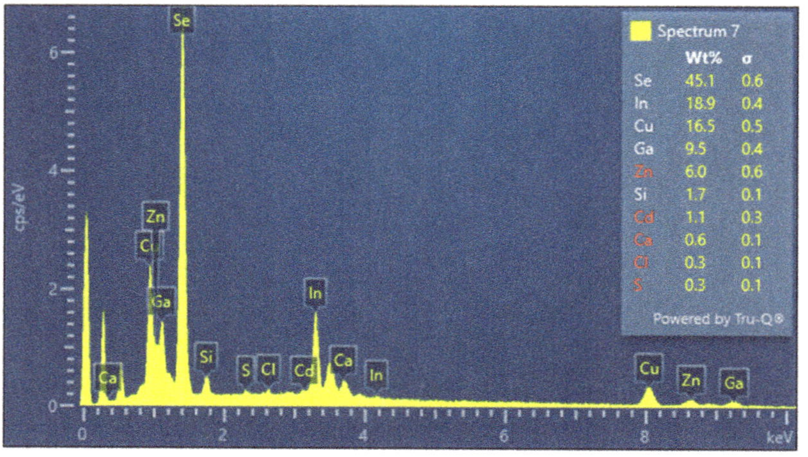

Figure 11. The result of the composition analysis at Spectrum point 7 for the sample of the absorbing layer of the CIGS panel.

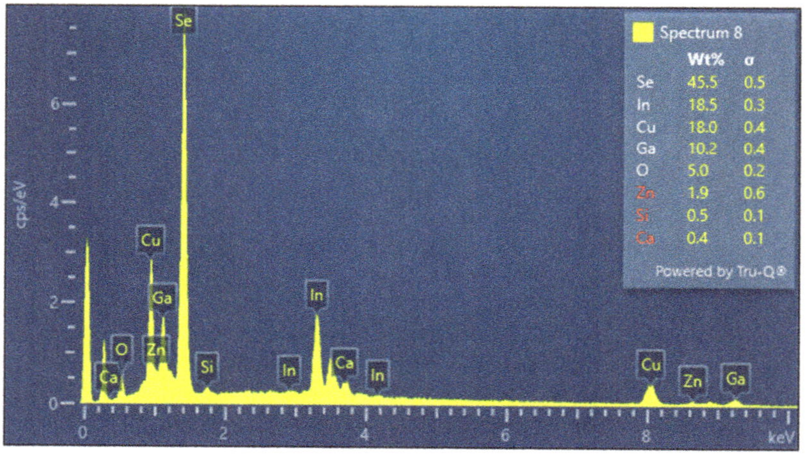

Figure 12. The result of the composition analysis at Spectrum point 8 for the sample of the absorbing layer of the CIGS panel.

The analysis of the electron microscope images and elemental analysis results presented in Figures 11 and 12 indicates that in the examined points of the absorbing layer of the panel, selenium, indium, copper, and gallium were present, and their distribution on the surface at the examined points was homogeneous (in the examined points, the surface % content of selenium, indium, gallium, and copper was marked at comparable levels, approximately 45%, 18–19%, 16–18%, and 9–10%, respectively).

These results confirm the assumption that metals such as In, Ga, and Cu are concentrated in the absorbing layer of the panel. The remaining part of the panel mainly consists of glass, which should be mechanically separated at the outset during the recycling process.

The absorbing layer sample from the panel was additionally subjected to analysis to determine its indium content. The acid-mineralized sample underwent total indium content determination using the ICP-MS method. The results of these analyses for the absorbing layer samples (designated as PA1–PA6 in this publication) of the CIGS panel are summarized in Table 3 and visually depicted in Figure 13.

Table 3. Summary of In content results in the absorbing layer of the panel (samples PA1-PA4). Analysis conducted using ICP-MS on acid-mineralized samples.

The Sample Designation	Indium [mg/g of CIGS Panel]
PA1	0.039
PA2	1.757
PA3	1.145
PA4	2.293
The minimum value	0.039
The maximum value	2.293
The median	1.451
The arithmetic mean	1.309

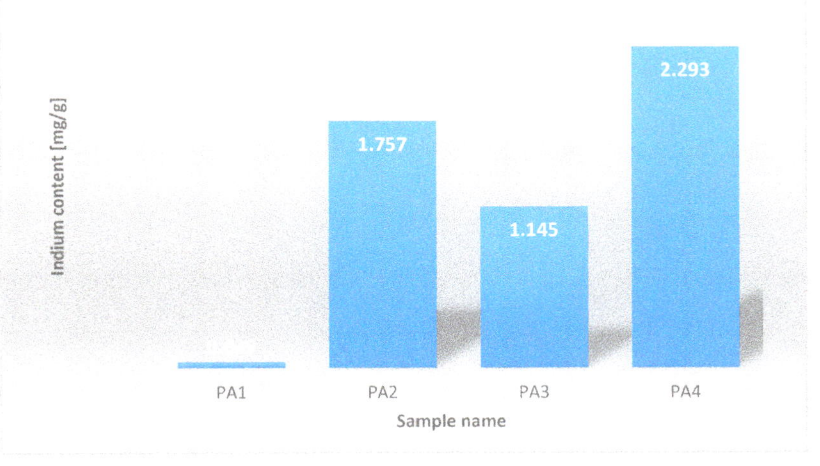

Figure 13. Summary of indium content results in the absorbing layer of the panel (samples PA1–PA4). Analysis conducted using ICP-MS on acid-mineralized samples.

The analysis of the data presented in Table 3 and depicted in Figure 13 revealed that in the absorbing layer of the panel, the determined contents for indium ranged from 0.039 to 2.293 mg/g (with a median of 1.45 mg/g). The results indicate incomplete homogenization of the sample. Nonetheless, the aim was to streamline the panel preparation process to minimize costs and energy consumption. Outliers were retained, and an estimated range of indium concentrations in samples from the absorbing layer of the panel was provided.

The indium contents in the absorbing layer of the panel were higher than those in the samples resulting from averaging the entire panel. Confirmation for the elemental analysis conducted using the ICP-MS method is provided by the results of the elemental analysis obtained from electron microscope images. The point analysis using electron microscopy also indicates that indium was concentrated in the absorbing layer of the panel.

These results allow us to conclude that a CIGS-type photovoltaic panel, where the absorbing part is encapsulated within a glass housing/structure, should be mechanically crushed during the recycling process to separate the glass from the absorbing part of the panel, which predominantly contains indium, a target for recycling.

After the mechanical separation of the panel parts concentrated with indium, an effective extraction and recovery method should be developed, as described in Section 3.2.

3.2. Preparation of Panel Samples for Electrolysis

The metal recovery process from a photovoltaic panel using electrolysis requires preliminary preparation. The photovoltaic panel was crushed and ground using a planetary mill, and the entirety was averaged (as described in Section 2.1). A representative sample was taken for analysis. A known mass of the sample was treated with a 3 M HCl solution, and indium were leached/washed using an ultrasonic bath at 80 °C. The leaching/washing process lasted for 4 h, with 30% H_2O_2 added to the mixture in four portions during washing. After the washing process was completed, the eluate was filtered into a 500 mL volumetric flask and topped up with deionized water. The indium content in the solution was determined using ICP-MS. Additionally, the pH of the solution was checked, ranging from 2 to 3.

To verify the efficiency of extraction conducted according to the procedure, extracts were prepared simultaneously from the CIGS panel and from the CIGS panel without (R1) and with the addition of a known amount of indium (R2), (Indium, beads, diam. 2–5 mm, 99.999% trace metals basis, Sigma Aldrich). A summary of key information regarding the preparation of metal extracts according to the procedure described for described method is provided in Table 4.

Table 4. The compilation of solutions prepared according to the described procedure is provided along with the results.

Solution Symbol	Preparation of the Solutions
R1	15 g of CIGS panel sample 460 mL of 3 M HCl 4 × 10 mL of H_2O_2 Mixed in an ultrasonic bath at 80 °C Washing time: 4 h The eluate was transferred to a 500 mL flask
R2	15 g of CIGS panel sample 1 g of indium 460 mL of 3 M HCl 4 × 10 mL of H_2O_2 Mixed in an ultrasonic bath at 80 °C Washing time: 4 h The eluate was transferred to a 500 mL flask

The obtained research results showed that the extraction process using 3 M HCl and H_2O_2 was effective. The studies demonstrated that as a result of this process, it is possible to extract up to 97% of indium from the panel into the solution.

3.3. Results for Samples Deposited on the Electrode

In Figure 14, photographs of the electrode with extracted indium from solutions prepared according to method described in Section 3.2 from a metal-enriched panel (solution R2) are presented. Additionally, electrolysis was conducted, resulting in the extraction of indium from the solution, which was the eluate from the CIGS panel (solution R1). In Figure 15, a photograph of the carbon electrode with extracted indium from solution R1 prepared from the CIGS panel is shown.

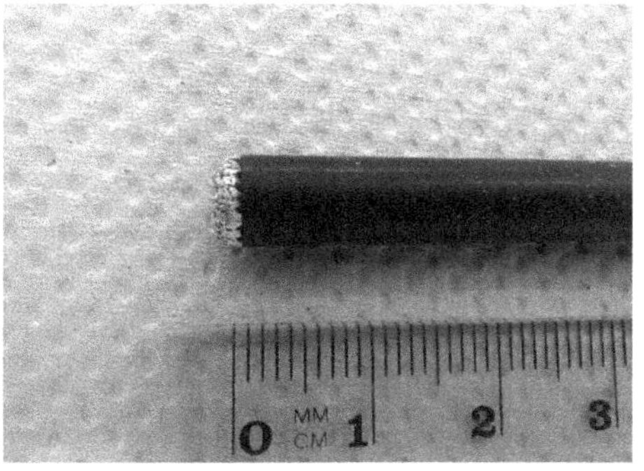

Figure 14. The carbon electrode with indium recovered from the solution during the electrolysis process. Optimization test results for the electrolysis parameters of indium from the solution with the addition of a known amount of indium (R2).

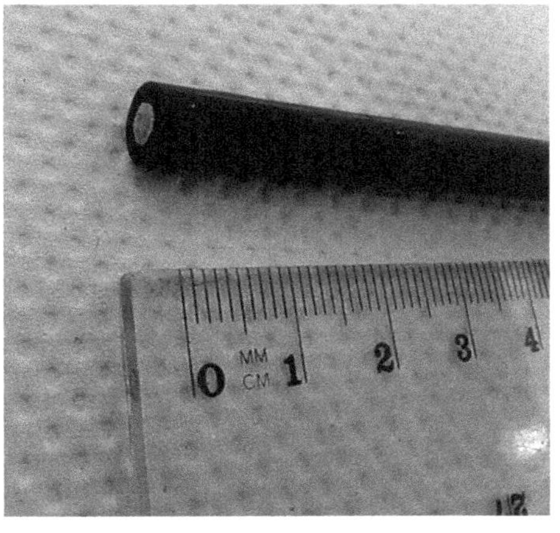

Figure 15. Carbon electrode with indium recovered from the solution during the electrolysis process without the addition of a known amount of indium (from the solution R1).

The samples deposited during the electrolysis process were examined using a scanning electron microscope. Figures 16–19 depict the results for solution R2, while Figures 20–22 show images taken for solution R1.

Figure 16. A photo from the electron microscope for the sample containing indium recovered from the solution with the addition of a known amount of indium during the electrolysis process (R2).

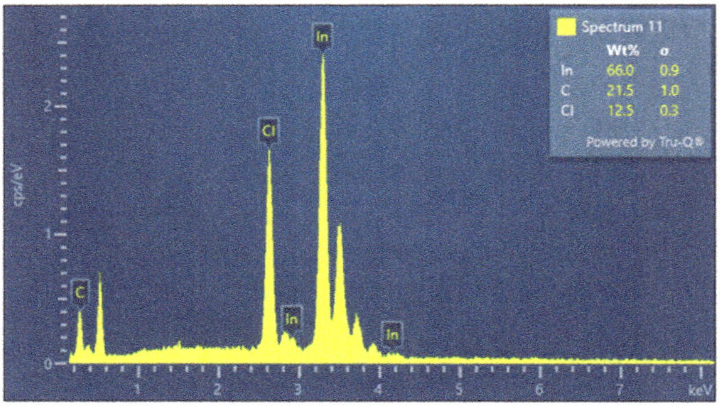

Figure 17. Result of the composition analysis at Spectrum 11 for the sample containing indium, recovered during the electrolysis process from the solution with the addition of a known amount of indium (R2).

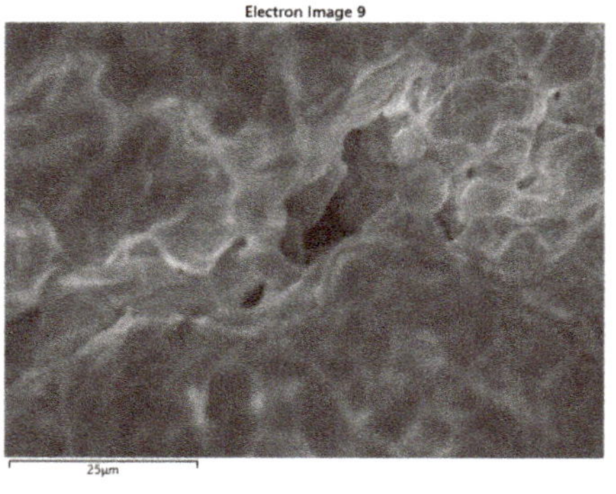

Figure 18. The image from the electron microscope at Spectrum 12 for the sample containing indium recovered during the electrolysis process from the solution with the addition of a known amount of indium (R2).

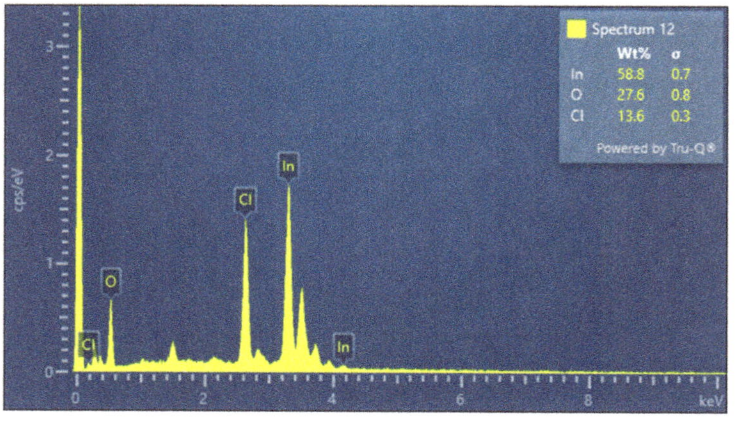

Figure 19. The result of the composition analysis at Spectrum 12 for the sample containing indium recovered during the electrolysis process from the solution with the addition of a known amount of indium (R2).

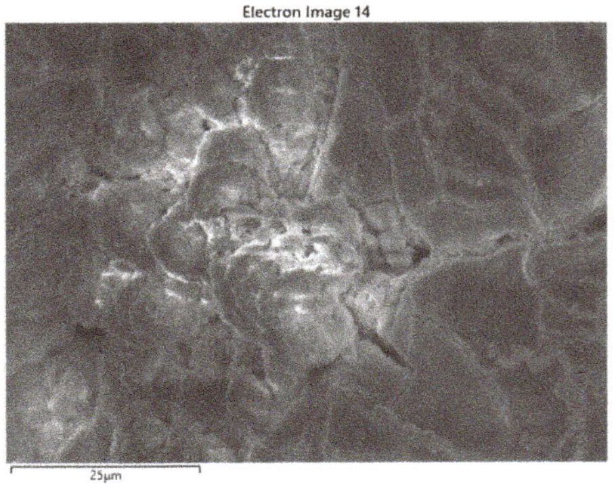

Figure 20. A scanning electron microscope image of the sample containing indium recovered from the solution without the addition of a known amount of indium during the electrolysis process (R1).

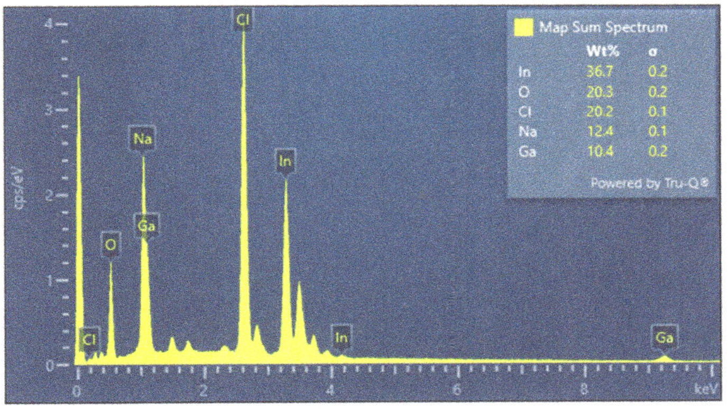

Figure 21. The result of the composition analysis for the sample containing indium recovered from the solution without the addition of a known amount of indium during the electrolysis process (R1).

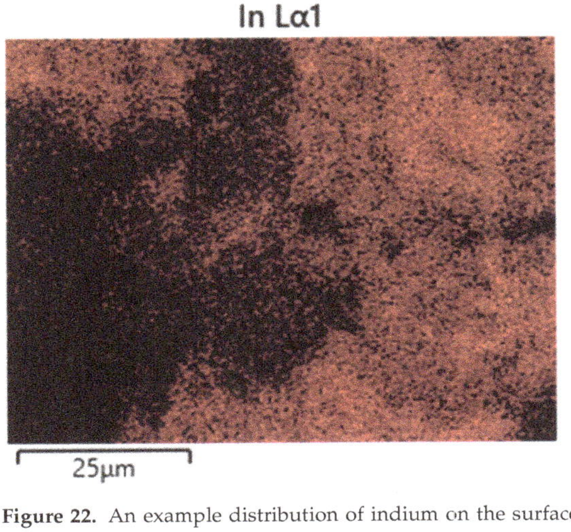

Figure 22. An example distribution of indium on the surface for the sample containing indium recovered from solution R1 during the electrolysis process.

The results presented in Figures 16–22 confirm that the electrolysis process employed successfully yielded indium (as shown in Figures 17, 19 and 21). Point analysis of the surface demonstrates that it constituted over 50% of the sample's composition at the examined point, with the remaining portion of the surface containing carbon, chlorine, and oxygen. Additionally, confirmation of the effectiveness of indium recovery in the electrolysis process is depicted in Figure 22, where an even distribution of indium in the examined sample obtained in process R1 can be observed.

Based on the obtained results, it can be concluded that the electrolysis method used in the metal recovery process is effective, allowing for selective recovery of metals (indium) from the solutions generated during the extraction process from the CIGS panel.

The electrolysis process for indium recovery from solution R1, which constitutes the liquid extract from CIGS panels prepared according to the procedure described in Section 3.2, was conducted under laboratory conditions at a temperature of 20–22 °C, as outlined in Table 1. The process lasted for 12–48 h, during which its effectiveness was monitored continuously. To confirm the effectiveness of the electrolysis process, samples were periodically taken from the solution undergoing electrolysis during the process. These samples were then analyzed using ICP-MS to assess the metal content. The obtained results are presented in Figure 23.

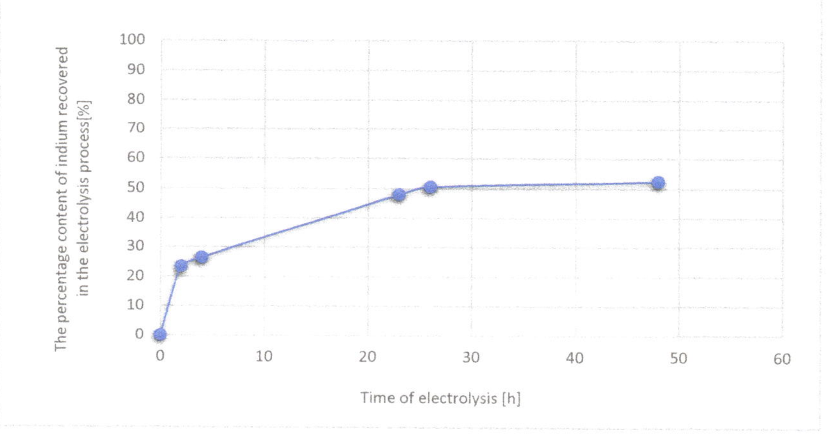

Figure 23. The graph depicting the relationship between indium recovery efficiency and electrolysis time.

Analysis of the results presented in Figure 23 indicates that in the electrolysis process applied in the study, up to 52% of the indium present in the solution was recovered within 48 h. The process was dynamic during the first 24 h, during which 50% of the indium was recovered from the solution. In the subsequent 24 h, only an additional 2% of the metal was recovered.

4. Conclusions

The dynamic advancement of photovoltaic technology has led to increased production, but it also raises concerns about the potential accumulation of waste. While photovoltaic panels are designed to operate efficiently for over 25 years, their lifespan eventually concludes, contributing to waste generation. Despite the environmentally friendly nature of photovoltaic energy production, capable of achieving zero CO_2 emissions during operation, the production process and disposal of panels can still pose environmental risks [31].

Recycling materials from used photovoltaic panels is paramount to mitigating their environmental impact. Indium, a crucial component in various photovoltaic panel types, particularly thin-film technologies like CIGS, plays a vital role in enhancing solar cell efficiency and performance. Moreover, indium is relatively scarce, with limited global reserves, underscoring the importance of its conservation and recycling to maintain a sustainable supply chain for photovoltaic production. Efficient recycling of indium from used photovoltaic panels serves to not only preserve this valuable resource but also reduce the environmental consequences associated with its extraction and processing. By implementing recycling processes for indium, reliance on new indium mining can be minimized, thereby mitigating environmentally disruptive practices. In essence, indium recycling is pivotal to advancing the sustainability of the photovoltaic industry and reducing its overall environmental footprint [32,33]. Based on the obtained results, it can be confirmed that the use of electrolysis in the metal recovery process demonstrates high efficiency, enabling selective recovery of these metals from the solutions obtained during the extraction from CIGS panels. In the conducted study, electrolysis allowed for the recovery of up to 52% of the indium present in the solution within 48 h. The dynamics of the process are particularly noticeable in the first 24 h, during which 50% of the indium was recovered from the solution. In the subsequent stage, i.e., the next day, an additional 2% of metal recovery was observed. These results suggest that electrolysis is an effective method for extracting metals from solutions obtained in the extraction process while enabling controlled selective recovery. This study presents a preliminary investigation into the electrolytic recovery of indium from CIGS photovoltaic panels using a simplified, low-energy process that does not require prior thermal treatment or the use of aggressive chemicals. Unlike many existing studies that rely on multi-step leaching and high-temperature pretreatment, the proposed method enables selective indium recovery through direct electrolysis in a diluted acidic medium. This approach minimizes chemical waste, reduces operational complexity, and demonstrates potential for scalability. While the findings are based on a single sample, they suggest that this method could offer a cost-effective and environmentally friendly alternative for indium recovery, particularly suited for small-scale or decentralized recycling operations. Most existing research on CIGS panel recycling adopts a holistic approach, aiming to recover multiple valuable elements such as copper, gallium, selenium, and components of the glass substrate. In contrast, this study focuses exclusively on indium due to its high economic value, limited natural availability, and strategic importance in the manufacturing of new thin-film photovoltaic devices. The primary objective of this work was to assess the feasibility of a selective electrolytic recovery process for indium as a first step toward a broader, modular multi-stage recovery system. Future research will aim to expand the method to include the recovery of additional critical metals and materials, with the ultimate goal of developing an integrated, environmentally responsible recycling pathway for the entire CIGS module.

Author Contributions: Conceptualization, M.G. and A.K.; methodology, M.G. and A.K.; writing—original draft preparation, M.G., A.K. and J.H.-R. writing—review and editing, T.K.; visualization, M.G.; supervision, J.H.-R. and E.K.-Z. All authors have read and agreed to the published version of the manuscript.

Funding: The research received funding as part of the statutory work commissioned by the Ministry of Education and Science; order no. 0073/GE/23, archival number: DK-4100-56/23.

Data Availability Statement: The data supporting the findings of this study are available at the Laboratory of Climate and Resources Protection, located at ul. Bagrowa 1, Kraków, Poland. Data are available upon reasonable request.

Conflicts of Interest: The authors declare no conflicts of interest.

Abbreviations

The following abbreviations are used in this manuscript:

CIGS	Copper indium gallium selenide panel
ICP-MS	Inductively Coupled Plasma Mass Spectrometry
SEM	Scanning Electron Microscopy
SE	Secondary Electrons detector
BSE	Backscattered Electrons detector
TED	Transmission Electron Detector
LCDs	liquid crystal displays
LEDs	light-emitting diodes
RF	radio frequency

References

1. Li, X.; Ma, B.; Wang, C.; Hu, D.; Lü, Y.; Chen, Y. Recycling and recovery of spent copper—Indium—Gallium—Diselenide (CIGS) solar cells: A review. *Int. J. Min. Met. Mater.* **2023**, *30*, 989–1002. [CrossRef]
2. Shamim, S.M.; Islam, S.; Huq, M.F.; Al Jobair, M. Design, performance analysis and efficiency optimization of copper indium gallium selenide (CIGS) solar cell. *Eur. Sci. J. ESJ.* **2015**, *11*. Available online: https://eujournal.org/index.php/esj/article/view/5133 (accessed on 11 April 2025).
3. Kuczyńska-Łażewska, A.; Klugmann-Radziemska, E.; Witkowska, A. Recovery of Valuable Materials and Methods for Their Management When Recycling Thin-Film CdTe Photovoltaic Modules. *Materials* **2021**, *14*, 7836. [CrossRef] [PubMed] [PubMed Central]
4. Isherwood, P. Reshaping the Module: The Path to Comprehensive Photovoltaic Panel Recycling. *Sustainability* **2022**, *14*, 1676. [CrossRef]
5. Petroli, P.A.; Camargo, P.S.S.; de Souza, R.A.; Veit, H.M. Assessment of Toxicity Tests for Photovoltaic Panels: A Review. *Curr. Opin. Green Sustain. Chem.* **2024**, *47*, 100885. [CrossRef]
6. PW Consulting. CIGS Solar Panels Market; PW Consulting Chemical & Energy Research Center: 9 February 2025. Available online: https://pmarketresearch.com/chemi/cigs-solar-panels-market/ (accessed on 11 April 2025).
7. European Commission. *Critical Raw Materials Resilience: Charting a Path Towards Greater Security and Sustainability*; COM(2023) 221 final; European Commission: Brussels, Belgium, 2023. Available online: https://ec.europa.eu/info/publications/critical-raw-materials-resilience-charting-path-towards-greater-security-and-sustainability_en (accessed on 11 April 2025).
8. Amato, A.; Beolchini, F. End-of-Life CIGS Photovoltaic Panel: A Source of Secondary Indium and Gallium. *Prog. Photovolt. Res. Appl.* **2018**, *27*, 229–236. [CrossRef]
9. Philipps, S.; Warmuth, W. Photovoltaics Report; Fraunhofer ISE and PSE Projects GmbH: Last Updated: 29 July 2024. Available online: https://www.ise.fraunhofer.de/en/publications/studies/photovoltaics-report.html (accessed on 9 April 2025).
10. Zhao, C.; Yu, S.; Tang, W.; Yuan, X.; Zhou, H.; Qi, T.; Zheng, X.; Ning, D.; Ma, M.; Zhu, J.; et al. Advances in CIGS Thin Film Solar Cells with Emphasis on the Alkali Element Post-Deposition Treatment. *Mater. Rep. Energy* **2023**, *3*, 100214. [CrossRef]
11. Theocharis, M.; Tsakiridis, P.E.; Kousi, P.; Hatzikioseyian, A.; Zarkadas, I.; Remoundaki, E.; Lyberatos, G. Hydrometallurgical Treatment for the Extraction and Separation of Indium and Gallium from End-of-Life CIGS Photovoltaic Panels. *Mater. Proc.* **2021**, *5*, 51. [CrossRef]
12. Rocchetti, L.; Beolchini, F. Recovery of Valuable Materials from End-of-Life Thin-Film Photovoltaic Panels: Environmental Impact Assessment of Different Management Options. *J. Clean. Prod.* **2015**, *89*, 59–64. [CrossRef]
13. Li, M.; Widijatmoko, S.D.; Wang, Z.; Hall, P. A Methodology to Liberate Critical Metals in Waste Solar Panel. *Appl. Energy* **2023**, *337*, 120900. [CrossRef]
14. Li, X.; Ma, B.; Hu, D.; Zhao, Q.; Chen, Y.; Wang, C. Efficient Separation and Purification of Indium and Gallium in Spent Copper Indium Gallium Diselenide (CIGS). *J. Clean. Prod.* **2022**, *339*, 130658. [CrossRef]
15. White, S.J.O.; Shine, J.P. Exposure Potential and Health Impacts of Indium and Gallium, Metals Critical to Emerging Electronics and Energy Technologies. *Curr. Environ. Health Rep.* **2016**, *3*, 459–467. [CrossRef]
16. Mufti, N.; Amrillah, T.; Taufiq, A.; Sunaryono; Aripriharta; Diantoro, M.; Zulhadjri; Nur, H. Review of CIGS-Based Solar Cells Manufacturing by Structural Engineering. *Sol. Energy* **2020**, *207*, 1146–1157. [CrossRef]
17. Moradi, M.; Teimouri, R.; Saadat, M.; Zahedifar, M. Buffer layer replacement: A method for increasing the conversion efficiency of CIGS thin film solar panels. *Optik* **2017**, *136*, 222–227. [CrossRef]
18. Fthenakis, V. Sustainability of Photovoltaics: The Case for Thin-Film Solar Cells. *Renew. Sustain. Energy Rev.* **2009**, *13*, 2746–2750. [CrossRef]

19. Wang, J.; Feng, Y.; He, Y. Advancements in Recycling Technologies for Waste CIGS Photovoltaic Modules. *Nano Energy* **2024**, *128*, 109847. [CrossRef]
20. Available online: https://tradingeconomics.com/commodity/ (accessed on 11 April 2025).
21. Gustafsson, A.M.K. Recycling of CIGS Solar Cell Waste Materials. Ph.D. Thesis, Chalmers University of Technology, Gothenburg, Sweden, 2014. Available online: https://publications.lib.chalmers.se/records/fulltext/203661/203661.pdf (accessed on 11 April 2025).
22. Pang, S.; Yan, Y.; Wang, Z.; Wang, D.; Li, S.; Wei, K. Enhanced Separation of Different Layers in Photovoltaic Panel by Mi-crowave Field. Sol. Energy Mater. *Sol. Cells* **2021**, *230*, 111213. [CrossRef]
23. Gustafsson, A.M.; Björefors, F.; Steenari, B.M.; Ekberg, C. Investigation of an Electrochemical Method for Separation of Copper, Indium, and Gallium from Pretreated CIGS Solar Cell Waste Materials. *Sci. World J.* **2015**, *2015*, 494015. [CrossRef] [PubMed] [PubMed Central]
24. Gu, S.; Fu, B.T.; Dodbiba, G.; Fujita, T.; Fang, B.Z. Promising approach for recycling of spent CIGS targets by combining electrochemical techniques with dehydration and distillation, ACS Sustain. *Chem. Eng.* **2018**, *6*, 6950.
25. Liu, F.-W.; Cheng, T.-M.; Chen, Y.-J.; Yueh, K.-C.; Tang, S.-Y.; Wang, K.; Wu, C.-L.; Tsai, H.-S.; Yu, Y.-J.; Lai, C.-H.; et al. High-yield recycling and recovery of copper, indium, and gallium from waste copper indium gallium selenide thin-film solar panels. *Sol. Energy Mater. Sol. Cells* **2022**, *241*, 111691. [CrossRef]
26. Teknetzi, I.; Holgersson, S.; Ebin, B. Valuable metal recycling from thin film CIGS solar cells by leaching under mild conditions. *Sol. Energy Mater. Sol. Cells* **2023**, *252*, 112178. [CrossRef]
27. Hu, D.; Ma, B.; Li, X.; Lv, Y.; Zhang, W.; Chen, Y.; Wang, C. Efficient separation and recovery of gallium and indium in spent CIGS materials. *Sep. Purif. Technol.* **2022**, *282*, 120087. [CrossRef]
28. Marchetti, B.; Corvaro, F.; Giacchetta, G.; Polonara, F.; Cocci Grifoni, R.; Leporini, M. Double Green Process: A Low Environmental Impact Method for Recycling of CdTe, a-Si, and CIS/CIGS Thin-Film Photovoltaic Modules. *Int. J. Sustain. Eng.* **2018**, *11*, 173–185. [CrossRef]
29. Teknetzi, I.; Click, N.; Holgersson, S.; Ebin, B. An Environmentally Friendly Method for Selective Recovery of Silver and ITO Particles from Flexible CIGS Solar Cells. *Sustain. Mater. Technol.* **2024**, *39*, e00844. [CrossRef]
30. Padoan, F.C.S.M.; Altimari, P.; Pagnanelli, F. Recycling of End-of-Life Photovoltaic Panels: A Chemical Prospective on Process Development. *Sol. Energy* **2019**, *177*, 746–761. [CrossRef]
31. Bae, D. A Statistical Approach to Scaling Up of CIGS PV Cells: Quantitative Analysis of Composition Uniformity within and between the Samples. *Mater. Sci. Semicond. Process.* **2023**, *164*, 107626. [CrossRef]
32. Bulińska, S.; Sujak, A.; Pyzalski, M. Sustainable Management of Photovoltaic Waste Through Recycling and Material Use in the Construction Industry. *Materials* **2025**, *18*, 284. [CrossRef]
33. Gerold, E.; Antrekowitsch, H. Advancements and Challenges in Photovoltaic Cell Recycling: A Comprehensive Review. *Sustainability* **2024**, *16*, 2542. [CrossRef]

recycling

MDPI

Article

Extraction of Rare Earth Elements from Organic Acid Leachate Using Formo-Phenolic-like Resins

Evan Lelong [1], Julien Couturier [2,3], Clément Levard [2], Stéphane Pellet-Rostaing [1] and Guilhem Arrachart [1,*]

1 ICSM, Univ Montpellier, CEA, CNRS, ENSCM, 30207 Marcoule, France
2 Aix-Marseille Univ, CNRS, IRD, INRAE, CEREGE, 13545 Aix-en-Provence, France; levard@cerege.fr (C.L.)
3 European Synchrotron Radiation Facility (ESRF), 38000 Grenoble, France
* Correspondence: guilhem.arrachart@umontpellier.fr; Tel.: +33-4-66-79-15-68

Abstract

Formo-phenolic-like resins were synthesized by replacing phenol with phloroglucinol, a biobased and biocompatible compound, and using different aldehydes, such as biomass-derived furaldehyde and glyoxal. Studies on the adsorption of rare earth elements from an aqueous organic acid solution indicate that these resins follow the Langmuir isotherm model, with maximum adsorption capacities ranging from 0.38 to 0.75 mmol/g. Adsorption was temperature-independent but strongly influenced by pH, with an up to fourfold increase between pH 2 and 5. Extraction kinetics were rapid, reaching equilibrium within two hours. Complete metal recovery was achieved within ten minutes using a 1 mol/L HCl desorption solution. Selectivity also varied with pH; glyoxal- and furfural-based resins showed superior separation performance at pH 2–3 and 3–4, respectively. The application of this method to real-world samples, including permanent magnet and red mud organic acid leachates, demonstrated effective extraction of rare earth elements and promising selectivity over iron (Fe), cobalt (Co), and nickel (Ni).

Keywords: rare earth elements; organic acid leachate; formo-phenolic resins; sorbents

1. Introduction

Rare earth elements (REEs)—fifteen lanthanides, scandium, and yttrium—are currently classified as critical materials due to the steady growth in their consumption because of new technologies [1,2]. The extraction process of REEs from mines is dominated by a few countries, and their cost and environmental impact will continue to increase. The pursuit of new sources of REEs has become increasingly critical in addressing supply chain vulnerabilities and reinforcing strategic resource independence [3,4]. While considerable research has focused on recycling REE-rich materials such as urban mine waste and end-of-life permanent magnets, other promising avenues remain underexplored. Notably, certain types of mining byproducts—such as bauxite residue—have been shown to contain significant concentrations of REEs, reaching up to 2700 mg/kg [5]. These underutilized materials represent a viable secondary resource for REE recovery and warrant further investigation as part of a diversified supply strategy [6–8].

Among hydrometallurgical processes, strong acidic leaching (HNO_3, H_2SO_4, HCl) followed by liquid–liquid extraction is considered the reference process for REE extraction from mines and waste [9]. However, the use of strong acids and large volumes of organic solvents raise health and environmental concerns.

The leaching of REEs from ores or waste materials using organic acids has emerged as an environmentally friendly alternative to traditional inorganic acid leaching [10,11]. Indeed, some organic acids can be produced from biomass, and solid adsorbents can be reused to avoid the problems associated with organic solvents, such as volatility, third-phase formation, and decomposition.

The leaching of REEs from NdFeB magnets has been demonstrated to be an efficient process when using organic acids (acetic, citric and tartaric acids, for example), minimizing the leaching of iron and other matrix components [12–14]. In the context of bauxite residue (red mud), organic acids like citric or lactic acid can effectively solubilize REEs through chelation and acid dissolution mechanisms [15,16]. Other secondary sources, including phosphate residues [17], coal fly ash [18], and electronic waste [19–21] also show promising REE leaching behavior with organic acids, offering selective recovery with reduced environmental impact.

Furthermore, solid-phase extraction is easier to implement than liquid–liquid processes. Various organic materials have been reported in the literature for the solid–liquid extraction of REEs [22], including polymers [23–25], phenolic polymers [26–31], modified silica [32–34], carbon particles [35–37], and impregnated resins [38–40]. The main challenges in developing organic adsorbents are combining high adsorption capacity with fast kinetics, ensuring good selectivity towards the targeted elements, preventing aggregation during the process, enabling the easy back-extraction of metals, and ensuring reusability.

Therefore, an interesting approach to recovering REEs involves leaching with organic acids followed by solid–liquid extraction. This appears to be an effective and suitable method [41,42].

With this in mind, the present study investigated the preparation of formo-phenolic-like resins for the extraction of REEs from organic acid leachate. In this process, phenol was substituted with phloroglucinol, a less toxic [43] and more biocompatible compound noted for its antispasmodic properties and high reactivity in polymerization reactions. Moreover, this triol is currently synthesized using a chemical process, but it could be produced using a greener process [44]. To discuss the effect of the aldehyde on the extraction properties, four aldehydes were used: formaldehyde, furfural, glyoxal, and terephthalaldehyde. The selection of these aldehydes was made with the intention of modulating the spacing between phloroglucinol units and the level of crosslinking [44–46]. Furthermore, furfural and glyoxal can be produced from renewable biomass resources, such as biobased xylose or ethylene glycol, respectively [45,46]. After a thorough study of the adsorption properties in acetic acid, the prepared materials were employed for metal extraction from the leachate of permanent magnets in acetic acid at a concentration of 1.6 mol/L as well as bauxite residue in acetic, citric, or lactic acid.

2. Results and Discussion

2.1. Resin Synthesis and Characterizations

The resins were obtained using a standard polymerization process with alkaline catalysis by means of sodium hydroxide [31]. Phenol, which is commonly used and classified as CMR 2, was substituted with **Ph**loroglucinol (Figure 1). This triol was pre-polymerized using **F**ormaldehyde (3 eq), **Fu**rfural (2 eq), **Gly**oxal (1 eq), or terephthalaldehyde (**TPA**, 1 eq) in a water/ethanol mixture (1/1) and then cured at 130 °C for 48 h. The four resulting insoluble materials, Ph-F, Ph-Fu, Ph-Gly, and Ph-TPA, were used for characterization and extraction tests. Solid-state ^{13}C NMR, FT-IR, and TGA analyses were performed and are presented in the Supplementary Materials (Figures S1–S3). Examination of the crushed resins by SEM showed that their morphology is typical of unmodified bulk formo-phenolic resins

(Figure S4). The material appears as a densely packed structure composed of particles with a rough surface texture. The absence of any visible pore openings or interconnected porous network indicates that the sample likely has a low surface area and minimal inherent porosity. An analysis of the data has indicated no significant disparities in terms of morphology when comparing the various sorbents. This observation suggests that the nature of the aldehyde does not appear to exert any influence on the observed outcomes.

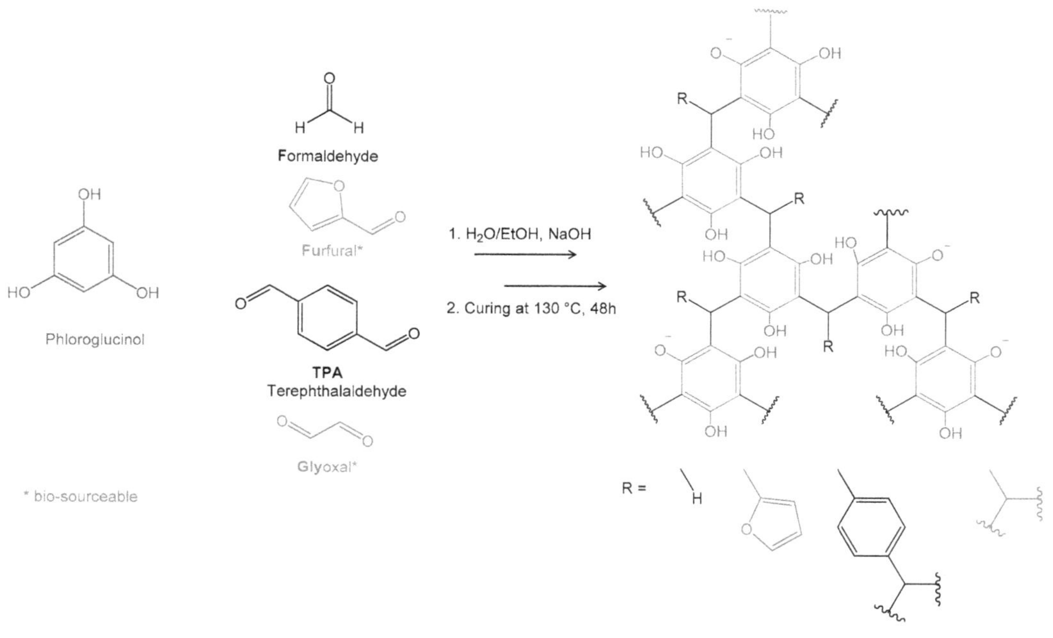

Figure 1. Synthesis pathway of phenolic-like resins.

2.2. Sorption/Desorption Experiments

The properties of resins toward REE adsorption were studied in 0.1 M acetic acid by varying the initial cation concentration, contact time, temperature, pH, and competing metal. Desorption was studied using a hydrochloric acid solution, one of the most suitable acid for REE refining [47], by varying the contact time and acid concentration.

2.2.1. Equilibrium Sorption Isotherm

Sorption isotherms were performed using a mixture of La and Yb in 0.1 M acetic acid to compare the behavior of the resins toward light and heavy REEs. The concentrations of cations varied from 2.5 to 200 mg/L. For a clearer representation, the isotherm has been plotted as a function of the initial concentration, as shown in Figure 2. The isotherm plotted as a function of the equilibrium concentration is shown in Figure S5. There are no significant differences between light and heavy REEs. The resins Ph-TPA and Ph-Fu showed lower capacities than Ph-F and Ph-Gly, with 30, 33, 50, and 65 mg/g for both cations. A decrease in Qe was observed for the higher initial concentrations, which can be attributed to potential factors such as surface site saturation, competitive adsorption, or aggregation effects at higher concentrations, as have already been observed for such materials [31]. Also, X-EDS analyses (Figure S6) confirmed the presence of cations in the resins after the extraction experiments.

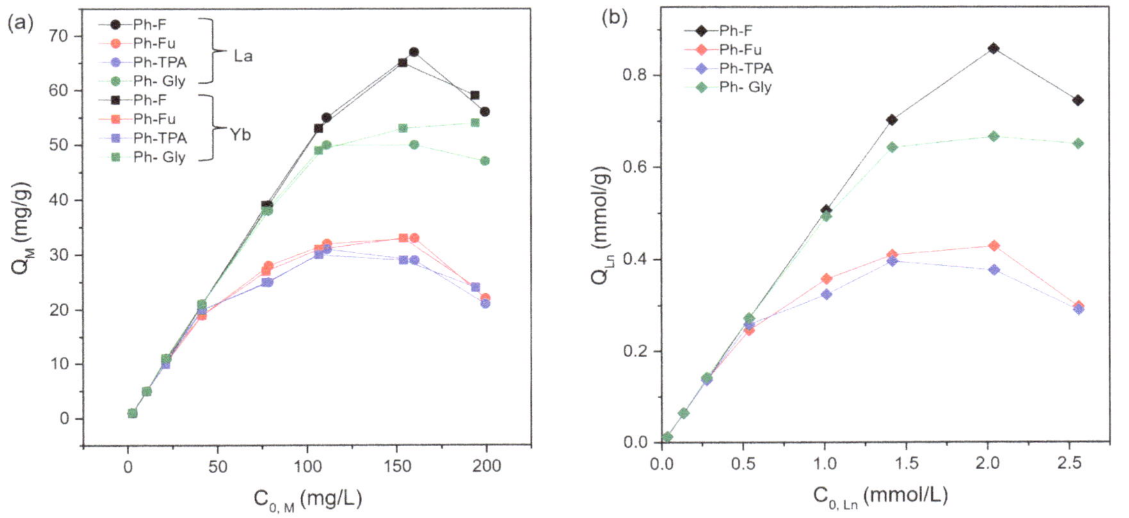

Figure 2. The adsorption isotherms of the resins (Ph-F, Ph-Fu, Ph-TPA, and Ph-Gly) for a mixture of La and Yb in acetic acid (0.1 M, pH = 2.8) at V/m = 0.5 for 16 h at 25 °C. (**a**) The capacity for each metal (M = La or Yb) expressed in mg/g; (**b**) the capacity for both metals (Ln = La and Yb) expressed in mmol/g.

The adsorption behavior of the lanthanides (Lns), specifically La and Yb, either individually or in combination, onto the various resins was evaluated using Langmuir (Figures S7 and S8) and Freundlich (Figure S9) isotherm models. A comparative analysis of the model fittings (Figures S8 and S9) indicated that the Langmuir isotherm provided a more suitable fit. This suggests that it is the most appropriate model for describing the adsorption of La and Yb onto the studied resins. This conclusion is supported by the high linearity and correlation coefficients (R^2) observed in the Langmuir plots, in contrast to the non-linear behavior exhibited by the Freundlich isotherms.

As shown in Figure 2, the Ln adsorption profiles across all resins closely followed the Langmuir model, characterized by rapid uptake at low concentrations, followed by a plateau, with a slight decrease at higher concentrations. This behavior is consistent with mechanisms involving monolayer adsorption on a homogeneous surface [48].

The maximum adsorption capacity of the resins (Q_{max}) was determined from the Langmuir plots (Figure S4) using the slope and intercept of the linearized isotherm equations. The calculated Q_{max} values for Ln ranged from 0.38 to 0.75 mmol/g in the following order: Ph-F (0.75) > Ph-Gly (0.66) > Ph-Fu (0.43) > Ph-TPA (0.38) (Table S1).

The adsorption performance of the synthesized sorbents (Ph-F, Ph-Fu, Ph-TPA, and Ph-Gly) was evaluated and compared with data reported for a variety of commercial resins (Table 1). Although the maximum adsorption capacities obtained in this study are lower than those of high-capacity sorbents, such as D113 resin [49] and gel-type weak acid resin [50], these results must be interpreted in light of the significantly different experimental conditions under which the adsorption studies were performed. The experiments in this study were carried out at a strongly acidic pH of 2.8. In contrast, most comparative studies were conducted at favorable pH levels ranging from 4.38 to 6.0. Under near-neutral conditions, the availability of binding sites for metal ions increases due to the lower concentration of competing protons (H^+). However, in highly acidic media, proton competition for active sites becomes more intense, often leading to decreased metal ion uptake.

Table 1. Comparison of the obtained results with commercial sorbents for La(III) and Yb (III) sorption from organic acid medium.

Sorbent	Acid Medium	Max Capacity (mg/g)	Reference
D113 resin	Acetate buffer (pH = 6)	273.3 (La^{3+})	Shu et al. [49]
Gel-type weak acid resin (110)	Acetate buffer (pH = 5.50)	265.8 (Yb^{3+})	Zheng et al. [50]
Diphonix Resin® Purolite S957	Citric acid (pH = 6)	60.01 (La^{3+}) 58.35 (La^{3+})	Araucz et al. [51]
Amberlite IRC86	Acetic-acid-buffered media (pH = 4.38)	40.56 (La^{3+})	Bezzina et al. [52]
Ph-F	Acetic acid (pH = 2.8)	56.91 (La^{3+}) 59.35 (Yb^{3+})	This work
Ph-Fu		33.14 (La^{3+}) 32.84 (Yb^{3+})	
Ph-TPA		29.04 (La^{3+}) 29.06 (Yb^{3+})	
Ph-Gly		47.68 (La^{3+}) 54.02 (Yb^{3+})	

Despite the challenging pH conditions, the synthesized sorbents exhibited significant adsorption capacities. For example, Ph-F and Ph-Gly displayed maximum capacities of 56.91 and 47.68 mg/g for La^{3+}, respectively, and 59.35 and 54.02 mg/g for Yb^{3+}, respectively. Ph-Fu and Ph-TPA also exhibited consistent performance. Notably, these values are similar to those reported for commonly used commercial sorbents, such as Diphonix Resin® (60.01 mg/g for La^{3+}), Purolite S957 (58.35 mg/g for La^{3+}) [51], and Amberlite IRC86 (40.56 mg/g for La^{3+}) [52]. These commercial sorbents were tested under less acidic conditions.

Thus, while the maximum capacities reported in this study are not the highest, the effectiveness of the synthesized sorbents at pH 2.8 highlights their potential as viable alternatives for metal ion recovery in highly acidic environments. Their robustness under such stringent conditions makes them promising candidates for practical applications.

2.2.2. Kinetic Studies

Kinetic studies were performed for the sorption and desorption of a mixture of La and Yb, as previously performed for the isotherms. The adsorption process was meticulously monitored by observing the immediate effect of the acetic acid solution on the activated resins. This observation continued for a period of one day, as depicted in Figure 3a. All materials exhibited rapid kinetic exchange for cations (Na to Ln), yet certain observations merit attention. It was observed that Ph-F and Ph-Gly achieved equilibrium within ten minutes. However, Ph-Fu exhibited a slight slowdown prior to reaching the plateau, while Ph-TPA demonstrated a modest decline in adsorption capacity. Nevertheless, equilibrium was achieved for all materials after two hours (black dashed line), with no significant differences observed in the La and Yb concentrations.

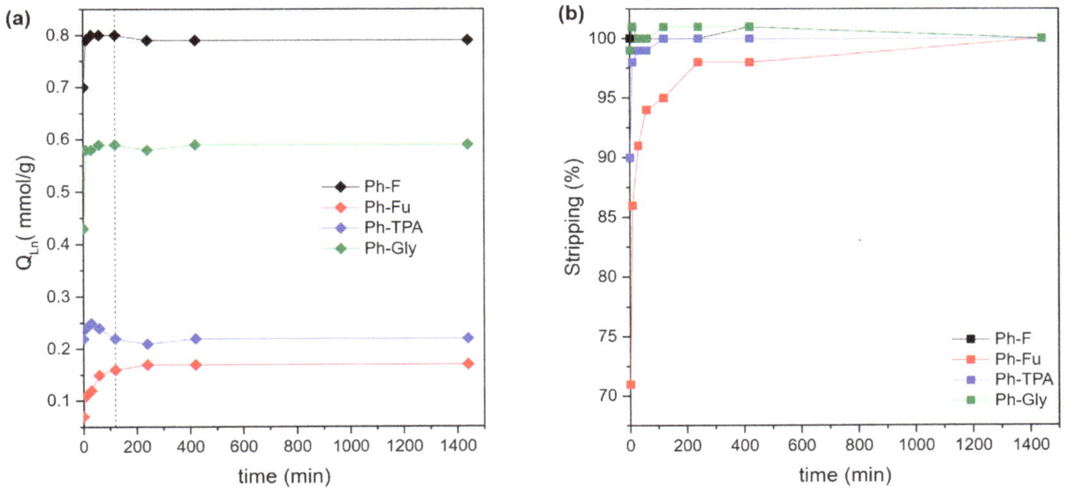

Figure 3. Kinetic analysis of the following: (**a**) The adsorption of a mixture of La (152 ppm) and Yb (196 ppm) in 0.1 M acetic acid (pH = 2.8); V/m = 0.5; t = 1, 10, 30, 60, 120, 240, 420, 1440 min; at 25 °C. (**b**) The back-extraction of La and Yb in a 1 M HCl solution, V/m = 0.5.

The rapid rate of adsorption observed in this experiment does not permit the utilization of the pseudo-first- or pseudo-second-order model. However, the rapid saturation of the adsorbent, which results in a precise line-up, suggests a chemisorption process in which all exchange sites are identical (pseudo-second-order).

Subsequent to the adsorption stage, identical samples were utilized to perform desorption kinetic experiments using a 1 M HCl solution (Figure 3b). A small quantity of solution was extracted from the batch experiments in order to quantify the amount of REEs released. As was the case with the adsorption process, the kinetics of release are extremely rapid for all materials. Full recovery of cations was observed after ten minutes, with the exception of Ph-Fu, which achieved 95% after two hours.

Furthermore, the efficiency of the back-extraction process was examined by varying the HCl concentration, with the objective of investigating the feasibility of implementing a less acidic solution (Figure S10). The absence of desorption when using H_2O exclusively indicates the feasibility of resin washing prior to REE recovery. The solution at 0.01 M facilitated the recovery of 43 to 56% of the adsorbed cations, 10-1 M resulted in 75–88% recovery, and 0.5 M yielded BE > 90%, while quantitative recovery was obtained with 1 M HCl. As demonstrated in Figure S11, the sorption and stripping capabilities of the sorbent remained high. This was observed even after two sorption/stripping cycles had been completed.

2.2.3. Influence of Temperature

In order to assess the influence of temperature on the adsorption process, extraction experiments were conducted with a mixture of La and Yb in acetic acid at temperatures ranging from 20 °C to 50 °C (Figure 4). After a two-hour period, no substantial alterations were detected, except in the case of Phy-Gly, where a slight increase as temperature rises in terms of adsorption capacity or selectivity between light and heavy REEs was observed.

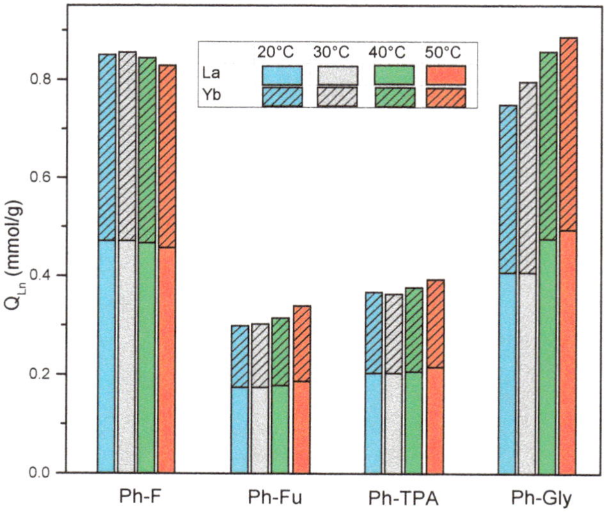

Figure 4. The adsorption capacity of the sorbents for a mixture of La and Yb in acetic acid 0.1 M (pH = 2.8) as a function of temperature. V/m = 0.5, contact time = 2 h.

Adsorption is a multifaceted process involving two distinct mechanisms: (i) the desorption of solvent molecules (e.g., water, acetic acid) and (ii) the adsorption of adsorbate on the surface (in this case, an exchange with Na^+) [53]. According to the Gibb free energy (ΔG0) equation, the spontaneity of a process can be determined by its relationship to enthalpy (ΔH0), entropy (ΔS0), and temperature (T) [54]. The experimental findings suggest that the sorption process is spontaneous and does not require energy input, at least at temperatures below 20 °C. Accordingly, in the context of Gibbs free energy, the value is negative, indicating that the exothermicity of cation exchange exceeds the energy required to denude the metal solvent sheath in solution. This phenomenon results in an adsorption capacity that remains temperature-independent.

2.2.4. Selectivity Towards Competing Metals

An investigation was conducted into the selectivity of the process towards scandium (Sc), iron (Fe), and aluminum (Al), given their prevalence as competitors in leaching solutions [55]. The presence of all three metals was identified in red mud leachates, with Fe emerging as the predominant competitor in permanent magnet leachate. In the course of these experiments, the effect of pH solution on selectivity was also investigated. Experiments were conducted at pH values of 2, 3, 4, and 5 (Figure 5). At a pH level below 2, there is a significant decrease in adsorption capacity. Conversely, at a pH level above 5, the process of metal hydroxide precipitation may interfere with the desired outcome. The extraction results demonstrated significant variations in the behavior of the materials, with a notable impact of pH on both adsorption capacity and selectivity.

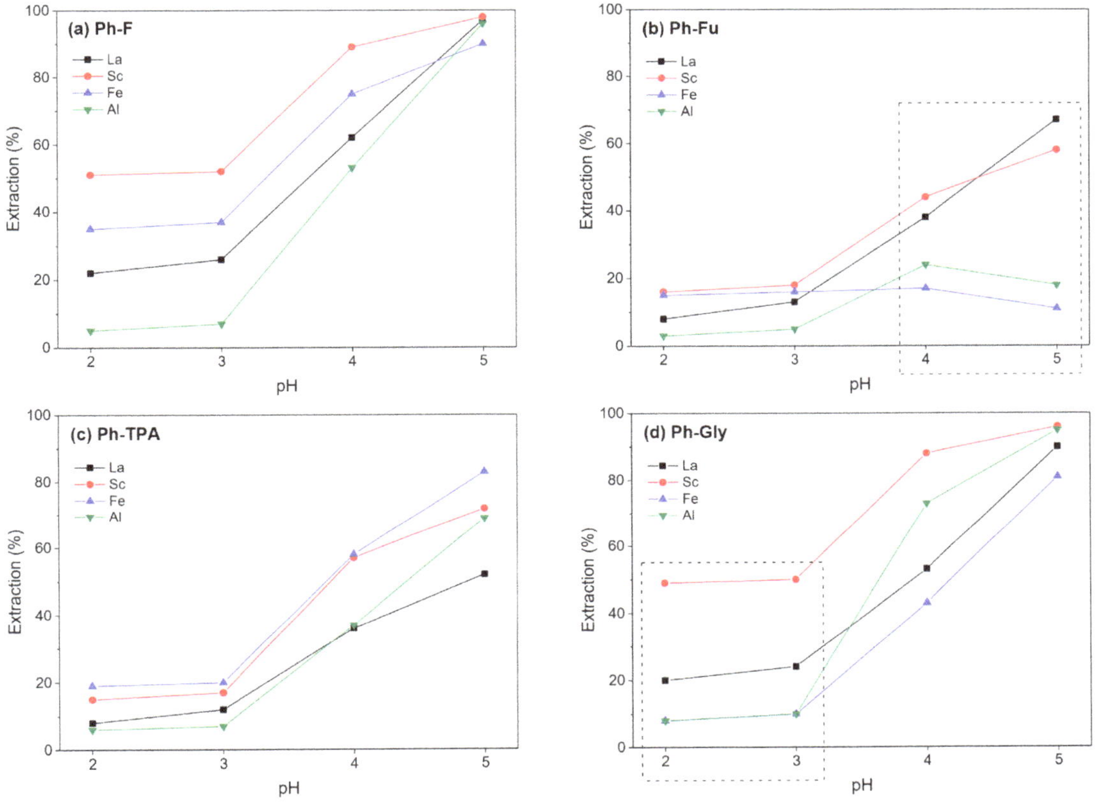

Figure 5. Extraction efficiencies (%) of sorbents for La, Sc, Fe, and Al from acetic acid solution as function of pH. Conditions: Al, Fe, Sc, La (≈50 ppm), acetic acid solution (pH adjusted with HCl or NaOH), V/m = 0.5, contact time = 4 h at 25 °C.

The observed disparities in selectivity among the various sorbent materials can be primarily attributed to differences in the accessibility and distribution of hydroxyl functional groups, which arise from the specific structural arrangements formed during the polymerization of phloroglucinol. These structural variations are, in turn, strongly influenced by the nature and reactivity of the aldehyde co-monomers employed in the synthesis. The type of aldehyde dictates not only the crosslinking density and spatial orientation of the polymer network, but also modulates the exposure of hydroxyl sites. These sites are critical for selective interactions with target analytes during sorption processes.

As was previously demonstrated in the sorption isotherm analysis at a pH of 2.8, the extraction percentage exhibited the following sequence: Ph-F ≥ Ph-Gly >> Ph-TPA ≥ Ph-Fu. Extraction at pH values of 4 and 5 resulted in a substantial enhancement of extraction, as illustrated in Figure S12. It has been demonstrated that all of the resins exhibit the capacity to extract metal at a rate that is twice that of Ph-Fu and up to four times that of the other materials at pH = 5. This trend in extraction capacity can be attributed primarily to the decline in H^+ concentration, leading to a reduction in the competition between the protonation of activated alcoholate function and metal chelation. Furthermore, acetic acid has a pKa constant of 4.8. Consequently, a mixture of acid and acetate exists at a pH of

4–5. This significantly impacts the selectivity between the four metals in solution (Table S2). The observed alterations can be attributed to the thermodynamic constant of metal–acetate complexes, which exhibit significant variation. This variation has a substantial impact on the selectivity of the system when the pH of the solution is altered. Notably, two resins have been identified as being particularly noteworthy within the range of pH values examined (Table 2). The Ph-Gly adsorbent demonstrates notable efficacy in extraction operations conducted at pH values ranging from 2 to 3, exhibiting a separation factor of approximately 3. This performance is particularly evident in the context of $SF_{La/Fe}$, which is observed to be lower in comparison to alternative resins. The adsorbent Ph-Fu exhibits remarkable efficacy in the separation process at pH values ranging from 4 to 5, exhibiting a substantially higher SF compared to other adsorbents, particularly $SF_{La/Fe}$, with an SF value exceeding 16 at a pH of 5.

Table 2. Separation factor (SF) for Ph-Gly at pH = 2, 3 and Ph-Fu at pH = 4, 5. All values available in Table S2.

	Ph-Gly				Ph-Fu		
pH	$SF_{Sc/La}$	$SF_{La/Fe}$	$SF_{La/Al}$	**pH**	$SF_{Sc/La}$	$SF_{La/Fe}$	$SF_{La/Al}$
2	3.7	2.9	2.8	4	1.3	3.0	1.9
3	3.2	2.9	2.8	5	0.7	16.6	9.4

2.3. Extraction in Real Leachates

This section of the study evaluates the efficiency and selectivity with which the sorbents extract metal ions in real leachates, such as those found in permanent magnets or red mud. The study particularly focuses on their ability to discriminate between REEs (Nd, Pr, and Dy) and common transition metals, particularly Fe.

An aqueous solution was obtained through the lixiviation of permanent magnet powder by 1.6 mol/L acetic acid. This solution was then contacted with four sorbents (Ph-F, Ph-Gly, Ph-Fu, and Ph-TPA) under identical extraction conditions (pH, volume-to-mass ratio, time, and temperature).

The extraction efficiencies and selectivities obtained with the different sorbents are summarized in Figure 6. The sorbent Ph-F demonstrated high extraction efficiencies for REEs (approximately 70%), along with strong selectivity for lanthanides over cobalt and nickel, with selectivity factors ($SF_{Ln/Co}$ and $SF_{Ln/Ni}$) of 11.1 and 10.9, respectively. Additionally, it exhibited an iron selectivity of $SF_{Ln/Fe}$ =3.7. Conversely, Ph-Fu exhibited moderate extraction of lanthanides, with efficiencies approaching 30%, and notably diminished selectivity, as evidenced by $SF_{Ln/Co}$ and $SF_{Ln/Ni}$ values of 2.0 and 2.1, respectively. However, its selectivity towards iron is much more pronounced, with a value of 8.9. The Ph-Gly sorbent exhibited balanced performance, achieving moderate REE extraction (approximately 50%) and exhibiting good selectivity over iron, with a selectivity factor $SF_{Ln/Fe}$ of 5.2. The extraction efficiencies of Ph-TPA were found to be marginally lower than those of Ph-Gly. However, Ph-TPA demonstrated acceptable selectivity, particularly in the context of cobalt (SFLn/Co = 3.9). Across all sorbents, iron consistently exhibited lower extraction, thereby underscoring the observed sorbents' preference for lanthanides over transition metals. These findings highlight the potential of the sorbents in developing efficient separation process for REEs from complex metal mixtures such as an acetic acid leachate of permanent magnets.

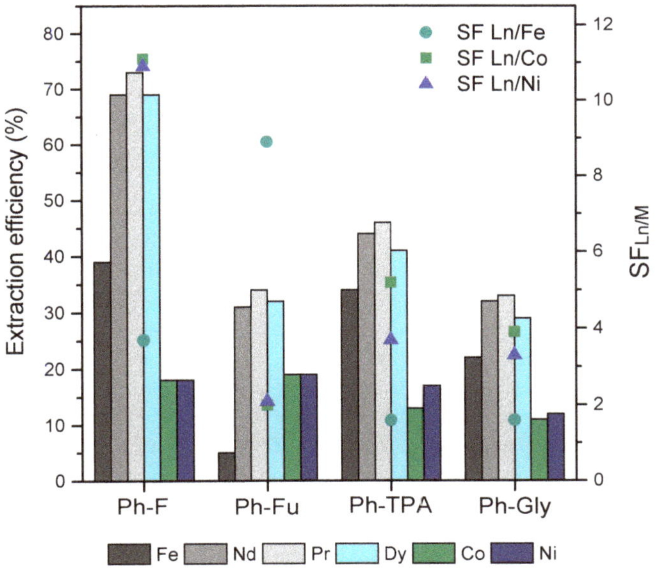

Figure 6. Extraction efficiency (%) of sorbents for Fe, Nd, Pr, Dy, Co, and Ni, as well as corresponding selectivity factors: $SF_{Ln/Fe}$, $SF_{Ln/Co}$, $SF_{Ln/Ni}$. Conditions: Fe^{3+} = 1536 mg/L, Nd^{3+} = 1006 mg/L, Pr^{3+} = 155 mg/L, Dy^{3+} = 58 mg/L, Co^{3+} = 56 mg/L, Ni^{2+} = 12 mg/L in acetic acid (1.6 mol/L, pH = 3), V/m = 0.1, contact time = 2 h at 25 °C.

A subsequent series of experiments was conducted to assess the extraction performance of sorbents in the context of an organic acid red mud leachate (Table S3). Red mud leachates were prepared using citric, lactic, and acetic acids as leaching agents to simulate environmentally relevant extraction conditions. These leachates were subsequently subjected to sorption experiments employing functionalized resins: Ph-F in both hydroxyl (-OH) and sodium (-ONa) forms, and Ph-Gly in the sodium form (-ONa). The Ph-Gly resin was evaluated under two different volume-to-mass ratios, specifically V/m = 0.2 and 0.5, to assess the influence of solid-to-liquid contact conditions on extraction efficiency.

The nature of the acid employed in the leaching process plays a crucial role in determining the extent to which metals form complexes in solution, which directly influences their availability for subsequent sorption. Among the acids studied—citric, lactic, and acetic—there are notable differences in their complexing strengths due to their molecular structures.

Citric acid, with its tricarboxylic structure, has been shown to form strong complexes with trivalent metal ions, including Fe^{3+}, Al^{3+}, and rare earth elements (REEs). This high affinity is attributed to its multiple coordination sites (three carboxylic acid groups and one hydroxyl group), which allow it to chelate metals efficiently and stabilize them in solution [56]. Lactic acid, in contrast, is classified as an alpha-hydroxy acid, bearing a carboxylic group and a hydroxyl group. It has been observed to form moderately strong complexes with metal ions, thereby offering a certain degree of stabilization, although not to the same extent as citric acid. Its simpler structure results in a limited number of coordination interactions that can be established with metal ions. Acetic acid demonstrates the weakest complexation behavior among the three acids studied. The presence of a single carboxyl group confers limited binding capacity, leading to a greater proportion of "free" metal ions remaining in solution. This lower complexation strength can influence the sorption process, as metals are more readily available for direct interaction with sorbents. The general trend in complexation strength with REEs is citric acid > lactic acid > acetic acid, which aligns with their molecular structures and number of coordination sites [57]. Consequently, the availability of metals for the sorption process is reversed. This observation is corroborated by the results presented in Figure S13, which demonstrate extraction efficiency, with maximum extraction percentages of approximately 10% for citric acid, 40% for lactic acid, and 60% for acetic acid.

It is evident across the three leaching systems that metal speciation in solution is the primary determinant of sorption efficiency. While citric acid is an excellent leachant, it creates metal complexes that are too stable for recovery via conventional resins. Lactic acid occupies an intermediate position, whereas acetic acid produces leachates that are rich in "free" or "labile" metal ions, which are optimal for sorption. These considerations must be taken into account in the context of selectivity for REEs, as this is more pronounced when sorption is performed from a lactic acid leachate. From a materials standpoint, the sodium-exchanged form of the phenolic resin (Ph-F-ONa) exhibited high and consistent performance, thanks to its efficient ion-exchange capacity. In contrast, the glyoxal-crosslinked variants (Ph-Gly-ONa) only demonstrated increased extraction under favorable conditions (acetic acid leachate). This suggests that crosslinking enhances physical properties, such as stability and swelling, rather than directly contributing to chemical affinity for metal ions. Over-crosslinking can reduce extraction efficiency by limiting diffusion and site availability.

The data on extraction efficiency obtained in citric, lactic, and acetic acid media demonstrate a distinct selectivity of the synthesized sorbents towards REEs over Al^{3+}, despite potential concerns regarding aluminum leaching. In particular, lactic acid solutions yielded the most pronounced differentiation, with REE extraction efficiencies reaching up to 40%, while Al^{3+} uptake remained consistently below 10% across all sorbents. These findings suggest that Al^{3+} does not exhibit competitive inhibition of REE sorption under the experimental conditions, thereby validating the sorbents' affinity for REEs in environments that are weakly to moderately complexing.

3. Materials and Methods

3.1. Chemicals

All obtained chemicals were analytically pure from Sigma–Aldrich (Saint Quentin Fallavier, France) or Thermo Fisher (Illkirch, France) and used without further purification: acetic acid (99%), phloroglucinol (99%), furfural (99%), terephthalaldehyde (98%), glyoxal (40% in water), and formaldehyde (37% in water). The ICP standards were purchased from SCP Science (Villebon-sur-Yvette, France).

3.2. General Procedure for Resin Synthesis

Phloroglucinol and sodium hydroxide (NaOH) pellets (0.5 eq.) were dissolved in a mixture of H_2O/EtOH (v/v with 50 eq. H_2O). Aldehyde (3 eq. Formaldehyde, 2 eq. Furfural, 1 eq. Glyoxal, or 1 eq. Terephthalaldehyde TPA) was added, and the reaction mixture was stirred overnight. Then, it was transferred to a crystallizer and cured in a ventilated oven at 130 °C for 48 h. The resulting solid, tough resin was recovered and ground using a ball mill for 15 min at 25 Hz. It was then washed with NaOH (1 M), HCl (1 M), and H_2O to obtain the protonated -OH form. The washed resins were then dried in a ventilated oven at 80 °C for 24 h, quickly dispersed by ball milling, and stored for extraction experiments. Before each experiment, the resins were activated using 0.1 M NaOH (0.5 mL/mg), then rinsed once with H_2O.

3.3. Characterization Techniques for Resins

The ^{13}C solid-state MAS NMR spectra were recorded at a rotation speed of 12 kHz (using 4 mm outer-diameter rotors) with a Bruker Advance 400 MHz spectrometer (Brucker, Wissembourg, France). The resins were mechanically ground at 25 Hz for 15 min using a Retsch mixer mill MM 200 (Retsch, Eragny sur Oise, France) with a Zr ball. Thermogravimetric analysis (TGA) was performed using a Mettler Toledo instrument (Mettler Toledo, Viroflay, France)with an air or nitrogen flow and a heating rate of 10 °C/min from 25 to 950 °C. An isothermal treatment of 30 min at 100 °C was performed under air to determine the resins' moisture regain from storage. Fourier transform infrared (FTIR) absorption spectroscopy was carried out on a PerkinElmer 100 spectrometer (Perkin Elmer, Villebon S/Yvette, France) between 615 and 4000 cm^{-1} using an attenuated total reflectance (ATR) crystal with a resolution of 4 cm^{-1}. Scanning electron microscopy (SEM) was performed using a Quattro ESEM instrument (ThermoScientific, Les Ulis, France) coupled with a XFlash6 / 100 Energy Dispersive X-ray Spectrometer with Quantax 400 software (Bruker, Wissembourg, France) (SEM/EDS) to study the morphology of the resins. Metal ion concentrations before and after sorption were determined using inductively coupled plasma/atomic emission spectroscopy (ThermoScientific, ICAP 7000 series; ThermoScientific, Les Ulis, France). The wavelengths used for Al, Fe, Sc, La, and Yb measurements were chosen to avoid spectral interference between the elements. The red mud leachates were quantified before and after sorption using inductively coupled plasma mass spectrometry (ICP-MS, PerkinElmer 300X Quadrupole; Perkin Elmer, Villebon S/Yvette, France).

3.4. Batch Extraction

A series of metal solutions were prepared for the sorption experiments, including a solution of La, a solution of Yb, and a solution of La and Yb in 0.1 mol/L acetic acid, which was prepared from the salts in their nitrate form, as well as a multicomponent solution containing Al, Fe, Sc, and La (50 mg/L of each element) in 0.1 mol/L acetic acid, with pH adjusted to 2, 3, 4, or 5 using HCl or NaOH.

A permanent magnet leaching solution was prepared from NdFeB magnet powder in acetic acid, as previously described in the literature [12,58].

A red mud leaching solution was prepared according to the literature protocol [15,16], employing organic acids including citric, lactic, and acetic acids. Typically, the leaching of 800 mg of a karstic bauxite residue from France [5] in 40 mL of organic acid was carried out in a microwave at 70 °C for 30 min. The resulting leachate was then filtered through a 0.22 µm filter to remove any remaining bauxite residue particles.

The ability of the thermosetting resins to extract REEs and competing metals was assessed by extraction tests, which were carried out in batch experiments using aqueous solutions. In these experiments, a known amount of resin was activated using a 0.1 M NaOH solution (V/m = 1), then washed with water and dried under air at 100 °C for 12 h. Then the resins were brought into contact with a volume of solution containing metal cations, typically 4 to 6 mg of resin brought into contact with 2 to 3 mL of aqueous solution (V/m = 0.5).

The mixture was stirred at 250 rpm at controlled temperature for a set number of hours, typically at 25 °C for 12 h. After contact, the mixture was centrifuged. The supernatant was filtered through a 0.22 µm cellulose acetate membrane and then diluted with 1% HNO_3 to achieve a suitable concentration. The amounts of various cations in the filtrate were analyzed using ICP-OES or ICP-MS.

The measured values led to the following:

- The cation uptake capacity Q_{ads} (mg/g) was calculated using the following equation:

$$Q_{ads} = \left(C_i - C_f\right) \cdot \frac{V}{m} \tag{1}$$

with C_i being the initial concentration of the metal ion in solution, C_f the residual metal ion concentration, and V/m the ratio of solution volume by resin-mass.
- The adsorption efficiency E (%) was calculated according to the following equation:

$$E = \frac{C_i - C_f}{C_i} \cdot 100 \tag{2}$$

- The distribution coefficient (K_D), expressed in mL/g, represents the ratio of the number of cations that the resin absorbs to the number of cations that remain in the solution after extraction. It was determined using the following formula:

$$K_D = \frac{C_i - C_f}{C_f} \cdot \frac{V}{m} \cdot 1000 \tag{3}$$

- The separation factor ($FS_{M1/M2}$) is defined as the ratio of the distribution coefficient of metal M1 to that of metal M2, the competing metal ion. This factor quantitatively expresses the selectivity of the resin for metal M1 in the presence of a competing species. The separation factor was calculated using the following equation:

$$FS_{M1/M2} = \frac{K_{D_{M1}}}{K_{D_{M2}}} \tag{4}$$

- The back-extraction or stripping efficiency S (%) is defined by the following equation:

$$S = \frac{Q_e - Q_f}{Q_e} \cdot 100 \tag{5}$$

with Q_e being the concentration of the metal ion loaded into the polymer and Q_f the residual metal ion concentration in the polymer after the release.

Langmuir isotherm parameters [59,60] were estimated by the following equation:

$$\frac{C_e}{Q_e} = \frac{1}{K_L Q_{max}} + \frac{1}{Q_{max}} \cdot C_e \tag{6}$$

with C_e being the concentration of the metal ion in solution at equilibrium, Q_e the amount of metal ion adsorbed at equilibrium, Q_{max} the maximum amount of metal ion adsorbed at equilibrium, and K_L the Langmuir constant. The maximum sorption capacity could be estimated from this isotherm when a plateau was obtained [61].

Freundlich isotherm [62] parameters were estimated by the following equation:

$$\log Q_e = \log K_F + \frac{1}{n} \cdot \log C_e \tag{7}$$

with C_e being the concentration of the metal ion in solution at equilibrium, Q_e the amount of metal ion adsorbed at equilibrium, and K_F and n Freundlich parameters.

3.5. Adsorption Kinetic

The kinetics data were fitted using pseudo-first-order and pseudo-second-order kinetic models [63,64]. The pseudo-first-order model, which is based on solid sorption capacity and was suggested by Lagergren, describes the adsorption phenomenon as a diffusion-controlled process. In the concentration range studied, this model suggests the formation of monolayer coverage of the adsorbate on the surface of the adsorbent [65,66]. This model was used to describe the sorption kinetics [67–69].

The pseudo-first-order rate is given as

$$\log\left(Q_e - Q_t\right) = \log Q_e - \frac{K_1}{2.303}\cdot t \tag{8}$$

where Q_e is the equilibrium metal ion concentration in the solid phase (mg/g), Q_t is the equilibrium metal ion concentration in the solid phase at time t (mg/g), K_1 is the pseudo-first-order equilibrium rate constant (min^{-1}), and t is the time (min).

The pseudo-second-order model is also based on solid-phase adsorption capacity but controlled by the sorption/desorption process and diffusion toward chelating sites [70–72].

The pseudo-second-order rate is given as

$$\frac{t}{Q_t} = \frac{1}{Q_e}\cdot t + \frac{1}{K_2 Q_e^2} \tag{9}$$

where Q_e is the equilibrium metal ion concentration in the solid phase (mg/g), Q_t is the equilibrium metal ion concentration in the solid phase at time t (mg/g), K_2 is the pseudo-first-order equilibrium rate constant (min^{-1}), and t is the time (min).

4. Conclusions

A series of four formo-phenolic-type resins were synthesized by replacing the phenol group with a phloroglucinol group, a biosourceable and biocompatible compound. The synthesis involved the use of four aldehydes or dialdehydes, with two of these aldehydes being produced from biomass, specifically furaldehyde and glyoxal. The isotherms study demonstrated that the resins comply with the Langmuir model, characterized by monolayer chemisorption on equal sites. The maximum adsorption capacity for Ln exhibited a range from 0.38 to 0.75 mmol/g, with Ph-F > Ph-Gly > Ph-Fu > Ph-TPA. The extraction capacity under consideration is independent of temperature (a minimum of 20 °C) but exhibits minimal dependence on the pH of the extraction medium. The extraction capacity is up to fourfold greater when the pH range is from 2 to 5. The rate of extraction is fast, with equilibrium largely reached after two hours. Furthermore, back-extraction is even faster, with the total recovery of metal after ten minutes in HCl 1M.

For instance, the selectivity of adsorption capacity exhibits a substantial variation in accordance with the pH level. It is noteworthy that the two resins demonstrating optimal separation properties are those polymerized using biosourceable aldehydes, specifically glyoxal at a pH of 2–3 and furfural at a pH of 3–4.

In order to assess the efficacy of resins in a real-life environment, an extraction process was conducted on permanent magnet leachate in 1.6 M acetic acid. This extraction process yielded highly promising results, with lanthanides being extracted from 30% to 70% and the selective extraction of lanthanides over Fe, Co, and Ni with from 2 up to 10 depending on the sorbent and the metal considered. This comparative study demonstrates that Ph-F and Ph-Gly are promising candidates for selective REE recovery from complex metal mixtures such as acetic acid leachates of permanent magnets.

A final experiment was conducted on red mud leachate, which demonstrated the efficacy of resins in diluted media. The results of this study demonstrate the crucial interplay between leachate chemistry and sorbent functionality in determining the efficiency of metal recovery from red mud. While the use of acetic acid resulted in the significant extraction of REEs due to the weak coordination environment it creates, the application of stronger complexing agents such as citric and lactic acid led to poor recovery, particularly for REEs. This suggests that standard phenolic-based sorbents, even when crosslinked with glyoxal, lack sufficient affinity to displace stable organic complexes in solution. To address this limitation, future work should explore the incorporation of selective chelating ligands directly into the polymeric matrix of the sorbent. Ideally, these ligands should exhibit higher complexation stability constants for REEs and transition metals than common organic acids, enabling competitive displacement and capture. Finally, combining this ligand-driven approach with smart regeneration strategies, such as pH switching, competitive stripping, and selective elution using mild chelators, could enable closed-loop, sustainable processes for recovering critical metals from red mud and other bauxite residues.

Supplementary Materials: The following supporting information can be downloaded at: https://www.mdpi.com/article/10.3390/recycling10040165/s1/, Figure S1: ^{13}C solid-state NMR of resins; Figure S2: FT-IR spectra of resins; Figure S3: Thermogravimetric analysis of resins; Figure S4: SEM image of resin Ph-Gly; Figure S5: Adsorption isotherms of the resins; Figure S6: X-EDS spectra of resin Ph-Gly after extraction; Figure S7: Plot of the Langmuir isotherm equation for Ln (La+Yb) adsorption; Figure S8: Plot of the Langmuir isotherm equation for La or Yb adsorption; Figure S9: Plot of Freundlich isotherm equation for La or Yb adsorption; Figure S10: Back-extraction efficiency (%), loaded sorbent; Figure S11: Sorption and stripping cycle; Figure S12: Adsorption capacity for the sorbent as function of pH; Figure S13: Extraction efficiencies (%) of the sorbents for red mud leachates; Table S1: Maximum adsorption capacity and Langmuir constant; Table S2: Separation factor (SF) for the sorbent as a function of pH; Table S3: Composition of red mud leachates.

Author Contributions: E.L. (Conceptualization, Methodology, Investigation, Data Curation, Writing—Original Draft), J.C. (Conceptualization, Methodology), C.L. (Conceptualization, Methodology, Project Administration, Funding Acquisition), S.P.-R. (Conceptualization, Methodology, Supervision), G.A. (Conceptualization, Methodology, Writing—Review and Editing, Supervision, Project Administration, Funding Acquisition). All authors have read and agreed to the published version of the manuscript.

Funding: The authors acknowledge financial support from the CNRS through the MITI interdisciplinary program (PRIME 2020: ExtraMet project) and the French Agence Nationale de la Recherche through the RECALL project (ANR-20-CE04-0007).

Data Availability Statement: The original contributions presented in this study are included in the article/supplementary material. Further inquiries can be directed to the corresponding author.

Acknowledgments: The authors would like to thank C. Rey, J. Lautru, B. Baus-Lagarde, and B. Angeletti for their assistance in conducting TGA, SEM-EDS, ICP-OES, and ICP-MS analyses, respectively.

Conflicts of Interest: The authors declare that they have no known competing financial interests or personal relationships that could have appeared to influence the work reported in this paper.

References

1. Reisman, D.; Weber, R.; Mckernan, J.; Northeim, C.; U.S. Environmental Protection Agency. *Rare Earth Elements: A Review of Production, Processing, Recycling and Associated Environmental Issues*; United States Environmental Protection Agency: Washington, DC, USA, 2013.
2. Grohol, M.; Veeh, C. *Study on the Critical Raw Materials for the EU 2023—Final Report*; European Commission; Publications Office of the European Union: Luxembourg, 2023.
3. Binnemans, K.; Jones, P.T.; Blanpain, B.; Van Gerven, T.; Yang, Y.; Walton, A.; Buchert, M. Recycling of Rare Earths: A Critical Review. *J. Clean. Prod.* **2013**, *51*, 1–22. [CrossRef]
4. Jowitt, S.; Werner, T.; Weng, Z.; Mudd, G. Recycling of the Rare Earth Elements. *Curr. Opin. Green Sustain. Chem.* **2018**, *13*, 1–7. [CrossRef]

5. Couturier, J.; Oularé, P.T.; Collin, B.; Lallemand, C.; Kieffer, I.; Longerey, J.; Chaurand, P.; Rose, J.; Borschneck, D.; Angeletti, B.; et al. Yttrium Speciation Variability in Bauxite Residues of Various Origins, Ages and Storage Conditions. *J. Hazard. Mater.* **2024**, *464*, 132941. [CrossRef]
6. Deady, É.; Mouchos, E.; Goodenough, K.; Williamson, B.; Wall, F. Rare Earth Elements in Karst-Bauxites: A Novel Untapped European Resource? In Proceedings of the ERES 2014: 1st conference on European Rare Earth Resources, Milos, Greece, 4–7 September 2014.
7. Liu, Y.; Naidu, R. Hidden Values in Bauxite Residue (Red Mud): Recovery of Metals. *Waste Manag.* **2014**, *34*, 2662–2673. [CrossRef] [PubMed]
8. Akcil, A.; Akhmadiyeva, N.; Abdulvaliyev, R.; Abhilash; Meshram, P. Overview On Extraction and Separation of Rare Earth Elements from Red Mud: Focus on Scandium. *Miner. Process. Extr. Metall. Rev.* **2018**, *39*, 145–151. [CrossRef]
9. Hidayah, N.N.; Abidin, S.Z. The Evolution of Mineral Processing in Extraction of Rare Earth Elements Using Liquid-Liquid Extraction: A Review. *Miner. Eng.* **2018**, *121*, 146–157. [CrossRef]
10. Zhao, M.; Zhang, H.; Han, H.; Jiang, X.; Yang, Y.; Li, T. Differential Leaching Mechanisms and Ecological Impact of Organic Acids on Ion-Adsorption Type Rare Earth Ores. *Sep. Purif. Technol.* **2025**, *362*, 131701. [CrossRef]
11. Banerjee, R.; Chakladar, S.; Mohanty, A.; Chakravarty, S.; Chattopadhyay, S.K.; Jha, M.K. Review on the Environment Friendly Leaching of Rare Earth Elements from the Secondary Resources Using Organic Acids. *Geosyst. Eng.* **2022**, *25*, 95–115. [CrossRef]
12. Belfqueh, S.; Seron, A.; Chapron, S.; Arrachart, G.; Menad, N. Evaluating Organic Acids as Alternative Leaching Reagents for Rare Earth Elements Recovery from NdFeB Magnets. *J. Rare Earths* **2023**, *41*, 621–631. [CrossRef]
13. Gergoric, M.; Barrier, A.; Retegan, T. Recovery of Rare-Earth Elements from Neodymium Magnet Waste Using Glycolic, Maleic, and Ascorbic Acids Followed by Solvent Extraction. *J. Sustain. Metall.* **2019**, *5*, 85–96. [CrossRef]
14. Gergoric, M.; Ekberg, C.; Foreman, M.R.S.J.; Steenari, B.-M.; Retegan, T. Characterization and Leaching of Neodymium Magnet Waste and Solvent Extraction of the Rare-Earth Elements Using TODGA. *J. Sustain. Metall.* **2017**, *3*, 638–645. [CrossRef]
15. Couturier, J.; Levard, C.; Collin, B.; Chaurand, P.; Vidal, V.; Mathon, O.; Duvivier, A.; Angeletti, B.; Borschneck, D.; Rose, J.; et al. Dissolution of Rare Earth Elements: Exploring the Ability of Deep Eutectic Solvents and Organic Acid Solutions, the Case of Lactic Acid. *Sep. Purif. Technol.* **2025**, *366*, 132740. [CrossRef]
16. Lallemand, C.; Ambrosi, J.; Borschneck, D.; Angeletti, B.; Chaurand, P.; Campos, A.; Desmau, M.; Fehlauer, T.; Auffan, M.; Labille, J.; et al. Potential of Ligand-Promoted Dissolution at Mild pH for the Selective Recovery of Rare Earth Elements in Bauxite Residues. *ACS Sustain. Chem. Eng.* **2022**, *10*, 6942–6951. [CrossRef]
17. Nayl, A.; Arafa, W.; Abd-Elhamid, A.; Elkhashab, R. Studying and Spectral Characterization for the Separation of Lanthanides from Phosphate Ore by Organic and Inorganic Acids. *J. Mater. Res. Technol.* **2020**, *9*, 10276–10290. [CrossRef]
18. Zhang, L.; Chen, H.; Pan, J.; Yang, F.; Long, X.; Yang, Y.; Zhou, C. Rare Earth Elements Recovery and Mechanisms from Coal Fly Ash by Column Leaching Using Citric Acid. *Sep. Purif. Technol.* **2025**, *353*, 128471. [CrossRef]
19. Golzar-Ahmadi, M.; Bahaloo-Horeh, N.; Pourhossein, F.; Norouzi, F.; Schoenberger, N.; Hintersatz, C.; Chakankar, M.; Holuszko, M.; Kaksonen, A. Pathway to Industrial Application of Heterotrophic Organisms in Critical Metals Recycling from E-Waste. *Biotechnol. Adv.* **2024**, *77*, 108438. [CrossRef]
20. Pavón, S.; Fortuny, A.; Coll, M.T.; Sastre, A.M. Rare Earths Separation from Fluorescent Lamp Wastes Using Ionic Liquids as Extractant Agents. *Waste Manag.* **2018**, *82*, 241–248. [CrossRef]
21. Kumari, A.; Sinha, M.K.; Pramanik, S.; Sahu, S.K. Recovery of Rare Earths from Spent NdFeB Magnets of Wind Turbine: Leaching and Kinetic Aspects. *Waste Manag.* **2018**, *75*, 486–498. [CrossRef] [PubMed]
22. Sharifian, S.; Wang, N.-H.L. Resin-Based Approaches for Selective Extraction and Purification of Rare Earth Elements: A Comprehensive Review. *J. Environ. Chem. Eng.* **2024**, *12*, 112402. [CrossRef]
23. Dave, S.R.; Kaur, H.; Menon, S.K. Selective Solid-Phase Extraction of Rare Earth Elements by the Chemically Modified Amberlite XAD-4 Resin with Azacrown Ether. *React. Funct. Polym.* **2010**, *70*, 692–698. [CrossRef]
24. Zereen, F.; Yilmaz, V.; Arslan, Z. Solid Phase Extraction of Rare Earth Elements in Seawater and Estuarine Water with 4-(2-Thiazolylazo) Resorcinol Immobilized Chromosorb 106 for Determination by Inductively Coupled Plasma Mass Spectrometry. *Microchem. J.* **2013**, *110*, 178–184. [CrossRef]
25. Karadaş, C.; Kara, D.; Fisher, A. Determination of Rare Earth Elements in Seawater by Inductively Coupled Plasma Mass Spectrometry with Off-Line Column Preconcentration Using 2,6-Diacetylpyridine Functionalized Amberlite XAD-4. *Anal. Chim. Acta* **2011**, *689*, 184–189. [CrossRef]
26. Ebraheem, K.A.K.; Mubarak, M.S.; Yassien, Z.J.; Khalili, F. Chelation Properties of Poly(8–Hydroxyquinoline 5,7–Diylmethylene) Crosslinked with Bisphenol–A Toward Lanthanum(III), Cerium(III), Neodimium(III), Samarium(III), and Gadolinium(III) Ions. *Sep. Sci. Technol.* **2000**, *35*, 2115–2125. [CrossRef]

27. Draye, M.; Czerwinski, K.; Favre-Réguillon, A.; Foos, J.; Guy, A.; Lemaire, M. Selective Separation of Lanthanides with Phenolic Resins: Extraction Behavior and Thermal Stability. *Sep. Sci. Technol.* **2000**, *35*, 1117–1132. [CrossRef]
28. Al-Rimawi, F.; Ahmad, A.; Khalili, F.; Mubarak, M. Chelation Properties of Some Phenolic-Formaldehyde Polymers toward Some Trivalent Lanthanide Ions. *Solvent Extr. Ion Exch.* **2004**, *22*, 721–735. [CrossRef]
29. Ameta, R.; Patel, V.; Joshi, J. Polychelates of Phenolic Ion-Exchange Resin: Synthesis and Characterization. *Iran. Polym. J.* **2007**, *16*, 615–625.
30. Arrambide, C.; Arrachart, G.; Berthalon, S.; Wehbie, M.; Pellet-Rostaing, S. Extraction and Recovery of Rare Earths by Chelating Phenolic Copolymers Bearing Diglycolamic Acid or Diglycolamide Moieties. *React. Funct. Polym.* **2019**, *142*, 147–158. [CrossRef]
31. Oye Auke, R.; Arrachart, G.; Tavernier, R.; David, G.; Pellet-Rostaing, S. Terephthalaldehyde-Phenolic Resins as a Solid-Phase Extraction System for the Recovery of Rare-Earth Elements. *Polymers* **2022**, *14*, 311. [CrossRef]
32. Liu, Z.; Liu, Y.; Gong, A. Preparation of Diglycolamide Polymer Modified Silica and Its Application as Adsorbent for Rare Earth Ions. *Des. Monomers Polym.* **2019**, *22*, 1–7. [CrossRef]
33. Wilfong, W.C.; Ji, T.; Duan, Y.; Shi, F.; Wang, Q.; Gray, M.L. Critical Review of Functionalized Silica Sorbent Strategies for Selective Extraction of Rare Earth Elements from Acid Mine Drainage. *J. Hazard. Mater.* **2022**, *424*, 127625. [CrossRef]
34. Ginot, L.; El Bakkouche, A.; Giusti, F.; Dourdain, S.; Pellet-Rostaing, S. Hydrophobic Porous Liquids with Controlled Cavity Size and Physico-Chemical Properties. *Adv. Sci.* **2024**, *11*, 2305906. [CrossRef]
35. Brown, A.; Balkus, K. Critical Rare Earth Element Recovery from Coal Ash Using Microsphere Flower Carbon. *ACS Appl. Mater. Interfaces* **2021**, *13*, 48492–48499. [CrossRef] [PubMed]
36. Cardoso, C.E.D.; Almeida, J.C.; Lopes, C.B.; Trindade, T.; Vale, C.; Pereira, E. Recovery of Rare Earth Elements by Carbon-Based Nanomaterials—A Review. *Nanomaterials* **2019**, *9*, 814. [CrossRef]
37. Ashour, R.; Abdelhamid, H.; Abdel-Magied, A.; Abdel-Khalek, A.; Ali, M.; Uheida, A.; Muhammed, M.; Zou, X.; Dutta, J. Rare Earth Ions Adsorption onto Graphene Oxide Nanosheets. *Solvent Extr. Ion Exch.* **2017**, *35*, 91–103. [CrossRef]
38. Matsunaga, H.; Ismail, A.; Wakui, Y.; Yokoyama, T. Extraction of Rare Earth Elements with 2-Ethylhexyl Hydrogen 2-Ethylhexyl Phosphonate Impregnated Resins Having Different Morphology and Reagent Content. *React. Funct. Polumers* **2001**, *49*, 189–195. [CrossRef]
39. İnan, S.; Tel, H.; Sert, Ş.; Çetinkaya, B.; Sengül, S.; Özkan, B.; Altaş, Y. Extraction and Separation Studies of Rare Earth Elements Using Cyanex 272 Impregnated Amberlite XAD-7 Resin. *Hydrometallurgy* **2018**, *181*, 156–163. [CrossRef]
40. Mondal, S.; Ghar, A.; Satpati, A.K.; Sinharoy, P.; Singh, D.K.; Sharma, J.N.; Sreenivas, T.; Kain, V. Recovery of Rare Earth Elements from Coal Fly Ash Using TEHDGA Impregnated Resin. *Hydrometallurgy* **2019**, *185*, 93–101. [CrossRef]
41. Borra, C.R.; Pontikes, Y.; Binnemans, K.; Van Gerven, T. Leaching of Rare Earths from Bauxite Residue (Red Mud). *Miner. Eng.* **2015**, *76*, 20–27. [CrossRef]
42. Asadollahzadeh, M.; Torkaman, R.; Torab-Mostaedi, M. Extraction and Separation of Rare Earth Elements by Adsorption Approaches: Current Status and Future Trends. *Sep. Purif. Rev.* **2021**, *50*, 417–444. [CrossRef]
43. Singh, I.; Sidana, J.; Bansal, P.; Foley, W. Phloroglucinol Compounds of Therapeutic Interest: Global Patent and Technology Status. *Expert Opin. Ther. Pat.* **2009**, *19*, 847–866. [CrossRef] [PubMed]
44. Abdel-Ghany, S.; Day, I.; Heuberger, A.; Broeckling, C.; Reddy, A. Production of Phloroglucinol, a Platform Chemical, in Arabidopsis Using a Bacterial Gene. *Sci. Rep.* **2016**, *6*, 38483. [CrossRef]
45. Li, X.; Jia, P.; Wang, T. Furfural: A Promising Platform Compound for Sustainable Production of C4 and C5 Chemicals. *ACS Catal.* **2016**, *6*, 7621–7640. [CrossRef]
46. Mattioda, G.; Blanc, A. *Ullmann's Encyclopedia of Industrial Chemistry*; Wiley-VCH: Weinheim, Germany, 2011.
47. Khawassek, Y.M.; Eliwa, A.A.; Haggag, E.S.A.; Omar, S.A.; Abdel-Wahab, S.M. Adsorption of Rare Earth Elements by Strong Acid Cation Exchange Resin Thermodynamics, Characteristics and Kinetics. *SN Appl. Sci.* **2018**, *1*, 51. [CrossRef]
48. Sahoo, T.R.; Prelot, B. Chapter 7—Adsorption Processes for the Removal of Contaminants from Wastewater: The Perspective Role of Nanomaterials and Nanotechnology. In *Nanomaterials for the Detection and Removal of Wastewater Pollutants*; Bonelli, B., Freyria, F.S., Rossetti, I., Sethi, R., Eds.; Elsevier: Amsterdam, The Netherlands, 2020; pp. 161–222.
49. Shu, Z.; Xiong, C.; Shen, Q.; Yao, C.; Gu, Z. Adsorption Behavior and Mechanism of D113 Resin for Lanthanum. *Rare Met.* **2007**, *26*, 601–606. [CrossRef]
50. Zheng, Z.; Xiong, C. Adsorption Behavior of Ytterbium (III) on Gel-Type Weak Acid Resin. *J. Rare Earths* **2011**, *29*, 407–412. [CrossRef]
51. Araucz, K.; Aurich, A.; Kołodyńska, D. Novel Multifunctional Ion Exchangers for Metal Ions Removal in the Presence of Citric Acid. *Chemosphere* **2020**, *251*, 126331. [CrossRef]
52. Bezzina, J.P.; Ogden, M.D.; Moon, E.M.; Soldenhoff, K.L. REE Behavior and Sorption on Weak Acid Resins from Buffered Media. *J. Ind. Eng. Chem.* **2018**, *59*, 440–455. [CrossRef]

53. Fan, Q.; Tan, X.; Li, J.; Wang, X.; Wu, W.; Montavon, G. Sorption of Eu(III) on Attapulgite Studied by Batch, XPS, and EXAFS Techniques. *Environ. Sci. Technol.* **2009**, *43*, 5776–5782. [CrossRef] [PubMed]
54. Saha, P.; Chowdhury, S. Insight Into Adsorption Thermodynamics. In *Thermodynamics*; Mizutani, T., Ed.; IntechOpen: London, UK, 2011. [CrossRef]
55. Judge, W.; Azimi, G. Recent Progress in Impurity Removal during Rare Earth Element Processing: A Review. *Hydrometallurgy* **2020**, *196*, 105435. [CrossRef]
56. Apelblat, A. Dissociation Equilibria in Solutions with Citrate Ions. In *Citric Acid*; Apelblat, A., Ed.; Springer International Publishing: Cham, Switzerland, 2014; pp. 143–212, ISBN 978-3-319-11233-6. [CrossRef]
57. Wang, R.; Zheng, Z. Rare Earth Complexes with Carboxylic Acids, Polyaminopolycarboxylic Acids, and Amino Acids. In *Rare Earth Coordination Chemistry: Fundamentals and Applications*; John Wiley & Sons: Singapore, 2010; pp. 91–136. [CrossRef]
58. Belfqueh, S.; Chapron, S.; Giusti, F.; Pellet-Rostaing, S.; Seron, A.; Menad, N.; Arrachart, G. Selective Recovery of Rare Earth Elements from Acetic Leachate of NdFeB Magnet by Solvent Extraction. *Sep. Purif. Technol.* **2024**, *339*, 126701. [CrossRef]
59. Langmuir, I. The Adsorption of Gases on Plane Surfaces of Glass, Mica and Platinum. *J. Am. Chem. Soc.* **1918**, *40*, 1361–1403. [CrossRef]
60. Ghosal, P.S.; Gupta, A.K. Determination of Thermodynamic Parameters from Langmuir Isotherm Constant-Revisited. *J. Mol. Liq.* **2017**, *225*, 137–146. [CrossRef]
61. Limousin, G.; Gaudet, J.-P.; Charlet, L.; Szenknect, S.; Barthès, V.; Krimissa, M. Sorption Isotherms: A Review on Physical Bases, Modeling and Measurement. *Appl. Geochem.* **2007**, *22*, 249–275. [CrossRef]
62. Freundlich, H. Über Die Adsorption in Lösungen. *Z. für Phys. Chem.* **1907**, *57U*, 385–470. [CrossRef]
63. Azizian, S. Kinetic Models of Sorption: A Theoretical Analysis. *J. Colloid Interface Sci.* **2004**, *276*, 47–52. [CrossRef]
64. Simonin, J.-P. On the Comparison of Pseudo-First Order and Pseudo-Second Order Rate Laws in the Modeling of Adsorption Kinetics. *Chem. Eng. J.* **2016**, *300*, 254–263. [CrossRef]
65. Lagergren, S. Zur Theorie Der Sogenannten Adsorption Geloster Stoffe. *K. Sven. Vetenskapsakademiens. Handl.* **1898**, *24*, 1–39.
66. Ho, Y.S. Citation Review of Lagergren Kinetic Rate Equation on Adsorption Reactions. *Scientometrics* **2004**, *59*, 171–177. [CrossRef]
67. Cheung, C.W.; Porter, J.F.; McKay, G. Sorption Kinetics for the Removal of Copper and Zinc from Effluents Using Bone Char. *Sep. Purif. Technol.* **2000**, *19*, 55–64. [CrossRef]
68. Das, A.; Chandrakumar, K.R.S.; Paul, B.; Chopade, S.M.; Majumdar, S.; Singh, A.K.; Kain, V. Enhanced Adsorption and Separation of Zirconium and Hafnium under Mild Conditions by Phosphoric Acid Based Ligand Functionalized Silica Gels: Insights from Experimental and Theoretical Investigations. *Sep. Purif. Technol.* **2020**, *239*, 116518. [CrossRef]
69. Trieu, Q.A.; Pellet-Rostaing, S.; Arrachart, G.; Traore, Y.; Kimbel, S.; Daniele, S. Interfacial Study of Surface-Modified ZrO_2 Nanoparticles with Thioctic Acid for the Selective Recovery of Palladium and Gold from Electronic Industrial Wastewater. *Sep. Purif. Technol.* **2020**, *237*, 116353. [CrossRef]
70. Ho, Y.S.; McKay, G. Pseudo-Second Order Model for Sorption Processes. *Process Biochem.* **1999**, *34*, 451–465. [CrossRef]
71. Vadivelan, V.; Kumar, K.V. Equilibrium, kinetics, mechanism, and process design for the sorption of methylene blue onto rice husk. *J. Colloid Interface Sci.* **2005**, *286*, 90–100. [CrossRef] [PubMed]
72. Plazinski, W.; Dziuba, J.; Rudzinski, W. Modeling of Sorption Kinetics: The Pseudo-Second Order Equation and the Sorbate Intraparticle Diffusivity. *Adsorption* **2013**, *19*, 1055–1064. [CrossRef]

recycling

MDPI

Review

Eco-Friendly and Complex Processing of Vanadium-Bearing Waste for Effective Extraction of Valuable Metals and Other By-Products: A Critical Review

Ahmed H. Ibrahim [1,*], Xianjun Lyu [2,*], Hani E. Sharafeldin [1] and Amr B. ElDeeb [1]

[1] Mining and Petroleum Department, Faculty of Engineering, Al-Azhar University, Cairo 11884, Egypt; hanisharafeldin@azhar.edu.eg (H.E.S.); engbasuony87@yahoo.com (A.B.E.)

[2] College of Chemical and Biological Engineering, Shandong University of Science and Technology, Qingdao 266590, China

* Correspondence: ahmedkraiz@azhar.edu.eg (A.H.I.); lyuxianjun@163.com (X.L.); Tel.: +20-100-8519628 (A.H.I.); +86-532-86057101 (X.L.)

Abstract: Achieving the New World Sustainability Vision 2030 leads to enacting environmental restrictions, which aim to partially or totally reduce the negative impacts of different forms of waste and develop alternative technologies for eco-friendly and cost-effective utilization. Solid waste is a hazardous waste with many environmental and economic problems resulting from its storage and disposal. However, at the same time, these wastes contain many valuable elements. One of these solid wastes is heavy oil fly ash "HOFA" generated in power stations using heavy oil as fuel. HOFA is produced annually in massive amounts worldwide, the storage of which leads to the contamination of water resources by the contained heavy metals, resulting in many cancerogenic diseases. At the same time, these ashes contain many valuable metals in significant amounts, such as vanadium "V" and nickel "Ni" that can be extracted effectively compared to their low content and difficulty processing in their main ores. Hence, recycling these types of wastes reduces the environmental adverse effects of their storage and the harmful elements in their composition. This paper critically reviews the world resources of vanadium-bearing waste and various approaches described in the literature for recovering V, Ni, as well as other valuable metals from (HOFA) and other wastes, including pyro- and hydro-metallurgical processes or a combination. Hydro-metallurgical processes include alkaline or acidic leaching using different reagents followed by chemical precipitation, solvent extraction, and ion exchange to extract individual elements. The pyro-metallurgical processes involve the non-salt or salt roasting processes followed by acidic or alkaline leaching processes. The operational parameters and their impact on the efficiency of recovery are also discussed. The digestion mixtures of strong mineral acids used to dissolve metal ions in HOFA are also investigated. Bioleaching is a promising eco-friendly technology for recovering V and Ni through appropriate bacteria and fungi. Oxidation leaching is also a promising environmentally friendly approach and more effective. Among all these processes, the salt roasting treatment showed promising results concerning the cost, technological, and environmental effectiveness. The possibility of complex processing of HOFA has also been investigated, proposing innovative technology for completely utilizing this waste without any remaining residue. Effective zeolite for wastewater treatment has been formulated as a good alternative for conserving the available water resources.

Keywords: sustainability; solid wastes; heavy oil fly ash "HOFA"; vanadium; nickel; hydrometallurgy; pyrometallurgy; wastewater treatment; zeolite

https://doi.org/10.3390/recycling10010006

1. Introduction

Vanadium is regarded as a rare metal, and its swift and consistently growing demand is challenged by its low concentrations in natural minerals. Vanadium extraction depends on the type of feed materials, efficiency, and technologies applied in processing raw materials. Therefore, recovering V and other valuable metals from secondary industrial resources like steel slag and ash containing these metal ions is highly intriguing regarding ecological and financial aspects. The increased industrial applications of vanadium contribute to the rising demand for vanadium, which has driven up the price of this metal. Currently, the value of vanadium ranges between $20,000 and $40,000 per metric ton [1,2]. These industrial uses encompass vanadium redox flow batteries, viewed as a promising solution to the global issue of large-scale energy storage production. They also include applications in the automotive and construction sectors, corrosion inhibitors, fertilizers, petrochemicals, catalysts for gas processing and sulfuric acid production, ferrovanadium alloys, thermistors, and more [3–5].

Unlike copper, nickel, or zinc, vanadium does not form concentrated deposits. Vanadium is found in many minerals, replacing Fe^{3+} or Al^{3+} in the octahedral positions of crystal lattices. This is possible because the ionic radii of V^{3+} and Fe^{3+} are almost the same, which means that vanadium can be found all over the Earth's crust, accounting for approximately 0.014% [6]. Vanadium is found in nature in the form of oxides, sulfides, and phosphates, and it is often associated with other metals such as iron, titanium, aluminum, zinc, lead, and uranium. These compounds include vanadinite, titanomagnetite, descloizite, patronite, and carnotite. The concentrations generally have a V_2O_5 grade between 0.2 and 1%, while higher-grade deposits may have V_2O_5 levels exceeding 2% [7–9].

It is estimated that there are 26 million tons of vanadium reserves worldwide. The only vanadium mine-operating countries with the largest reserves are China with 9.5 million tons, Russia with 5 million tons, Australia with 6 million tons, and South Africa with 3.5 million tons. In contrast, the world's resources total over 63 million tons [10]. Nearly all of the world's vanadium production in 2022 came from China (68%), Russia (17%), South Africa (9%), and Brazil (6%). The total global production of vanadium this year was approximately 120 kilotons/year [8,10]. Over 88% of the total vanadium is primarily produced as co-products from slag during the smelting of titanomagnetite ores or steel refining, representing over 69% of the starting raw material and making them an essential source of vanadium. In contrast, approximately 19% of the vanadium was produced from primary vanadium ores and 12% from secondary sources (Figure 1) [1,8].

The world's vanadium consumption reached 102.1 kilotons, representing an increase of nearly 45% over 2011. By the end of 2023, consumption is predicted to increase by up to 120~125 kilotons at a 5% annual growth rate, with increased demand for high-strength steel serving as the primary driver of this growth [8].

Coulsonite, a mineral of FeV_2O_4, is commonly found alongside magnetite and forms a series with $FeCr_2O_4$, also known as chromite [11]. The mineral ilmenite can also host vanadium, possessing a minor quantity [12]. Likewise, clay containing vanadium in weathered mineral deposits is commonly found, with V^{3+} taking the place of Al^{3+} in phyllosilicates and Fe^{3+} in goethite, FeOOH. A variety of vanadium oxides have been discovered in weathered zones of vanadium titanomagnetite deposits, including chlorite, $(Fe, Mg, Al)_6(Si, Al)_4O_{10}(OH)_8$, and goethite. It is worth noting that a lot of uranium deposits also contain vanadium in the form of minerals like carnotite ($K_2(UO_2)_2V_2O_8.3H_2O$) and tyuyamunite ($Ca(UO_2)_2V_2O_8.5–8H_2O$) that have uranyl vanadate [13].

From ores rich in minerals like carnotite ($K_2(UO_2)_2V_2O_8 \cdot 3H_2O$), vanadium has been generated as a by-product of uranium mining at numerous mines in the southwestern regions of the USA [14]. After a moderate leaching process, a significant portion of the uranium deposits was dissolved. The remaining vanadium in the uranium-leaching residue was then extracted using a more aggressive leaching condition [1]. The processing of South Korean shale-hosted uranium/vanadium ore has been suggested using a similar strategy [15]. Liu [16] has also been found to have comparable deposits in Western Australia. Coal and bituminous shale also contain trace levels of vanadium [17]. In northern Queensland, an abundant oil shale deposit at Julia Creek has been found to include vanadium and trace quantities of molybdenum [18].

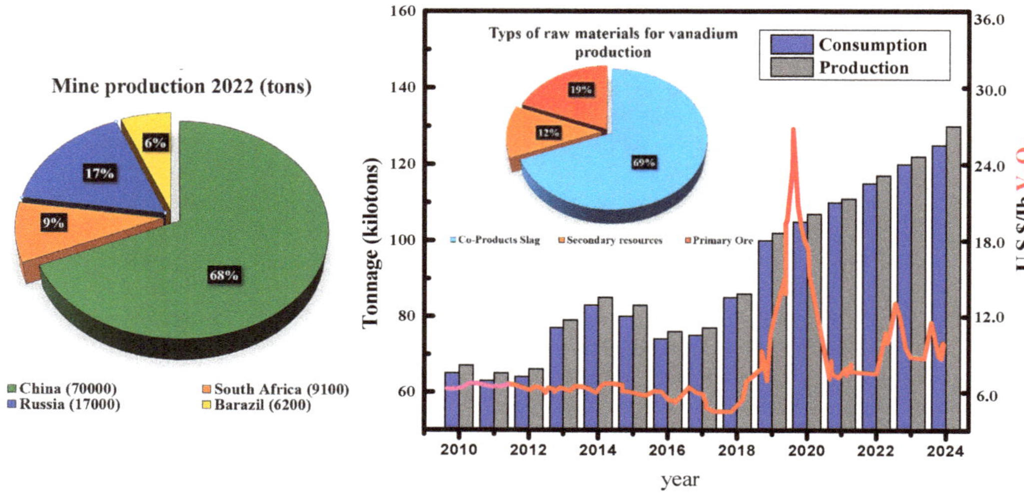

Figure 1. World mine production and global annual vanadium production and consumption [1,8]

Trace amounts of Vanadium have also been found in crude oil [19–22], among other potentially essential sources. Vanadium levels in crude oil are very high, reaching up to 0.05% in certain regions of Venezuelan oil [23,24]. During oil combustion, its vanadium content is transferred to the ash, which contains up to 10~18% of V_2O_5, making it an attractive material for vanadium recovery [25–27].

According to a report by White and Levy [28], the main sources of vanadium released into the environment are steel co-production, which accounts for 75% of the total, crude oil processing, which accounts for 10%, and mining, which accounts for less than 10%. In the Middle East and Africa, fuel oil and crude oil are easily accessible resources for power plants and desalination plants. For instance, it has been approximated that Saudi Arabia, with the world's largest oil reserves, consumes more than 40 million metric tons of crude oil and heavy fuel oil annually for thermal power plants and seawater desalination [29]. Egypt, Africa's largest non-OPEC oil producer, produces approximately 11 million metric tons of heavy fuel oil (HFO) annually, of which approximately 9 million metric tons are consumed domestically [30–32]. Petroleum and other liquids contribute 36%, and natural gas 57%; they were considered the most-consumed fuels in Egypt in 2020, according to BP's Statistical Review of World Energy 2021, in addition to renewable energy and coal, which accounted for 6% and 1% of the country's total energy consumption [33,34]. The country implemented measures to decrease its natural gas usage in power generation, which typically accounts for up to 90% of its total generation. As part of this effort, the country has significantly increased its use of oil-fired generation, resulting in an average fuel oil burn of over 100 kb/d in the first nine months of 2022, the highest level seen in four years [29].

To date, in many developing countries, most of the generated ashes do not have any industrial application but are only stored in landfills. Compared to other V resources, which are relatively limited and low-grade, these materials have no mining or operating costs. The disposal of ash or steel slag is a major environmental problem facing thermal power plants and iron production industries, and no monitoring is taken to prevent the contamination of groundwater and soil [30,35]. It is recommended to continuously monitor the metals of environmental impact, such as nickel, vanadium, and zinc, in ash samples due to carcinogenic and health hazards. Therefore, it is recommended to consistently monitor and recover valuable metals from industrial waste, as this is highly beneficial for both environmental protection and economic considerations. This review aims to offer a summary of advanced technologies used for vanadium recovery from various types of vanadium-bearing waste, as well as the comprehensive utilization of their solid residue.

2. Chemical and Mineralogical Composition of Vanadium-Bearing Waste

Table 1 shows the chemical composition of vanadium-bearing waste produced from different iron and steel production facilities in different locations worldwide. The chemical structure of vanadium oxides, such as V_2O_3 or V_2O_5, varies based on the type of feed materials introduced into the furnaces, the efficiency of the processing methods, and the desired final products. Typically, CaO, SiO_2, MgO, and Al_2O_3 contained in the vanadium-bearing slag are in the ranges 2~45%, 6~25%, 1~16%, and 2~10%, respectively. During reduction smelting, chromium and titanium oxide are also found in the vanadium-bearing slag, so obtaining pure vanadium requires critical separation of those metals after extraction.

Table 1. Chemical composition of vanadium-bearing slag generated from various industrial plants in different locations in the world.

Sources	Technology	Chemical Composition					Ref.
		V_2O_5/V_2O_3	FeO/Fe_2O_3	Cr_2O_3	TiO_2	MnO/MnO_2	
Panzhihua Iron and Steel	BF-BOF	8.5–14.3	32.9–42.1	2.1–4.3	11.1–12.9	5.9–9.1	[36]
Kapok Iron and Steel	BF-BOF	14.2–16.0	31.9–32.3	-	12.0	7.6–8.7	[37]
Chengde Xinxin Vanadium and Titanium Chemical	BF-BOF	8.5	51.1	3.5	10.5	5.2	[38]
Desheng Iron and Steel Group	BF-BOF	15.2–20.8	35.3–64.5	7.7–10.6	6.39–8.5	5.5–7.6	[39]
Chengde Iron and Steel	BF-BOF	10.2–13.4	36.7–49.1	1.7–4.2	6.8–11.1	5.2–7.2	[40]
Esfahan Steel	BF-BOF	1.5–1.9	17.3–17.6	-	1.0–1.5	4.2–4.5	[41]
Sichuan Weiyuan Iron and Steel	BF-BOF	14.3	24.8	4.4	7.4	8.5	[42]
Pan Steel	DR-EAF	8.9	24.3–25.2	2.0–8.7	3.3–14.7	1.6–13.8	[43]
CITIC Jinzhou Ferroalloy	DR-EAF	15.3	30.5	2.3	13.7	10.9	[44]
British Steel	BF-BOF	0.82	32.0	0.2	0.3	4.5	[45]

BF = blast furnace; BOF = basic oxygen furnace; DR = direct reduction; EAF; electric arc furnace.

The chemical structure of HOA reveals that it primarily consists of carbon, along with several metallic elements. Table 2 presents the chemical composition of ash samples generated from burning heavy oil in various power and desalination plants across different global locations. Industrial ashes primarily consist of vanadium (V) and nickel (Ni), along with toxic elements like chromium, arsenic, cadmium, lead, cobalt, and selenium. The maximum content of V and Ni in ash is 5 wt.% and 1.8 wt.%, respectively [35]. Heavy oil ash chemical composition also differs based on the makeup of the heavy fuel oil, parameters of operation, and chemicals used [35,46]. Two types of ash (boiler ash and fly ash) are generated by the combustion of HFO in power plants, which is expected to increase as energy consumption rises in the next few years. Around 3 kg of ash residue is produced when 1000 L of heavy oil is burned [47]. A higher level of V and Ni was found in heavy oil ash (HOA) produced by various power plants worldwide [48]. The ash fraction is too heavy to be introduced into the flue gas and hence deposited outside the boiler tube, which is known as the boiler ash. It is usually classified as high-grade ash which contains

(4.4~19.2%) V, (2.7~8.5%) Ni, and burned C [49,50]. The other type of ash is known as fly ash, which is fine combusted residue usually collected from the stack or electrostatic precipitators (ESPs). It is usually classified as low-grade ash which contains (0.3~1.5%) V, (0.2~0.5%) Ni, and unburned (30~80%) C [51]. The refining of crude oil often entails some kind of thermal cracking to produce a light oil fraction and a heavier residual oil. Most metals and a small amount of sulfur are still linked to the non-volatile elements and end up in a residual oil fraction. Several factors influence the elemental composition of heavy fuel oil, including the source of crude oil, the yield of HFO from crude oil, and the degree of processing [52].

Table 2. Chemical composition of vanadium-bearing waste reported in different studies.

Element (ppm)	References										
	Ref. [53], PP	**Ref. [53], DP**	**Ref. [21]**	**Ref. [54]**	**Ref. [55]**	**Ref. [56]**	**Ref. [57]**	**Ref. [58]**	**Ref. [35]**	**Ref. [59]**	**Ref. [25]**
Carbon (%)	90.40	56.70	85.56	51.86	-	67.4	-	77.40	-	14.73	18.65
Sulfur (%)	7.77	27.18	-	-	-	8.6	-	7.10	-	5.83	11.23
Vanadium	9072	31,044	2958	34,487	50,000	38,000	7670	12,900	3540	85,000	49,100
Nickel	2382	13,633	1762	11,852	15,400	16,000	18,000	6800	1055	52,500	24,200
Cadmium	1.65	3.7	3.28	1.59	-	-	-	-	60.4	-	920
Arsenic	2.54	1846	2.24	68.29	-	-	-	-	128.5	-	1.61
Cobalt	2.88	12.33	3.28	247.79	2200	-	-	-	60.7	-	2140
Chromium	36.79	113.09	4.06	107.60	8000	-	-	-	70.4	-	-
Selenium	1.00	6.81	11.60	13.20	-	-	-	-	7.25	-	-
Lead	17.09	13.94	11.00	116.10	-	-	-	-	27.85	7200	1900
Zinc	21.92	118	130	592.10	-	-	1110	4000	65.5	38,000	7600
Copper	10.44	50.40	170.4	120.30	-	-		17,000	57.12	290	1760
Iron	7210	8771	-	-	220,500	8000	59,800	1400	176.5	97,400	59,600
Magnesium	6971	94,608	-	-	2000	-	3430	14,100	3615	5920	3.6
Manganese	23.9	149.26	-	-	-	-	-		12.8	900	2.4
Calcium	582.3	4121.2	-	-	-	-	-	2300	3380	2100	2354
Sodium	1395	7555	-	-	-	19,000	-	-	1310	2045	1524
Aluminum	3541	1041.8	-	-	-	1040	2870	2500	642.4	7123	4252
Barium	7.42	49.68	-	-	-	-	-	-	69.0	-	5214
Others (ppm)	-	-	Hg, 0.25	-	Mo, 3500	Si, 8000	-	Si, 800; O, 93 200	-	Si, 39,000	Si, 36,800

PP = power plant; DP = desalination plant.

Table 3 presents the mineralogical compositions of vanadium-bearing slag produced by different iron and steel-making facilities and ashes. The spinel phases of [$(Fe, Mn)_2(V, Ti) O_4$] and [$(Fe, Mg, Mn) (V, Cr)_2 O_4$] contain vanadium, titanium, and chromium. The spinels are distributed throughout the olivine matrix $(Fe, Mg)_2SiO_4$, augite ($Ca(Fe, Mg)Si_2O_6$), fayalite (Fe_2SiO_4), and diopside phases ($Ca(Fe, Al)_2SiO_6$) [60–62]. Previously reported data indicated that V-Ti and V-Cr spinels are arranged in a layered core configuration. Chromium is primarily located in the central core of the grains, vanadium is prevalent in the outer layer, and titanium is situated along the spinel grain boundary. This distribution reflects the varying crystallization abilities of each spinel, ranked as Cr > V > Ti [43]. Other elements, such as manganese, are evenly dispersed within the spinel grain. Magnesium is concentrated alongside chromium, forming the minor phase of $MgCr_2O_4$ [60]. The $Ca_2V_2O_7$ phase of vanadium was also documented [41], which resulted from the substantial amount of limestone utilized in the iron and steel production process [63].

Additional oxides, originating from the fluxes added to the furnace charge, have also been identified. Dondi [64] states that X-ray diffraction patterns show that two types of fly ash (ash B from Apulia and ash F from Sardinia) produced from the combustion of orimulsion in Italian thermal power plants have a complex composition. These fly ashes contain high grades of V, Ni, S, and Mg and exhibit increased reflections, several line interferences, and overlaps. This complexity makes it difficult to make certain attributes. The interpretation of these patterns indicates that magnesium sulfate monohydrate ($MgSO_4.H_2O$) and hydrogen vanadyl sulfate [$(VO)_2H_2(SO_4)_3$] are predominated, along with magnesium monoxide (MgO), anhydrous, monohydrate, and pentahydrate vanadyl sulfate (a-$VOSO_4$,

$VOSO_4.H_2O$, $VOSO_4.5H_2O$). These phases are associated with minor amounts of calcium sulfate ($CaSO_4$) and nickel monoxide (NiO), and presumably sodium aluminum silicate ($NaAlSi_3O_8$) and traces of magnesium sulfate hexahydrate ($MgSO_4.6H_2O$). Other analyses of the insoluble residues indicated that in addition to magnesium, nickel oxides, and anhydrous vanadyl sulfate, other vanadium compounds, i.e., VO_2, B-V_2O_5, VOOH, and $VO(OH)_2$, were present in small amounts of untreated ash, which cannot be detected in the collective sample due to their low concentration. However, identifying these compounds suggests the presence of V^{3+} and V^{5+}, together with the monopolized vanadyl ion valence of V^{4+}. The existence of these vanadyl ions can vary depending on the oxidation conditions during the combustion of heavy oil or the cooling techniques used [65,66].

Table 3. Characterization of the minerals in vanadium-containing waste produced from various industrial plants in different locations in the world.

Source	Mineralogical Phases	Ref.
Panzhihua Iron and Steel	$(Fe,Mn)V_2O_4$,$(Fe,Mn)_2SiO_4$,$CaMn(SiO_3)_2$	[36]
Kapok Iron and Steel	(Fe, Mn)O, $(V,Ti)_2O_3$, 2(Fe,Mn)O, SiO_2, SiO_2	[37]
Chengde Xinxin Vanadium and Titanium Chemical	$(Mn_{0.84}Fe_{0.16})(Mn_{0.16}Fe_{1.34}Cr_{0.5})O_4$, $Fe_{2.1}Ti_{0.74}Mn_{0.02}V_{0.01}Ca_{0.01}Si_{0.01}Al_{0.05}Mg_{0.06}O_4$,$Fe_2SiO_4$, SiO_2	[38]
Desheng Iron and Steel Group	$(Mg,Fe)(V,Cr)_2O_4$, Fe_2SiO_4, $Ca(Fe,Mg)Si_2O_6$,Fe_2TiO_4	[39]
Chengde Iron and Steel	$(Mn,Fe)(V,Cr)_2O_4$, Fe_2SiO_4, SiO_2	[40]
Esfahan Steel	$Ca(OH)_2$,Ca_3SiO_5, $CaFeO_2$,$Ca_2V_2O_7$	[41]
Sichuan Weiyuan Iron and Steel	$(Mn,Fe)(V,Cr)_2O_4$,Fe_2SiO_4, $(Fe,Mg)Si_2O_6$,Mg_2TiO_4	[42]
Pan Steel	$(Mn,Fe)(V,Cr)_2O_4$,Fe_2TiO_4, $(Fe,Mn)_2SiO_4$	[43]
CITIC Jinzhou Ferroalloy	$(Mn,Fe)(V,Cr)_2O_4$,$CaFeAlSiO_6$,$Fe_{2.5}Ti_{0.5}O_4$	[44]
British Steel	MgO, Al_2O_3, β-Ca_2SiO_4, $Ca_2(Al,Fe)_2O_5$, FeO, $Mg(OH)_2$, C	[45]
Fiume Santo-Porto Torres	$MgSO_4.H_2O$, $(VO_2)_2H_2(SO_4)_3$, $VOSO_4.nH_2O$ $CaSO_4$, $NiSO_4.nH_2O$	[64]
El Kriymat Electric power station	$(VO_2)_2$, FeV_2O_4, Mg_2SiO_4, $Ca_2V_2O_7$, $(Ca(Fe,Al)_2SiO_6)$, $NiSO_4.H_2O_6$	[22,25]

3. Vanadium Processing Technologies

As previously mentioned, steel slag and oil ash are important secondary sources of valuable elements (V, Ni, Cr, and Zn). However, there are two main ways for processing industrial solid waste to extract vanadium and other useful metals. The first method, direct leaching, involves using acid (such as H_2SO_4, HCl, and HNO_3) or alkaline (such as NaOH and Na_2CO_3) solutions to dissolve the valuable metal from vanadium-bearing waste. The dissolved metals are then recovered from the solution using various techniques such as chemical precipitation, solvent extraction, or ion exchange. The second is the roasting-assisted leaching process, including salt-free, calcium, or salt roasting, in which the solid waste is heated to a high temperature to allow the separation of valuable vanadate materials. Acid, alkaline, and water are frequently employed as leaching agents, as outlined below [67–74]. A comprehensive flowchart for recovering V from solid waste is shown in Figure 2 [35]. After extracting vanadium and nickel as their respective chemical compounds (NH_4VO_3 and NiC_2O_4), calcination is conducted to produce V_2O_5 and NiO.

3.1. Direct Leaching Process

Direct acidic or alkaline leaching is used to dissolve almost all valuable metals into the solution from various types of solid waste and then extract these metals separately using adequate techniques [75–82].

3.1.1. Acid Leaching

H_2SO_4 is a cost-effective option for leaching procedures, which makes it a popular choice due to its economic benefits. The H_2SO_4 acid has the ability to dissolve several metallic cations like V, Ni, Fe, and Mg, among others, creating metal sulfates. The leaching of significant economic and nuclear elements such as uranium, hafnium, and zirconium, as well as vanadium and nickel, with or without oxidant agents (NaClO, MnO_2, or $KClO_3$), has been investigated by Abd El-Hamid [83]. It was noted that vanadium leaching efficiency of 95% at a concentration of 200 g/L H_2SO_4, 6% (solid/solid) manganese dioxide, 80 °C, 6 h, 1/10 solid/liquid ratio of −200 mesh ore granule size. According to Deng [84], the direct acid leaching technique could increase the vanadium leaching efficiency from 14 to 73% by adding an oxidant. Barik [85] created a technique for extracting vanadium (V) and nickel (Ni) from solid industrial wastes abundant in these metals. The procedure includes leaching with sulfuric acid (H_2SO_4), followed by solvent extraction, precipitation, and crystallization. It was noted that using 1.35 M H_2SO_4 at 40 °C for 90 min, both V and Ni were extracted with a 98% recovery rate. Vanadium is selectively extracted from the pregnant solution using 40% LIX 84-I at pH 0.5, and then by adding an aqueous solution of ammonia NH_3; it precipitates as ammonium metavanadate (NH_4VO_3). Ammonium oxalate $(NH_4)_2C_2O_4$ is used to selectively separate Ni that has been converted into Ni oxalate in the raffinate (NiC_2O_4). Finally, NH_4VO_3 can be heated to produce vanadium pentoxide (V_2O_5) [86]. On the other hand, nickel oxide is obtained by heating nickel oxalate at 450 °C for 2 h.

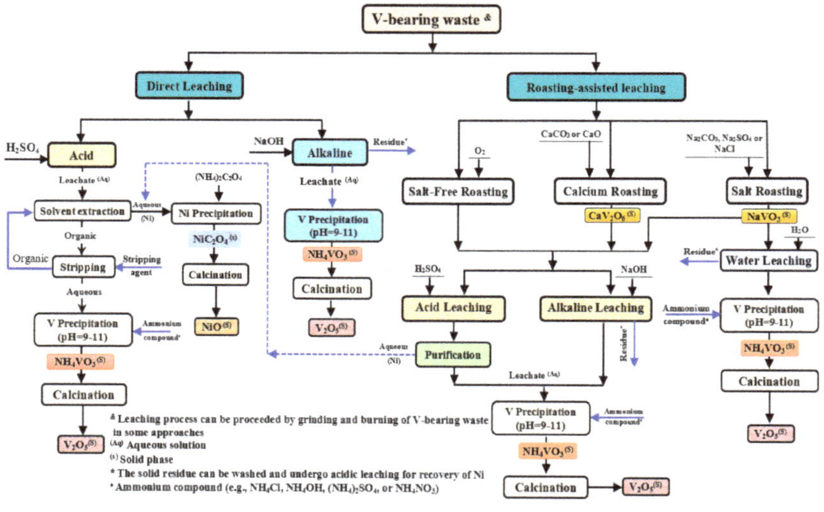

Figure 2. Overall flowchart showing the most common methods for the recovery of vanadium and nickel from industrial solid waste.

Nazari [87] optimized the parameters of the H_2SO_4 leaching process to maximize V and Ni recovery from fly ash samples. The maximum recovery of 94% V and 81% Ni has been achieved at the following optimal conditions: 19.5% H_2SO_4 concentration at 80 °C, 9.15% S/L ratio by weight, and for 2 h. Fly ash was subjected to a three-step procedure, which included leaching with H_2SO_4, oxidative precipitation, and washing to extract V as V_2O_5 from the fly ash, and about 90% of V was successfully recovered. Boiling 1 M H_2SO_4 for 30 min. at 3 mL/g L/S ratio is considered to be the operational parameter for the leaching process. During the oxidative precipitation process, sodium chlorate ($NaClO_3$) was added as an oxidizing agent of V^{4+} to V^{5+}. The pH dropped because of this oxidation of V, which also released H into the environment; hence, the appropriate pH was maintained by adding Na_2CO_3. In order to remove Na and other impurities, the V_2O_5 precipitate was then washed with an acidic solution [26]. Leaching experiments on HOA were performed by Navarro [88] in 0.5 M H_2SO_4 solution at 4 mL/g as L/S ratio at 200 rpm agitation speed for 24 h and at room temperature. After the leaching procedure, the material was washed for 1 h. in water at 4 mL/g L/S ratio. Ni recovery was only about 12% under these conditions, but V reached 98%.

Aburizaiza [89] detailed a multi-step approach for leaching and extracting V, Ni, and Fe from HOA, which involves the following steps: (i) Acidic leaching of fly ash in a solution containing mineral acids like perchloric acid ($HClO_4$), hydrochloric acid (HCl), and hydrogen peroxide (H_2O_2); (ii) Ammonium pyrrollidine dithiocarbamate ($C_5H_9NS_2.NH_3$) in chloroform ($CHCl_3$) is combined and agitated with the leachate during the organic extraction of the metals. Once equilibrium is reached, a layer of organic phase that contains the metal cations is formed. The findings demonstrate that the organic phase successfully extracts the targeted metal ions (V, Ni, and Fe); (iii) the V, Ni, and Fe metal ions are removed from the organic phase by combining them with 1 M HNO_3 that contains Hg^{2+} ions. Therefore, metallic cations (V^{3+}, Ni^{2+}, and Fe^{2-}) are present in the final aqueous nitric acid solution; (iv) The addition of dimethylglyoxime ($C_4H_8N_2O_2$) in chloroform allows the extraction of the total Ni from the stripped nitric acid solution while leaving the V^{3+} and Fe^{2+} ions dissolved in the aqueous nitric acid solution phase. In order to recover nickel in the form of NiO, the Ni-containing organic phase is separated and evaporated till it is dried; (v) Total Fe is extracted from the aqueous nitric acid solution phase produced in step (iv) by 4-pypyridyl treatment. The organic phase that contained Fe was separated, evaporated, and ignited to produce Fe in the form of FeO; (vi) vanadium as V^{3+} and other trace elements can be found in the aqueous solution that was left over. The study does not present an efficient technique for extracting V from the final aqueous solution. Additionally, this approach demands significant amounts of chemicals to handle a small quantity of fly ash.

Tokuyama [58] studied the dissolving of HOA in two steps. In the first step, Ni, Zn, Mg, Al, and Fe are dissolved in water, and in the second stage, V is dissolved in H_2SO_4, as shown in Figure 3. Moreover, it was found that HCl and H_2SO_4 have similar leaching efficacy. In the same study, leaching with concentrated NaOH might result in the selective leaching of V.

Tsygankova [48] concluded that the HOA leaching process depends on both its phase change and chemical composition. The outcomes of acidic leaching for three distinct types of fly ash, each with different chemical and phase compositions, have been examined. Each type of fly ash requires specific leaching conditions, involving 5 to 9% H_2SO_4 concentration, with temperatures set between 20 and 80 °C, and durations spanning from 30 to 60 min. Vanadium is precipitated as V_2O_5 by oxidizing the leaching solution with H_2O_2 at 20 °C for 30 min. After further heating to 95 °C and at a pH range from 1.8 to 2, precipitation is accomplished.

Khalafalla [90] examined the extraction of V, Ni, and Zn by leaching the heavy oil ash (HOA) in H_2SO_4 solution, as shown in Figure 4. High-tension Magnetic Separation was used to physically concentrate the fly ash, increasing the amounts of V_2O_5, ZnO, and NiO, to 18.1%, 8.11%, and 11.18%, respectively. Following that, leaching was performed using a 180 g/L H_2SO_4 solution with 4% MnO_2 as an oxidant for 10 h at 80 °C which resulted in the recovery of 96.5% of vanadium, 94.8% of nickel, and 99.1% of zinc. Ultimately, vanadium was extracted from the leach solution using 3% Alamine 336 in kerosene as the solvent, while nickel and zinc remain dissolved in the raffinate. After vanadium was extracted from the organic phase, $V_2O_5.H_2O$ was precipitated by adding H_2SO_4. In contrast, when Na_2S was added to the raffinate solution, Ni and Zn co-precipitated, forming a (Zn-Ni) sulfide cake. This cake could be suggested for hydro-metallurgical techniques to recover and separate Zn and Ni [91].

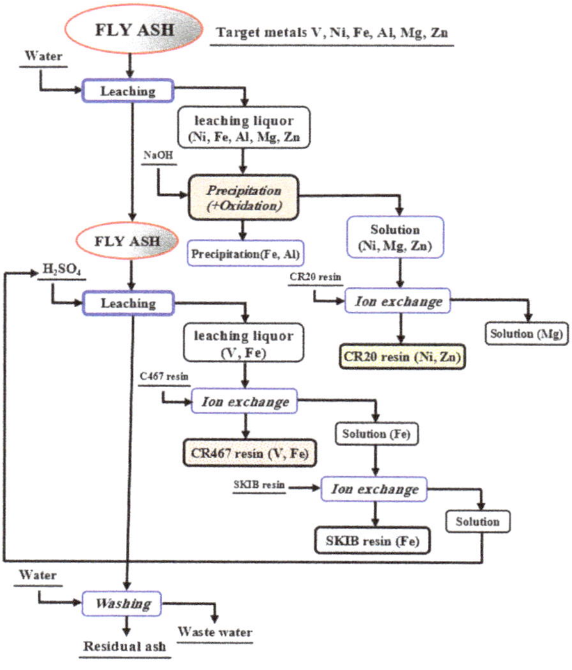

Figure 3. Flowsheet of processing fly ash for recovering metal ions [58].

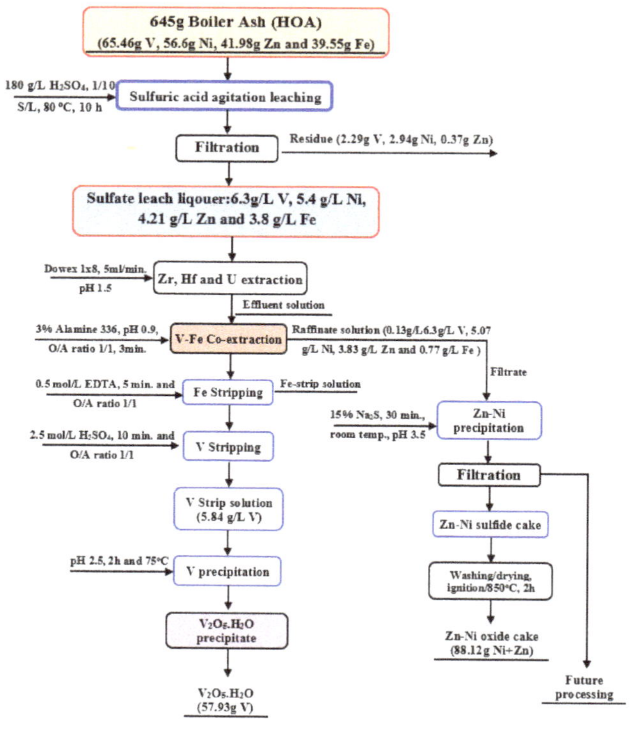

Figure 4. The proposed technical flowsheet for the chemical treatment of El-kuriemat boiler ash [90].

Rahimi [92] developed a distinctive technique for extracting vanadium from HOA utilizing lemon juice and organic acids. The citric acid (90 mg/g), malic acid (0.86 mg/g), and ascorbic acid (1.24 mg/g) are all present in lemon juice. The idea was to obtain organic acids by an ecologically acceptable method as an alternative to bioleaching, which produces such acids through extremely sluggish microbial development. The recommended conditions for the maximum recovery of 88.7% V were 2 h. ultrasonic leaching of HOA in a 27.9% lemon juice solution with 10% hydrogen peroxide at 35 °C and 0.01% S/L ratio. Sodium carbonate (Na_2CO_3) was added to the liquid to raise the pH level to 9 or 10, and the addition of $CaCl_2$ precipitates Al and Fe. Next, ammonium chloride (NH_4Cl) was added, leading to the precipitation of V as pure NH_4VO_3, then calcined at 500 °C to obtain vanadium pentoxide (V_2O_5). Eventually, the results revealed that assisting leaching agents, such as H_2O_2 and ultrasonic, is essential in accelerating V dissolution's kinetics.

High-pressure and high-temperature leaching demonstrates significant improvements compared to traditional acid leaching. For instance, according to Mu [93], 97.9% vanadium and 94.5% iron were leached out at a temperature of 140 °C, a concentration of 200 g/L H_2SO_4, a partial pressure of 0.5 MPa of oxygen, a period of 120 min, and an L/S ratio of 20:1. These are in completely agree with the findings published by Zhang [94–96]. Zhou [97] studied the dissolution kinetics of vanadium trioxide in sulfuric acid-oxygen medium. The highest dissolution rate was obtained with increased stirring speed before 800 rpm, oxygen partial pressure in the range of 0.6~1.4 MPa, and particle size decreased. Still, it was nearly independent of the sulphuric acid concentration from 0.4~2.0 mol/L and stirring speed over 800 rpm. Amer [98] recovered $V_2(SO_4)_3$ and $NiSO_4$ from Egyptian boiler ash by direct leaching using H_2SO_4 under oxygen pressure. The recommended leaching conditions have been determined to be 200 °C, 15 min, 15 bar of oxygen pressure, 60 g/L of H_2SO_4 concentration, and 1 S/L ratio. Adding an aqueous solution of ammonia NH_3 and neutralizing the acidic sulfate solution, V is precipitated as vanadium hydroxide ($V(OH)_3$). On the other hand, metallic Ni can be electrodeposited from the $NiSO_4$ electrolyte. The results of comprehensive studies employing acidic leaching methods on various vanadium resources are shown in Table 4.

Table 4. Results of different studies on direct acidic leaching of vanadium from various V-bearing waste.

Ref.	Lixiviant	Acid Leaching Conditions	Temp. °C	Time	S/L Ratio	Leaching Efficiency
[98]	H_2SO_4	−250 µm, 60 g/l H_2SO_4, 15 bar of O_2 pressure	200 °C	15 min	1:1	• V = 95%
[85]	H_2SO_4	1.35 M H_2SO_4, 300 rpm stirring	40 °C	90 min	1:5	• V = 98% • Ni = 98%
[87]	H_2SO_4	−75 µm, 19.5% H_2SO_4 Conc.	80 °C	120 min	1:9	• V = 94% • Ni = 90%
[26]	H_2SO_4	1 M H_2SO_4, 300 rpm stirring	100 °C	30 min	1:3	• V = 90%
[88]	H_2SO_4	−500 µm, 110 °C drying, 0.5 M H_2SO_4	Room Temp.	24 h	1:4	• V = 98% • Ni = 12%
[89]	H_2SO_4 + HCl + HNO_3	1.0 M Conc., 300 rpm stirring	RT	6 h	1:5	• V = 98% • Ni = 95% • Fe = 86%
[48]	H_2SO_4	−75 µm, 5~9% H_2SO_4 Conc.	20~80 °C	30~60 min	1:4	• V = 96%

Table 4. *Cont.*

Ref.	Lixiviant	Acid Leaching Conditions	Temp. °C	Time	S/L Ratio	Leaching Efficiency
[90]	H_2SO_4	180 g/L H_2SO_4, 4% MnO_2, 500 rpm stirring	80 °C	600 min	1:10	• V = 96.5% • Ni = 94.8%
[92]	Ultrasound, H_2SO_4	−75 μm, 27.9% lemon juice, 10% H_2O_2	35 °C	120 min	0.01%	• V = 88.7%
[99]	H_2SO_4	−100 μm, 0.5 N H_2SO_4, 400 rpm stirring	30 °C	120 min	1:5	• V = 65% • Ni = 60% • Fe = 42%

3.1.2. Alkaline Leaching

Several investigations on direct alkaline leaching of V-bearing waste have been carried out, indicating that Ni is insoluble in alkali solutions, which are more suitable for selective leaching of V to increase recovery. A two-stage leaching approach for HOA was proposed by Tsai [99]. The fly ash is leached in ammonia water containing ammonium sulfate $(NH_4)_2SO_4$, and the Ni is extracted from the solution in the first stage. The second step involves recovering V by leaching the ash residue in an alkaline (NaOH) solution. However, the method was not used in more recent research because of the decreased recovery of Ni (60%) in the ammonia/$(NH_4)_2SO_4$ solution, which also dissolves 8% of V in the first step. Furthermore, the study did not propose or conduct a method to recover V and Ni from the leaching solution. In a power plant, alkaline leaching and sodium hydroxide "NaOH" solution were used by Al-Zuhairi [100] to extract V with a high recovery yield (98%) from red crude oil residues. According to their discussion, 2 M NaOH concentration at 100 °C for 2 h is the ideal extraction environment for V. The effluent is then acidified to lower the pH and treated with aqueous solution of ammonia (NH_3), to precipitate V as NH_4VO_3.

According to Navarro [88], alkaline leaching of HOA by NaOH is favored over acidic leaching because it is more selective for V, which is less efficient than acidic leaching by H_2SO_4. Alkaline leaching with NaOH is carried out using a 2 M NaOH solution at 1/4 g/mL S/L ratio for 24 h at room temperature and stirring at 200 rpm. Leaching is followed by three times for 1 h water rinses. At these conditions, vanadium recovery achieves 90%, but repeated leaching and washing 6 times, each for 1 h, enhances V recovery to 98%. When the contact time approaches 12 h, it is observed that the temperature has a minimal impact on the effectiveness and kinetics of V leaching. In order to improve selectivity for V recovery while preventing Si leaching, sodium carbonate (Na_2CO_3) is used for leaching HOA. However, by employing 0.66 M Na_2CO_3 solution in the leaching procedure under the same conditions as for NaOH leaching, 80% of V is recovered. Leaching with Na_2CO_3 exhibits more excellent vanadium selectivity while having a lower leaching efficiency than leaching with NaOH.

Hakimi [101] established a flowsheet for extracting V with a high percent recovery from V-rich HOA using NaOH solution at 95 °C for 4 h. as shown in Figure 5. Leaching is followed by adding 4.5 M H_2SO_4 solution to neutralize the alkaline leachate. Al and Si are deposited once the pH is raised to 8, leaving V dissolved as sodium vanadate in the filtrate. After being treated with an ammonium compound, the latter precipitates V as NH_4VO_3, which is then heated at 450 °C for 1 h to form V_2O_5.

Akita [102] studied the recovery of nickel and vanadium from fly ash using a two-step leaching process. The first stage involves leaching Ni with NH_4Cl solution, while the second step involves leaching the remaining ash with Na_2CO_3 to dissolve the vanadium. Subsequently, vanadium is extracted using a solvent mixture of TOA and toluene, and then precipitated as NH_4VO_3 using NH_4Cl. However, nickel can be extracted as NiS from its leach solution by precipitating it with Na_2S, as illustrated in Figure 6.

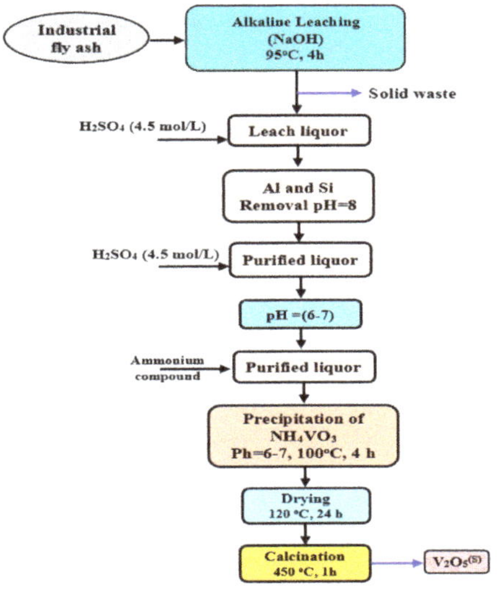

Figure 5. Flowchart of vanadium extraction from fly ash [101]. $^{(S)}$ Solid phase.

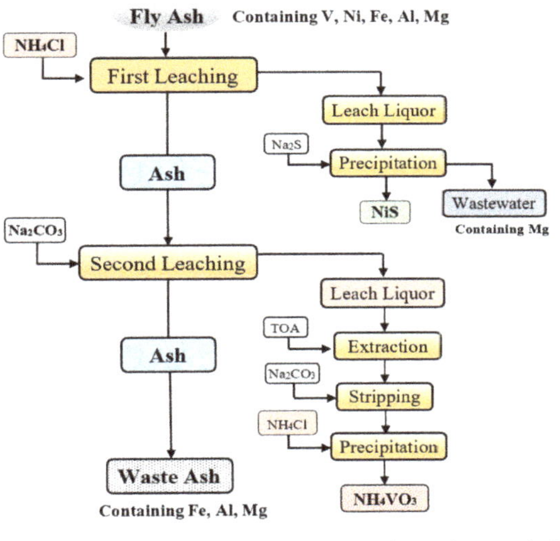

Figure 6. Process flow chart for recovery of vanadium and nickel from fly ash [102].

Vanadium, nickel, and molybdenum recovery from HOA was accomplished by Stas [55], using a two-stage leaching procedure, as shown in Figure 7. The first step is the alkaline leaching using NaOH solution to extract V and Mo. In the subsequent second step, acid leaching using H_2SO_4 leaches the residual fly ash to recover Ni. The maximum V recovery of 90% is obtained at 5 mL/g L/S ratio, at 100 °C, and for 3 h. The alkaline leaching solution containing V and Mo is gently agitated and cooled to 5 °C for one hour, which causes V to precipitate as sodium vanadate. This precipitate is then removed through filtration, leaving behind a filtrate that contains sodium molybdate. The later filtrate is heated to 90 °C and acidified with HNO_3 to precipitate Mo as H_2MoO_4. Conversely, sodium vanadate is dissolved again in a 5% HNO_3 solution (pH = 8), and $(NH_4)_2SO_4$ is introduced to the solution to precipitate vanadium as N_4HVO_3. The latter is converted to V_2O_5 after being calcined at 500 °C for 24 h. The remaining ash is leached using a 5 M H_2SO_4 solution for 3 h at 100 °C and a 4 mL/g L/S ratio, which leads to extraction of 80% in the second acidic leaching stage. The pH is subsequently increased by adding a NaOH solution to the acidic leachate in two stages, which results in the precipitation of Fe and the residual V. Finally, sodium carbonate is introduced to precipitate nickel in the form of nickel carbonate ($NiCO_3$).

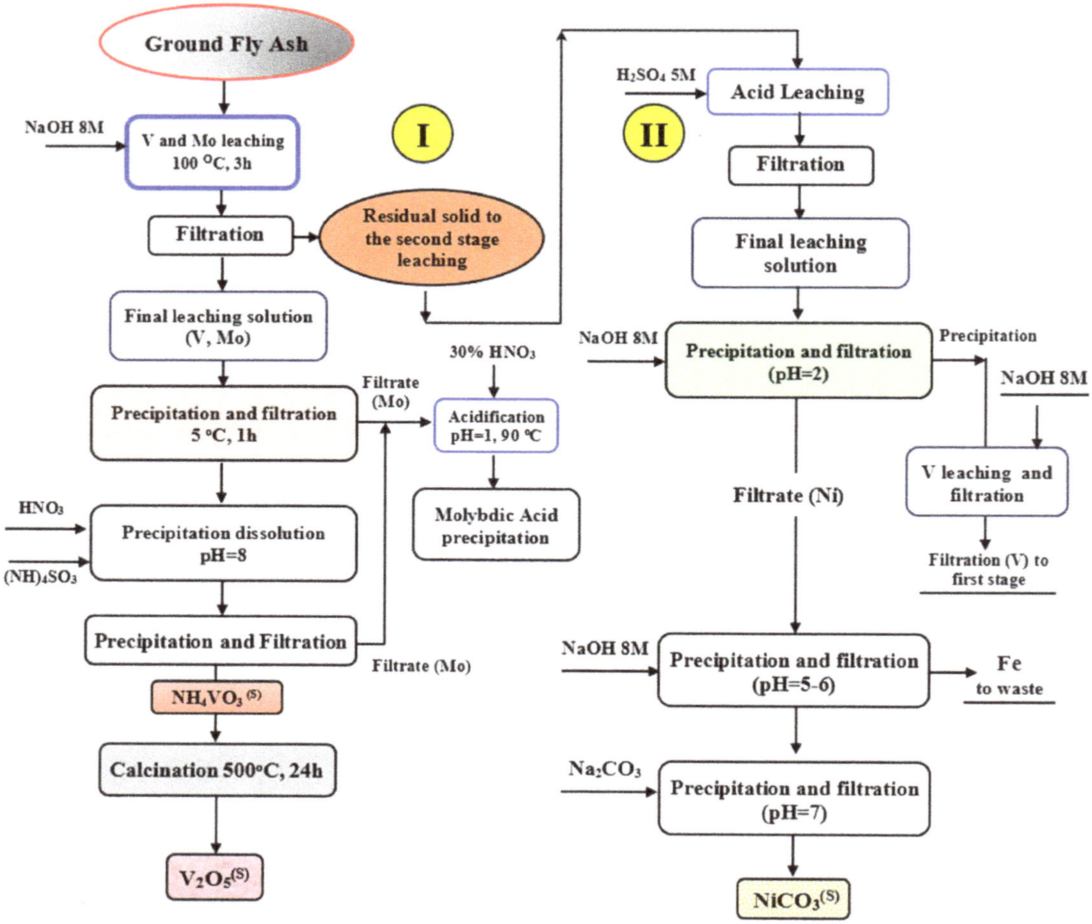

Figure 7. Two-stage leaching procedure to extract V, Mo, and Ni from fly ashes; (I) first stage alkaline leaching; (II) second stage acidic leaching [55]. $^{(S)}$ Solid phase.

Al-Ghouti [103] studied the recovery of vanadium and nickel from HOA by two leaching processes. Using NH_4Cl/NH_3 solution made up of 2 M NH_4CL and 2 M NH_3 at 50 °C, at 19 mL/g 1 L/S ratio, and for 6 h, Ni is selectively extracted in the first stage. After treating the leachate with sodium sulfide (Na_2S), nickel is precipitated as NiS with 56% recovery. In the following step, vanadium is extracted from the nickel-free remaining ash using agitation leaching in a 2 M Na_2CO_3 aqueous solution at 70 °C and a pH of 5.5 for a duration of 6 h. However, along with V, Fe, Mg, and Ca are also removed by

Na_2CO_3 solution. Triethylamine/toluene solution [0.1 M of $(CH_3CH_2)_3N$ in toluene] is added to the Na_2CO_3 solution to perform selective extraction of V. The organic layer is removed and then blended once again for 3 h with a fresh 2 M Na_2CO_3 solution to extract V from the aqueous layer. Finally, the resultant aqueous layer is mixed with 1.3 M NH_4Cl for 20 h to achieve maximum recovery of 45% by precipitating the extracted V. Alkaline pressure leaching was carried out by Shahnazi [104] with a maximum vanadium leaching of 90% under the concentration of 40 to 50% NaOH and a temperature range of 110 to 160 °C. In the same way, Qiu [105] examined how pure vanadium trioxide dissolves in an oxygen system under alkaline conditions. The findings indicated that higher temperatures (130 °C) and oxygen pressure (700 kPa) resulted in a faster rate of vanadium dissolution. Additionally, stirring speeds above 800 rpm, and sodium hydroxide concentrations at 1.0 M were found to maximize the dissolution rate of vanadium (90%). A novel approach utilizing NaOH-$NaNO_3$ binary melts has been suggested for treating vanadium slag [106]. This method enables the extraction of both vanadium and chromium during the leaching process. When the reaction conditions are optimized (1:1 NaOH-$NaNO_3$ solid ratio, 400 °C, 0.5 L/min atmospheric pressure for 2 h), vanadium and chromium recovery rates can reach 93.7% and 88.2%, respectively, within 6 h. The results indicate that the reaction temperature and KOH-to-ore mass ratio are more influential factors for the extraction of vanadium and chromium. Table 5 summarizes the results of in-depth investigations on different vanadium resources using the alkaline leaching technique.

Table 5. Results of different studies on the alkaline leaching of vanadium from various V-bearing wastes.

Ref.	Lixiviant	Alkaline Leaching Condition	Temp. °C	Time	S/L Ratio	Leaching Efficiency
[99]	NaOH	−250 μm, 2 N NaOH, pH 14, 400 rpm agitation speed	30 °C	120 min	1:5	• V = 80%
[99]	NH_4OH	−250 μm, 4 N NH_4OH, pH 10	30 °C	120 min	1:5	• V = 50% • Ni = 60%
	NH_4OH + $(NH_4)_2SO_4$ + NaOH	0.25 N NH_4OH, 4 N $(NH_4)_2SO_4$, pH 8.5, 2 M NaOH	30 °C	120 min	1:5	• V = 80% • Ni = 60%
[100]	NaOH	2 M H_2SO_4, 300 rpm stirring	100 °C	120 min	1:3	• V = 98%
[88]	NaOH	2 M NaOH+ H_2O, 200 rpm agitation speed	RT	24 h	1:4	• V = 98%
	Na_2CO_3	0.66 M Na_2CO_3, 200 rpm stirring	RT	24 h	1:4	• V = 80%
[101]	NaOH	−500 μm, 110 °C drying, 0.5 M H_2SO_4	100 °C	240 min	1:2.7	• V = 99.6%
[102]	HN_4Cl + NH_4OH	2 M HN_4Cl + NH_4OH	50 °C	300 min	1:5	• Ni = 59%
	Na_2CO_3	2 M Na_2CO_3	70 °C	240 min	1:4	• V = 63%
[103]	NH_4Cl + NH_3	2 M NH_4Cl+ 2M NH_3	50 °C	300 min	1:19	• Ni = 56%
	Na_2CO_3	2 M Na_2CO_3, 200 rpm stirring	70 °C	300 min	1:20	• V = 45%
[55]	NaOH + H_2SO_4	8 M NaOH 5 M H_2SO_4,	100 °C 100 °C	180 min	1:5 1:4	• V = 90% • Ni = 80%

Even though direct acid (H_2SO_4, HCl, and HNO_3) [107–109] and alkaline (NaOH, NH_4OH, $NaCO_3$, and NH_4Cl) [110–112] leaching methods are the cheapest operational and energy-efficient ways to extract valuable metal ions from heavy oil ash. They have several drawbacks, including a lack of selectivity in leaching the target metals, requiring large amounts of high acid concentration, using an oxidizing agent, polluting metals in solid ash residue, and producing leachate with hazardous elements that can complicate the precipitation and purification processes [4,113–115]. Due to the lack of an economically feasible alternative for processing vanadium-bearing waste, roasting is utilized to enhance the leaching process.

3.1.3. Bioleaching Process

Choosing the right bioleaching conditions is crucial for the microorganisms to grow ideally. The process has the advantage of being low in production cost and having a minimal environmental impact [116–120]. Bacteria like Acidithiobacillus, A. ferrooxidans, and Acidithiobacillus thiooxidans are well suited for chemolithoautotrophic applications in acidic conditions [121,122]. A study using A. ferrooxidans discovered that after 14 days, the roasted slag could leach out 83% of the chromium while only 20% of the vanadium was leached out [117]. Gomes's [123] results align with those obtained from a modified acidophilic culture of A. thiooxidans and A. ferrooxidans. Under alkaline conditions, high pH-tolerant organisms are being assessed [116,117]. These include Pseudomonas putida, a heterotrophic bacterium, and Aspergillus niger, a heterotrophic fungus. A study involving *P. putida* revealed that after 15 days, 75% of vanadium was leached from a roasted slag sample. Using the bacterium on the calcine increased dissolution to 90% [107,117].

3.1.4. Electro-Oxidation Leaching

In the leaching reactor, electrodes are inserted, and an electric field is used to convert low-valent vanadium to a higher oxidation state. Outstanding outcomes were obtained, with high recoveries of 95% for vanadium and 90% for chromium. The alkaline leaching system was applied with current densities ranging from 750 to 1000 A/m^2, NaOH concentration of 40~50 wt.%, and 90~120 °C temperature [124–126]. Both macro and micro perspectives allow for an understanding of the electrooxidation system's assistance in leaching. The schematic of both mechanisms is explained in more detail by Lee et al. [127]. Diffusion of reaction products is the focal point of the micro-viewpoints [124]. During leaching, metallic ions escape from the iron phase more readily in an electrical field, facilitating their directional movement. Macro-perspectives were correlated with direct and indirect oxidation processes [126]. During indirect oxidation, slag particles collide with cathodes and anodes, indirectly oxidizing vanadium and chromium as Fe^{3+} and H_2O_2 are generated.

3.2. *Roasting-Leaching Processes*

The roasting process of ash aims at the disintegration of the more stable and less soluble spinel phases, like Fe_2VO_4, as shown in Figure 8, and transforms them into more oxidized phases, enhancing the fact that they are more easily attacked by water/acid/alkaline solutions during the subsequent recovering process [128]. The roasting process is conducted with or without additives at temperatures ranging from 400 to 1000 °C, and advanced technologies have been utilized on fly ash to improve vanadium recovery [129,130]. To recover vanadium, the process involves oxidizing V^{3+} into acid-soluble V^{4+} and/or water-soluble V^{5+} compounds, which can then be dissolved using an appropriate reagent.

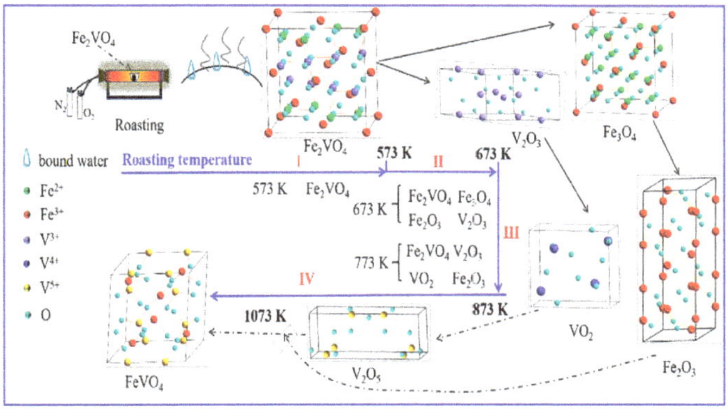

Figure 8. Schematic diagram of selective oxidation roasting of Fe_2VO_4 [131].

3.2.1. Non-Salt Roasting Assisted Leaching

This method results in the breakdown of the iron phases encasing the vanadium and chromium spinels, as explained in Equations (1) and (2) [62,108,132], allowing the formation of highly acidic leachable phases, including CaV_2O_6, $Ca_3V_2O_8$, MgV_2O_6, and $Mg_3V_2O_8$. As a result, the procedure involves the treatment of slags that have high levels of CaO/MgO.

$$4/5FeV_2O_4 + O_2 \rightarrow 2/5Fe_2O_3 + 4/5V_2O_5 \ \Delta G(T) = -293.45 + 0.15\ T\ (kJ/mol\ O_2) \quad (1)$$

$$4/5FeCr_2O_4 + O_2 \rightarrow 2/5Fe_2O_3 + 4/5Cr_2O_5 \ \Delta G(T) = -47.01 + 0.04\ T\ (kJ/mol\ O_2) \quad (2)$$

The olivine phase breaks down at 500 °C, whereas the spinels decompose at 800 °C, based on the results obtained earlier [36,110]. Wang [110] developed a method for direct extraction of low valence vanadium (LVV) from vanadium slag without producing poisonous waste, as shown in Figure 9. Application of roasting and leaching technique leads to the extraction of LVV from vanadium slag with a recovery rate of 69.37% at 800 °C roasting temperature for 1 h in an atmosphere of N_2/O_2 equals 10. Above this temperature or with longer roasting durations, V^{5+} was leached in the leaching solution. Smaller vanadium slag particles resulted in a higher vanadium recovery rate but lowered the amount of LVV.

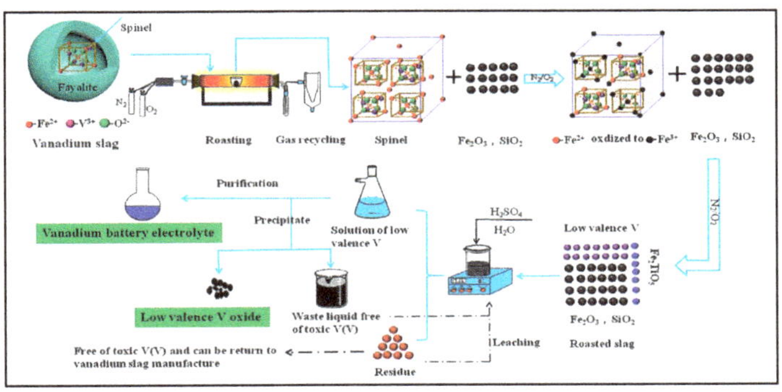

Figure 9. Novel process for extracting low valence vanadium from vanadium slag [110].

In the same context, Vitolo [56] created their three-step technique by first burning the ash under controlled conditions to lower the amount of carbon in the ash (Figure 10). The recovery of vanadium increased to 97% in the subsequent leaching process because of the reducing the carbon component in the fly ash. The best-recommended temperature for the burning carbon phase was found to be 850 °C and resulted in a higher total recovery of V (83%) as V_2O_5 with fewer impurities. Lower V recovery resulted from higher burning temperatures (over 950 °C), which caused V to fuse and volatilize as well as produce complex V-Ni refractory compounds.

According to Li [109,133,134], the oxidation process of vanadium spinel can be described as follows: (i) conversion of vanadium spinel into a solid solution of $Fe_2O_3.V_2O_3$ at the early stages of roasting, (ii) oxidation of this solid solution to $Fe_2O_3.V_2O_4$, followed by an incomplete reaction of V^{4+} with MgO to produce Mg_2VO_4, and (iii) extended roasting time leads to the further oxidation of vanadium, resulting in the formation of the highest-valence vanadates. In the non-salt roasting process, the chromium spinel could not convert to carcinogenic chromate salts to avoid the cost and disposal of chromium waste. Also, the vanadium may be directly leached out as NH_4VO_3, which can be separated after cooling and crystallization (Figure 11). The leaching efficiency of vanadium could be up to 90%.

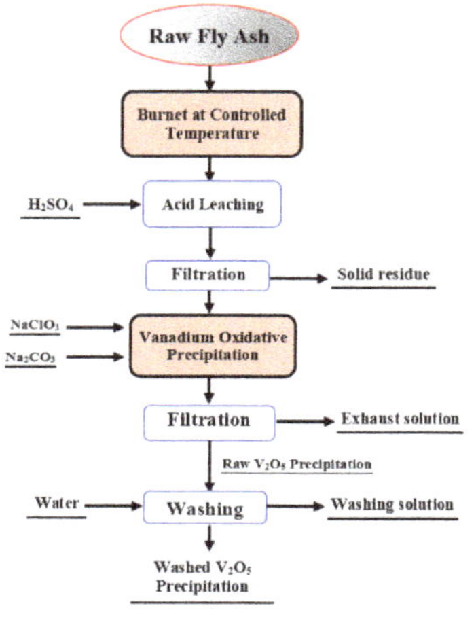

Figure 10. The block diagram for the carbonaceous fly ash combustion followed by acid leaching and oxidative precipitation of vanadium as V_2O_5 [56].

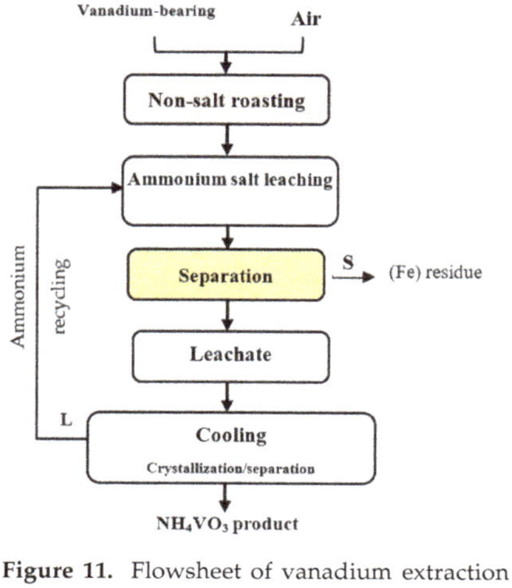

Figure 11. Flowsheet of vanadium extraction from vanadium slag using the non-salt roasting method [109].

To perform the thermodynamic analysis of this process, we attempted to create a diagram, illustrating the relationship between standard Gibbs free energy change and temperature (ΔG^0-T), and the data required for each reaction were obtained from HSC Chemistry 9.3.0/Fact-Sage software [127]. The ΔG^0 values for all reactions are negative throughout the temperature range of 400 to 1000 °C.

This indicates that direct oxidation of the spinels ($FeV_2O_4/FeCr_2O_4$) is thermodynamically feasible. The Gibbs free energy change for reaction (1.1) is much more negative than that of reaction (1.2), exhibiting that V-spinels can be preferentially oxidized and hence decomposed before that of Cr-spinels. Zhang [135] compared the different effects of microwave and conventional blank roasting on high-chromium vanadium slag's oxidation conduct, microstructure, and surface structure. The normal spinel was oxidized to inverse spinel at 400 °C which then decomposed at 600 °C. Carrying out the roasting process at high temperatures resulted in the minority of Cr^{3+} ions in the spinel phase being incorporated into VO_2 to form the $Cr_{0.07}V_{1.93}O_4$ or $CrVO_4$. Wang [136] found that the roasting process depends on the original state of vanadium in the coal stone. In the case of existing vanadium in amorphous phase form, the non-salt-roasting technology is excellent for leaching out vanadium. While the occurrence of vanadium in the vanadium-bearing waste is in a crystalline phase, the addition of a fluxing agent is necessary to achieve high vanadium leaching efficiency. Table 6 presents a summary of the extensive research conducted on vanadium resources through non-salt roasting and leaching procedures.

Table 6. Various studies on vanadium's non-salt roasting and leaching from various V sources.

Ref.	Non-Salt Roasting Conditions	Lixiviant	Leaching Conditions	Results
[56]	850 °C, −250 μm for 60 min.	H_2SO_4	2.0 M of H_2SO_4 Conc., at 100 °C; for 60 min.	• V = 97%
[36]	850 °C for 60 min.	Na_2CO_3	160.0 (g/L) of Na_2CO_3 Conc., and 10 (mL/g) L/S ratio at 95 °C for 150 min	• V = 90% • Cr remains in the residue
[109]	900 °C for 150 min.	$(NH_4)_2CO_3$	25% of Na_2CO_3 Conc., and 4 L/S ratio at 50 °C for 150 min.	• V = 92.8%
[110]	10 N_2/O_2, at 800 °C for 60 min.	H_2SO_4	2 M of H_2SO_4 Conc., at 90 °C; for 150 min.	• V = 69.37%
[133]	900 °C for 150 min.	$(NH_4)_2C_2O_4$	13% of $(NH_4)_2C_2O_4$ Conc., and 4 L/S ratio at 70 °C for 60 min.	• V = 90%

However, non-salt technology offers several advantages, such as reducing the carbon content of HOA, lowering its volume, and increasing the levels of V and Ni in the ash. There are some limitations, including the release of gas emissions, treating only waste with a high content of CaO or MgO, the chance of V compounds becoming volatile, the possibility of fusion and the creation of V-Ni refractory compounds at temperatures exceeding 900 °C, and alterations in the ash pH that might negatively impact the recovery of V and Ni during the leaching process.

3.2.2. Calcification Roasting Assisted Leaching

As an alternative to non-salt roasting, the clean technique of calcification roasting was used. The calcium roasting process involves combining limestone, lime, or other calcium compounds with vanadium wastes, grinding the mixture to an acceptable size, and then roasting the mixture in a vertical kiln [137,138], as illustrated in Figure 12. According to the chemical reactions listed in Table 7 and the previous studies of thermodynamic assessment, the oxidation processes involving CaO and V-spinels (Equations (3) and (4)) begin around 600 °C, whereas those using Cr-spinels (Equations (5) and (6)) start at 800 °C [139,140]. It is noted that the vanadium and calcium reaction during the roasting process leads to the formation of calcium vanadates (CaV_2O_6 and $Ca_3V_2O_8$); the compounds mentioned are not soluble in water (Equations (7)–(13)); therefore, they must be leached out through H_2SO_4 acid (Figure 13) [137,141,142] or alkaline leaching with ammonium carbonate solution (Figure 14) [139,143,144].

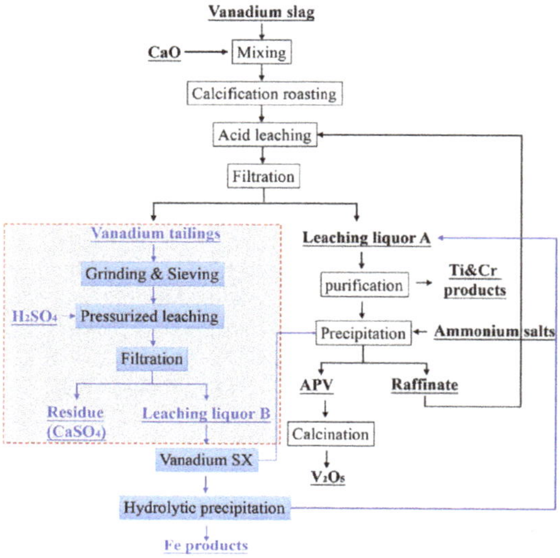

Figure 12. Flowchart of vanadium recovery from slag using calcification roasting-acid leaching technique [137].

Table 7. Reactions during calcium-roasting and leaching of vanadium-bearing waste.

Calcium Oxide (CaO)	ΔG (kJ/mol O_2)	Eqs.
$4/5FeV_2O_4 + 4/5CaO + O_2 \rightarrow 4/5CaV_2O_6 + 2/5Fe_2O_3$	$\Delta G(T) = -439.86 + 0.21T$ (kJ/mol O_2)	(3)
$4/5FeV_2O_4 + 12/5CaO + O_2 \rightarrow 4/5Ca_3V_2O_8 + 2/5Fe_2O_3$	$\Delta G(T) = -655.92 + 0.44T$ (kJ/mol O_2)	(4)
$4/7FeCr_2O_4 + 8/7CaO + O_2 \rightarrow 8/7CaCrO_4 + 2/7Fe_2O_3$	$\Delta G(T) = -255.77 + 0.15T$ (kJ/mol O_2)	(5)
$2/3MgCr_2O_4 + 4/3CaO + O_2 \rightarrow 4/3CaCrO_4 + 2/3MgO$	$\Delta G(T) = -468.37 + 0.32T$ (kJ/mol O_2)	(6)
Acid-leaching of calcified roasted product		
$CaV_2O_6 + 2H_2SO_4 \rightarrow (VO_2)_2SO_4 + CaSO_4 + 2H_2O$		(7)
$Ca_3V_2O_8 + 4H_2SO_4 \rightarrow (VO_2)_2SO_4 + 3CaSO_4 + 4H_2O$		(8)
$CaCrO_4 + H_2SO_4 \rightarrow H_2CrO_4 + CaSO_4$		(9)
Alkali-leaching of calcified roasted product		
$CaV_2O_6 + Na_2CO_3 \rightarrow 2NaVO_3 + CaCO_3$		(10)
$Ca_3V_2O_8 + 3Na_2CO_3 \rightarrow 2Na_3VO_4 + 3CaCO_3$		(11)
$Ca_3V_2O_8 + 3(NH_4)_2CO_3 \rightarrow 2(NH_4)_3VO_4 + 3CaCO_3$		(12)
$CaCrO_4 + (NH_4)_2CO_3 \rightarrow (NH_4)_2CrO_4 + CaCO_3$		(13)

Roasting and leaching behaviors of vanadium and chromium using calcification roasting-acid leaching from high-chromium vanadium slag were investigated [145]. The results indicated that more CaO combined with vanadium and reacted to form calcium vanadate ($Ca_3V_2O_8$), resulting in a high leaching efficiency of vanadium, approximately 91.14%, while the chromium remained in the leaching residue, and only 8.48% was leached. Gao [130] studied the roasting process of high chromium vanadium slag using CaO and microwave heating techniques. The effect of CaO dosage on vanadium-roasted products was investigated. At low m(CaO)/m(V_2O_5), CaV_2O_6 was formed, which was then converted to $Ca_2V_2O_7$ and $CaVO_3$ with increasing m(CaO)/m(V_2O_5) ratio. Results indicated that microwave radiation can reduce particle size and shorten roasting time, achieving 98.29% vanadium leaching efficiency under optimal conditions. Zhang [42] investigated the use of lime in roasting vanadium-bearing waste. Vanadium recovery was significantly influenced by the heating rate. Hence, reducing the heating rate resulted in high vanadium recovery. The formation of $Ca_2V_2O_7$ was more favorable for leaching vanadium with sulfuric acid. Xiang [146] applied mechanical activation treatment to enhance the extraction of vanadium from converter slag at a 1:1 mole ratio of vanadium to calcium. Then, the roasted slag was leached in a 15% sulfuric acid solution for 60 min at a temperature of 50 °C, stirring speed of 150 rpm, and S/L ratio of 1:20. The results demonstrated that the mechanical activation significantly decreased the optimum roasting temperature from 900 to 800 °C and increased the leaching efficiency of vanadium from 86.0 to 90.9%, respectively.

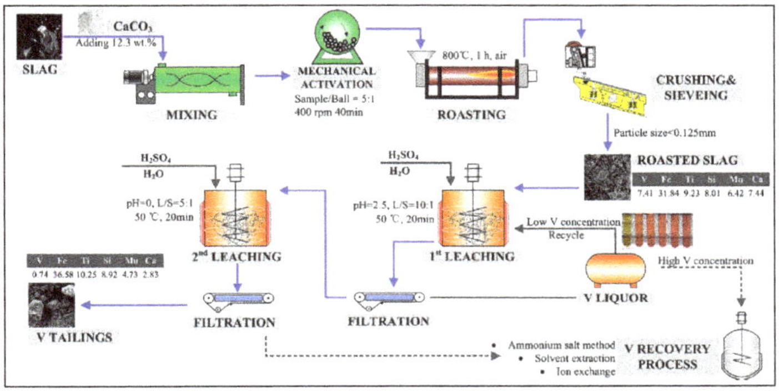

Figure 13. Flowsheet for the recovery of vanadium from vanadium slag by calcification roasting-two step leaching process [142].

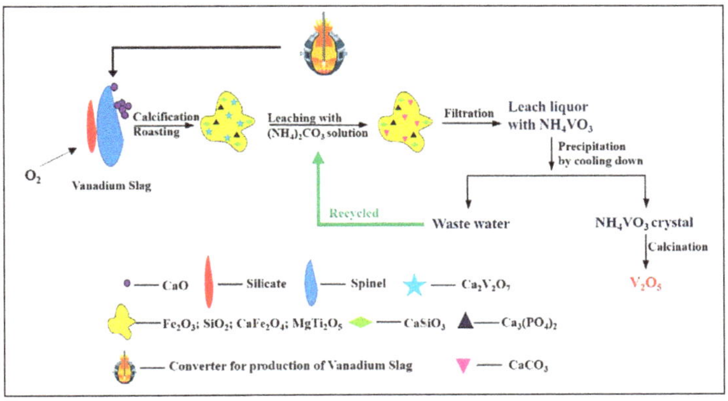

Figure 14. Flowsheet of ammonium carbonate leaching of vanadium slag [143].

Table 8 summarizes extensive research on different vanadium resources utilizing calcium roasting and leaching techniques. Utilizing this technology makes it possible to obtain high-purity vanadium leach solutions with minimal interference from elements such as silicon and phosphorus. The other advantage is that the leaching may proceed at a lower temperature when a particular ammonia salt is used. However, the leaching efficiency is much lower when using ammonium salts compared to the maximum 90% filtration of vanadium with sodium salts due to the precipitation of vanadium as NH_4VO_3. So, problems linked to ammonia volatilization must also be considered. Additionally, there are several drawbacks associated with this technology, such as the production of a large quantity of leaching residues containing sulfate compounds, equipment corrosion, and the emission of harmful gasses, which can cause environmental contamination.

Table 8. Results of different studies on vanadium's calcium roasting and leaching from various V sources.

Ref.	Roasting Salt	Roasting Conditions	Leaching Lixiviant	Leaching Conditions	Results
[143]	CaO	1:1.1 V/Ca molar ratio, 900 °C for 120 min.	$(NH_4)_2CO_3$	600 g/L of $(NH_4)_2CO_3$ Conc., 20 (mL/g) L/S, 80 °C, 70 min.	• V = 96% • P = 9.2%
[144]	CaO	900 °C for 180 min.	NH_4HCO_3	15% NH_4HCO_3, 75 °C, 180 min.	• V = 69.2%
[147]	CaO	1:16 CaO/solid ratio, 850 °C for 120 min.	H_2SO_4	15% H_2SO_4 Conc., 10 (mL/g) L/S ratio, 55 °C, 70 min.	• V = 93%
[42]	CaO	0.42 CaO/solid ratio, 850 °C for 150 min.	H_2SO_4	pH 2.5, 4 L/S ratio, 65 °C, 60 min.	• V = 91.5%
[148]	CaO	0.5 CaO/V_2O_3 molar ratio, 900 °C for 60 min.	H_2SO_4	20% H_2SO_4 Conc., 5 (mL/g) L/S ratio, 50 °C, 60 min.	• V = 99.4%
[149]	CaO	1 CaO/V_2O_5 molar ratio, 900 °C for 120 min.	$(NH_4)_2SO_4$ + H_2SO_4	250.0 g/L $(NH_4)_2SO_4$ Conc., 3.75 M H_2SO_4 conc., 10 (mL/g) L/S ratio, 20 °C, 60 min.	• V = 93.5% • Cr = 0.2%
[139]	CaO	0.5 CaO/V_2O_5 molar ratio, 900 °C for 60 min.	Na_2CO_3	160 (g/L) Na_2CO_3 Conc., 10 L/S, 80 °C, 60 min.	• V = 93.2% • Cr = 0.04%
[150]	$CaCO_3$	1 Ca/V molar ratio, 850 °C for 120 min.	H_2SO_4	15% H_2SO_4 Conc., 10 (mL/g) L/S, 10 (mL/g), 50 °C, 60 min.	• V = 83%
[151]	(1st) CaO (2nd) Na_2CO_3	1st stage: 5 CaO/V_2O_3 molar ratio, 780 °C for 60 min 2nd stage: 3.3 Na_2CO_3/Cr_2O_3 at 950 °C	1st stage: H_2SO_4 2nd stage: Water	1st stage: 3.75 M H_2SO_4 Conc., 5 L/S ratio, 70 °C, 60 min. 2nd stage: 3 L/S ratio, 3 (mL/g), 25 °C, 20 min.	1st stage: • V = 90% • Cr = 1% 2nd stage: • Cr = 87% • V = 99%

3.2.3. Sodium Salt Roasting Assisted Leaching Process

Traditional roasting technologies include sodium roasting, which can be traced back to Bleecker's first salt-roasting technology in 1912 [128]. Basically, a mixture of sodium salts (Na_2CO_3, Na_2SO_4, NaCl, and NaOH) as a source for alkalis was roasted with vanadium-bearing waste in a furnace at high temperatures, depending on the melting points of the sodium sources as follows: 1200~1250 °C for Na_2SO_4, 750~850 °C for NaCl, 800~1000 °C for Na_2CO_3, and 400~800 °C for NaOH [43,152,153]. A maximum amount of oxygen was used during roasting to oxidize the V^{3+} and convert it into soluble sodium vanadates ($NaVO_3$, $Na_4V_2O_7$) through the chemical reactions listed in Table 9 [127]. The vanadium conversion degree in these sodium sources follows the following order: NaOH > Na_2SO_4 > Na_2CO_3 > NaCl [154]. This order can be attributed to several reasons, including the diffusivity of the different sodium sources into the interior layers of the wastes, where they react with the V-spinels [155–157]. Despite all the limitations of this technology, which include the formation of harmful gaseous products, such as SO_2, Cl_2, or CO_2, fusion of NaCl at high temperatures, and high energy consumption while using sodium salts as the alkali source, it remains the best method in terms of selective vanadium extraction with low operating cost. The above serious environmental pollution greatly hinders the sustainable development of V extraction from V-bearing materials [158,159]. Consequently, efforts are still needed to develop an eco-friendly and effective method for extracting V from V-bearing materials. Attention has also been paid to recycling the Na^+ from the leach solution, using a membrane-assisted electrochemical cell to separate the cation from vanadium ions [160].

Table 9. Reactions during salt roasting and leaching of vanadium-bearing waste.

Sodium Sulfate (Na_2SO_4)	**ΔG (kJ/mol O_2)**	**Eqs.**
$4/3FeV_2O_4 + 4/3Na_2SO_4 + O_2 \rightarrow 8/3NaVO_3 + 2/3Fe_2O_3 + 4/3SO_2$	$\Delta G(T) = -155.62 - 0.08T$	(14)
$4FeV_2O_4 + 8Na_2SO_4 + O_2 \rightarrow 4Na_4V_2O_7 + 2Fe_2O_3 + 8SO_2$	$\Delta G(T) = 985.43 - 1.19T$	(15)
$FeCr_2O_4 + 2Na_2SO_4 + O_2 \rightarrow 2Na_2CrO_4 + 1/2Fe_2O_3 + 2SO_2$	$\Delta G(T) = 936.43 - 0.59T$	(16)
$2MgCr_2O_4 + 4Na_2SO_4 + O_2 \rightarrow 4Na_2CrO_4 + 2MgO + 4SO_2$	$\Delta G(T) = 1195.64 - 0.78T$	(17)
Sodium chloride (NaCl)		
$4/5FeV_2O_4 + 8/5NaCl + 4/5H_2O + O_2 \rightarrow 8/5NaVO_3 + 2/5Fe_2O_3 + 8/5HCl$	$\Delta G(T) = -245.36 + 0.04T$	(18)
$4/9FeV_2O_4 + 16/9NaCl + 4/9H_2O + O_2 \rightarrow 4/9Na_4V_2O_7 + 2/9Fe_2O_3 + 8/9HCl$	$\Delta G(T) = -113.65 - 0.05T$	(19)
$4/7FeCr_2O_4 + 16/7NaCl + 8/7H_2O + O_2 \rightarrow 8/7Na_2CrO_4 + 8/7Fe_2O_3 + 16/7HCl$	$\Delta G(T) = 79.07 - 0.01T$	(20)
$2/3MgCr_2O_4 + 8/3NaCl + 4/3H2O + O_2 \rightarrow 4/3Na_2CrO_4 + 4/3MgO + 8/3HCl$	$\Delta G(T) = 148.14 - 0.12T$	(21)
$4/7FeV_2O_4 + 8/7NaCl + O_2 \rightarrow 8/7NaVO_3 + 2/7Fe_2O_3 + 4/7Cl_2$	$\Delta G(T) = -186.26 + 0.09T$	(22)
$4/9FeV_2O_4 + 16/9NaCl + O_2 \rightarrow 4/9Na_4V_2O_7 + 2/9Fe_2O_3 + 8/9Cl_2$	$\Delta G(T) = -80.26 + 0.06T$	(23)
$4/7FeCr_2O_4 + 16/7NaCl + O_2 \rightarrow 8/7Na_2CrO_4 + 2/7Fe_2O_3 + 8/7Cl_2$	$\Delta G(T) = 55.83 + 0.02T$	(24)
$2/5MgCr_2O_4 + 8/5NaCl + O_2 \rightarrow 4/5Na_2CrO_4 + 2/5MgO + 4/5Cl_2$	$\Delta G(T) = 36.74 + 0.005T$	(25)
Sodium carbonate (Na_2CO_3)		
$4/5FeV_2O_4 + 4/5Na_2CO_3 + O_2 \rightarrow 8/5NaVO_3 + 2/5Fe_2O_3 + 4/5CO_2$	$\Delta G(T) = -345.3 + 0.04T$	(26)
$4/5FeV_2O_4 + 8/5Na_2CO_3 + O_2 \rightarrow 4/5Na_4V_2O_7 + 2/5Fe_2O_3 + 8/5CO_2$	$\Delta G(T) = -306.70 + 0.07T$	(27)
$4/7FeCr_2O_4 + 8/7Na_2CO_3 + O_2 \rightarrow 8/7Na_2CrO_4 + 2/7Fe_2O_3 + 8/7CO_2$	$\Delta G(T) = -94.64 - 0.04T$	(28)
$2/3MgCr_2O_4 + 4/3Na_2CO_3 + O_2 \rightarrow 4/3Na_2CrO_4 + 2/3MgO + 4/3CO_2$	$\Delta G(T) = -92.39 + 0.14T$	(29)
Sodium hydroxide (NaOH)		
$4/5FeV_2O_4 + 8/5NaOH + O_2 \rightarrow 8/5NaVO_3 + 2/5Fe_2O_3 + 4/5H_2O$	$\Delta G(T) = -458.66 + 0.11T$	(30)
$4/5FeV_2O_4 + 16/5NaOH + O_2 \rightarrow 4/5Na_4V_2O_7 + 2/5Fe_2O_3 + 8/5H_2O$	$\Delta G(T) = -533.41 + 0.08T$	(31)
$4FeCr_2O_4 + 8NaOH \rightarrow 4Na_2CrO_4 + 2Fe_2O_3 + 4H_2O$	$\Delta G(T) = -273.10 - 0.02T$	(32)
$2/3MgCr_2O_4 + 8/3NaOH + O_2 \rightarrow 4/3Na_2CrO_4 + 2/3MgO + 4/3H_2O$	$\Delta G(T) = -210.17 + 0.01T$	(33)
Water leaching of sodium roasted product		
$NaVO_3 + H_2O \rightarrow H_2VO(_4^-) + Na^+$		(34)
$Na_4V_2O_7 + H_2O \rightarrow 2HVO(_4^{2-}) + Na^+$		(35)
Acid-leaching of sodium roasted product		
$2NaVO_3 + 2H_2SO_4 \rightarrow (VO_2)_2SO_4 + Na_2SO_4 + 2H_2O$		(36)
$Na_4V_2O_7 + 3H_2SO_4 \rightarrow (VO_2)_2SO_4 + 2Na_2SO_4 + 3H_2O$		(37)
$Na_2CrO_4 + H_2SO_4 \rightarrow H_2CrO_4 + Na_2SO_4$		(38)
Alkaline-leaching of sodium roasted product		
$Na_2O.V_2O_5 + 4NaOH \rightarrow 2Na_3VO_4 + 2H_2O$		(39)
$2NaVO_3 + (NH_4)_2CO_3 \rightarrow 2NH_4VO_3 + Na_2CO_3$		(40)
$Na_4V_2O_7 + 2(NH_4)_2CO_3 \rightarrow (NH_4)_4V_2O_7 + 2Na_2CO_3$		(41)
$Na_2CrO_4 + 2(NH_4)_2CO_3 \rightarrow (NH_4)_4V_2O_7 + 2Na_2CO_3$		(42)

Anyway, the water-leaching process mostly applies to products undergoing sodium roasting, which takes place through Equations (34) and (35). Metal sulfates are formed by dissolving sodium vanadate compounds by H_2SO_4, according to Equations (36)–(38), and ammonium carbonate leaching is accomplished through Equations (39)–(42). However, the leaching effectiveness depends on the degree of metal converted during roasting, where sodium vanadate can be leached out. With regard to this case, there are certain common features for the processes that are used in the commercial plants of vanadium production from industrial wastes, such as the sodium salt roasting-water leaching-ammonia precipitation process utilized at Highveld Steel and Vanadium Corp. (*South Africa*), Chengde Iron and Steel Group (*China*), Nizhniy Tagil Iron and Steel Works (*NTMK, Russia*), etc. [127]. Li [43] reported that only 87.9% of V and 6.3% of Cr were recovered from calcined waste at Na/V molar ratio of 3.3 at 800 °C using water leaching, as shown in Figure 15. When the leaching residue was subjected to second sodium calcination at (Na/(V, Cr) molar ratio of 2.86 and 950 °C), about 90.7% of vanadium and 96.4% of chromium were leached, giving an overall recovery of vanadium and chromium of 98.9% and 96.4%, respectively.

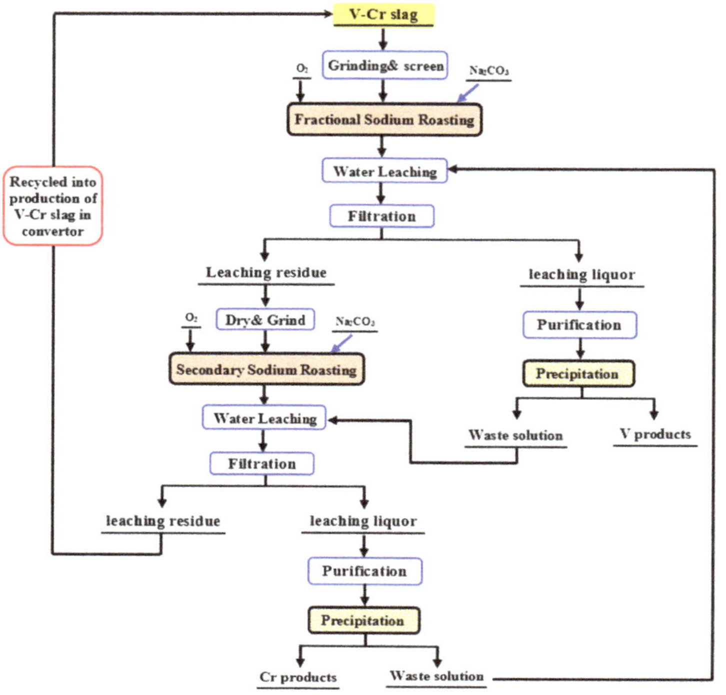

Figure 15. Novel process based on salt-roasting to extract vanadium and chromium from the V-Cr slag [43].

Wen [161] evaluated the leaching of the vanadium and chromium using sodium salts roasting (Na_2CO_3) and then leaching them with $(NH_4)_2SO_4$, showing that 94.6% vanadium and 96.5% chromium were leached. According to Zhao [162], vanadium can be leached from stone coal with the addition of 6 wt.% sodium chloride and 10 wt.% sodium sulfates during the roasting process. It was noted that the vanadium-bearing muscovite with quartz was converted to feldspar group minerals (albite, orthoclase, and anorthite). These results could produce some potential methods for vanadium recovery from stone coal. Puhong [129] roasted red cake obtained from multi-processing of stone coal with 22.5 g NaOH/25 g red cake at 170 °C for 1 h, then used water leaching at 98 °C for 60 min. with a solid/liquid ratio of 1:3.3 g/mL. It was noted that the V leaching efficiency was up to 97% after purification and calcination as V_2O_5, with a purity of 99.3%.

A three-step process was performed on HOA containing 85 wt.% unburned C and 2.2 wt.% V, beginning with carbon burning, followed by salt roasting and water leaching. It was noted that the 4 h. calcining step at 650 °C removes most of the carbon and thus increases the V content to 19 wt.%. Then, the burned enriched-V ash was roasted with sodium carbonate Na_2CO_3 at 650 °C for 4 h to convert V oxides into sodium metavanadate ($NaVO_3$), which is a water-soluble compound [163]. The calcined ash was leached with water for 4 h at 60 °C with an S/L ratio of 1:50 g/mL. Water leaching leads to the selective dissolving of V with about 92% recovery, leaving Fe and Ni compounds undissolved in the residue of the ash. By adding $(NH_4)_2SO_4$, V is precipitated as NH_4VO_3 from the leaching solution to separate V from the solution [101].

In the same context, Ibrahim et al. [59] discussed a recent technique for converting V compounds into water-soluble vanadates ($NaVO_3$) through roasting with NaCl, as shown in Figure 16. They found that over 95.5% of the vanadium leached out after the HOA was roasted with 20% NaCl for 2.5 h. Then, the vanadium was leached using distilled water with an L/S ratio of 10 mL.g^{-1}. The purity of the obtained NH_4VO_3 powder was estimated to be about 92% using direct precipitation from vanadium-pregnant solution by NH_4Cl addition, and the precipitation efficiency of NH_4VO_3 powder was estimated to be around 91.5%, and the total recovery rate of vanadium reached up to 87.60%. Then, the ammonium meta-vanadate was calcined at 550 °C for 3 h to produce V_2O_5 powder with an estimated 83% purity (Figure 17).

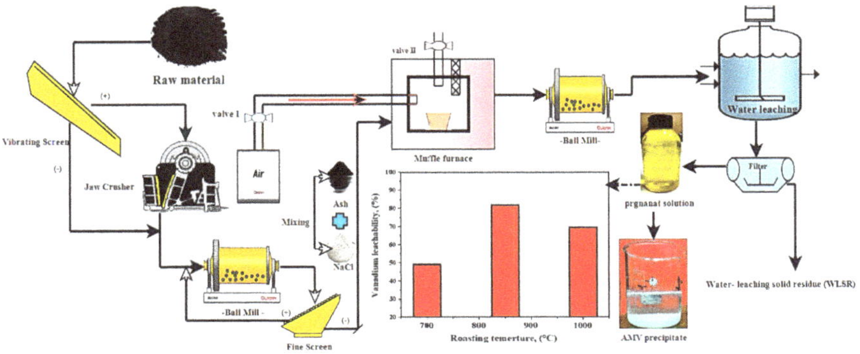

Figure 16. Schematic flow of salt-roasting and water-leaching process from boiler ash for vanadium extraction [59].

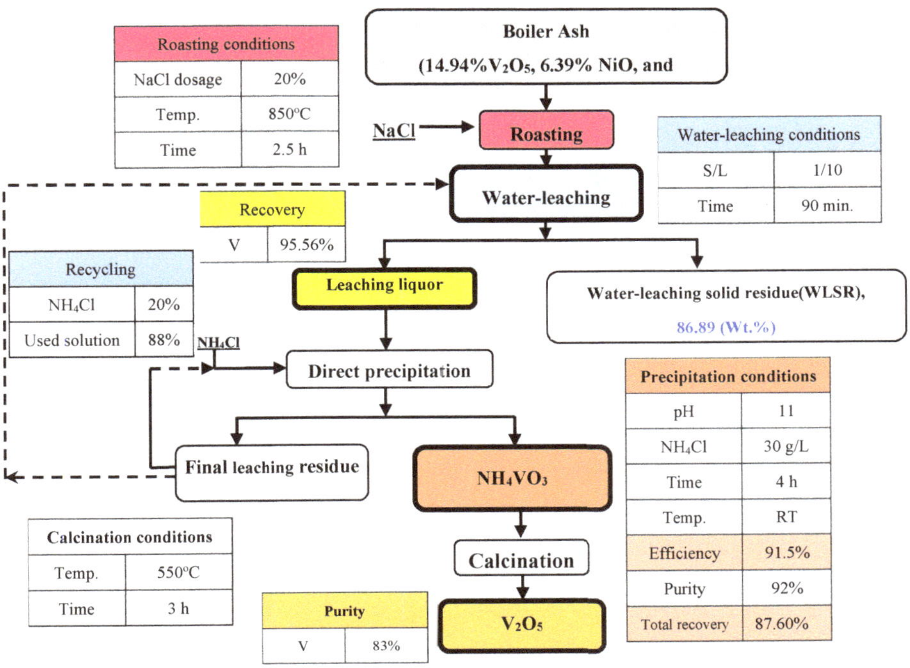

Figure 17. The quantitative flowsheet for V recovery from Egyptian boiler ash using sodium salt-roasting and water-leaching process [59].

The authors [59] comprehensively investigated the solid-state and phase transformation mechanisms that govern the selective extraction of vanadium from HOA. The

SEM images are displayed in Figure 18 for the roasted sample produced at 700 °C, and the corresponding EDS analysis for different points on the burnt ash is listed in Table 10. The results (Figure 18a) showed that the roasted particles are presented in an irregular shape with different sizes, and a few areas of spinel surface structure have been destroyed. According to the elemental mapping analyses shown in points S_1 and S_4, the decomposition of the spinel started to release the vanadium molecules, which indicates that sodium and vanadium are partially compatible (Figure 18b). The main information shown in this figure includes the following: (1) the particles are loose and dispersed, indicating that the interaction between particles is not significant under this temperature; (2) the distribution of sodium is relatively extensive and uniform, indicating that NaCl has not entirely reacted and formed a new sodium phase; and (3) the relative enrichment of vanadium appeared in some particles with high sodium content, indicating that some phases containing sodium and vanadium were formed.

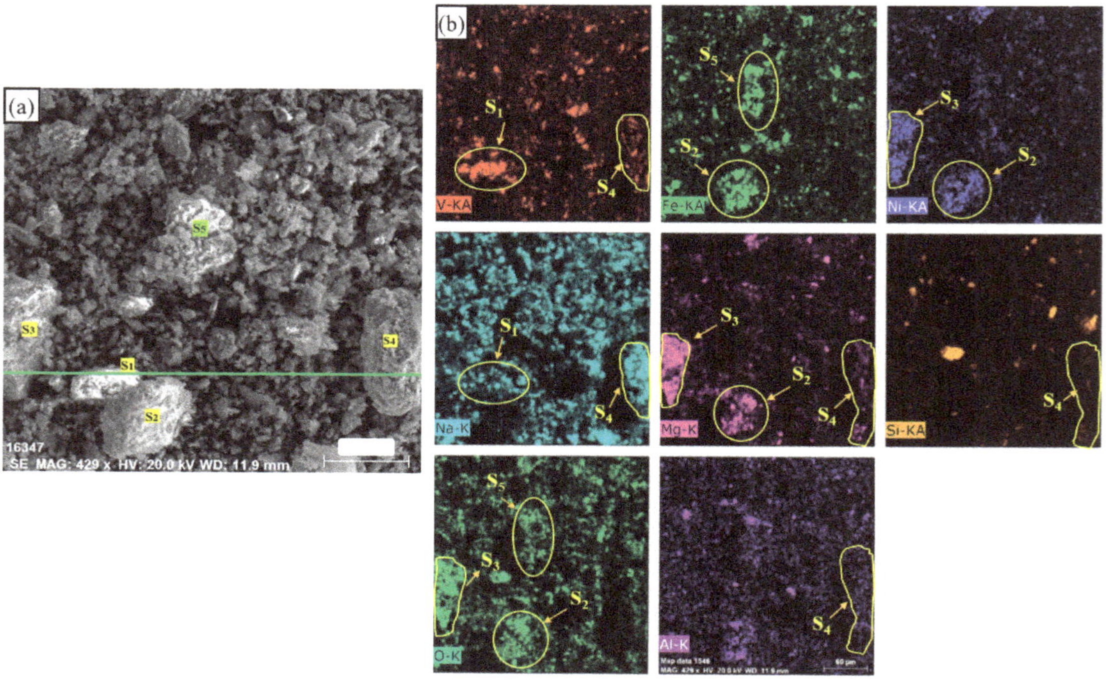

Figure 18. SEM images of roasted boiler ash sample: (**a**) at 700 °C for 2 h, and (**b**) elemental mapping images of burned surface.

Table 10. Spots EDS analysis of roasted boiler ash at 700 °C for 2 h and 20 wt.% of NaCl.

Spots	Composition (wt., %)							
	V	Ni	Fe	O	Na	Mg	Al	Si
S_1	18.57	2.93	5.88	28.07	25.39	15.12	3.03	1.01
S_2	2.24	7.72	4.47	26.34	23.21	31.59	3.15	1.28
S_3	1.2	10.39	17.58	20.56	18.05	22.88	5.61	3.73
S_4	10.08	4.61	6.01	30.5	28.24	13.5	6.02	1.04
S_5	3.94	3.14	25.2	24.6	19.83	11.42	9.34	2.53

Based on the EDS analysis (Table 10) of the specific locations (S_1 and S_4), it was clear that the chemical composition showed a moderate correlation between the levels of vanadium and sodium due to the limited reaction range of O_2 with vanadium spinel to release their oxides. So, not all quadruple vanadium (V^{4+}) was oxidized to the highest V^{5+} valency [64,164]. As a result, the reaction temperature at 700 °C was not sufficient to convert more of the V found in the vanadium compound into water-soluble sodium vanadates because almost all of V still exists inside the spinel, with less outer distribution than the central region and not enough opportunity to react with NaCl, resulting in decreasing the vanadium leachability [111]. To further explore, a line scan analysis was performed, as shown in Figure 19. Clearly, it was noted that the V and Na lines are aligned to some extent, suggesting an inconsiderable liberation of the vanadium element from the spinel surface. Furthermore, a large amount of free sodium remains along the line, meaning it did not enter into the chemical reaction.

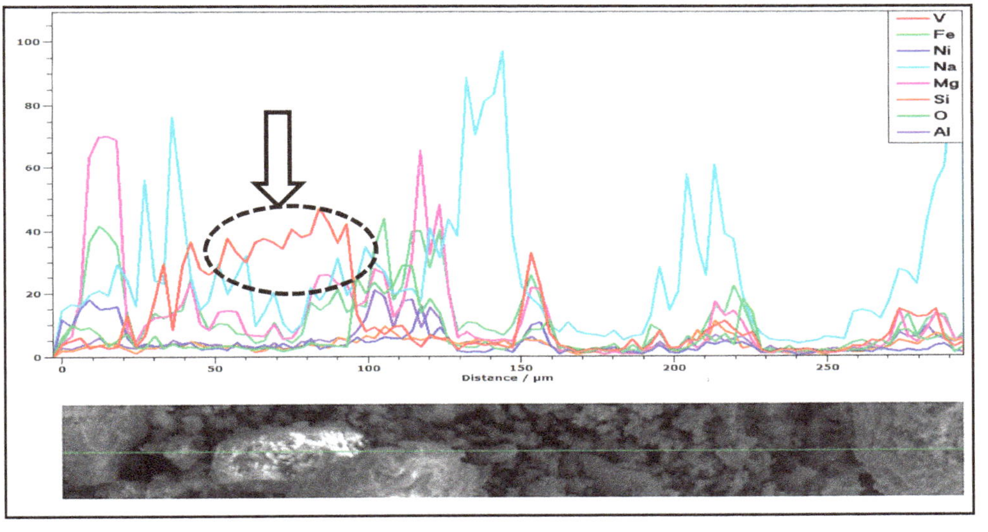

Figure 19. Line scan and distribution elements of the roasted boiler ash at 700 °C.

Their findings indicate that a roasting temperature of 850 °C (Figure 20) appears optimal for facilitating the O_2 reaction with vanadium-bearing minerals. This temperature provides an excellent opportunity for vanadium molecules to react with sodium chloride, forming the sodium metavanadate phase ($NaVO_3$). However, this phase can be easily extracted through a water-leaching process, resulting in the liberation of vanadium molecules, as shown in. Elemental scan lines demonstrated that the Na and V lines align, indicating that a considerable amount of the V element has been released from the spinel surface.

Elemental scan lines demonstrated the distribution of elements in the roasted ash at 850 °C is shown in Figure 21. Also, it was noted that there is an abroad alignment of the Na and V lines, suggesting that there is a significant liberation of the V element from the spinel surface. Also, along the line, there are no remains of free sodium, which means that all of it was entirely consumed in the process of forming a water-soluble sodium vanadate compound. This result confirms the hypothesis explained in the TGA data of the same study, which has been attributable to the complete thermal decomposition of vanadium spinel (FeV_2O_4) and vanadyl sulfates at 850 °C.

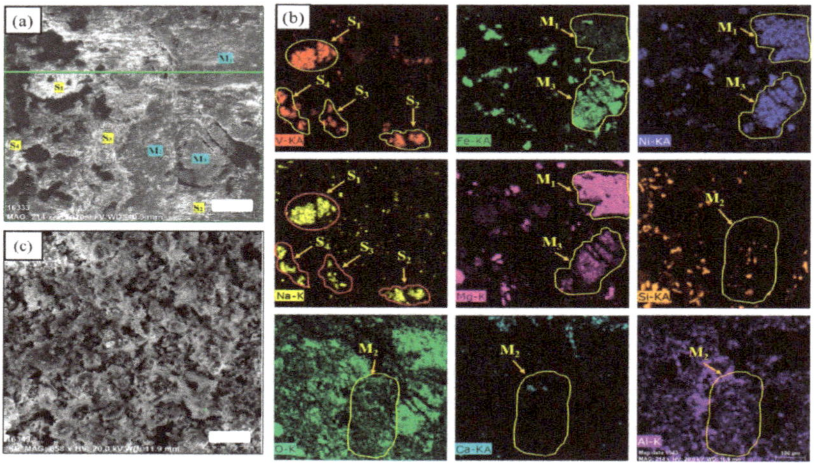

Figure 20. SEM images of roasted boiler ash sample: (**a**) at 850 °C for 2 h, (**b**) elemental mapping images of the surface, and (**c**) back-scattered morphology image [59].

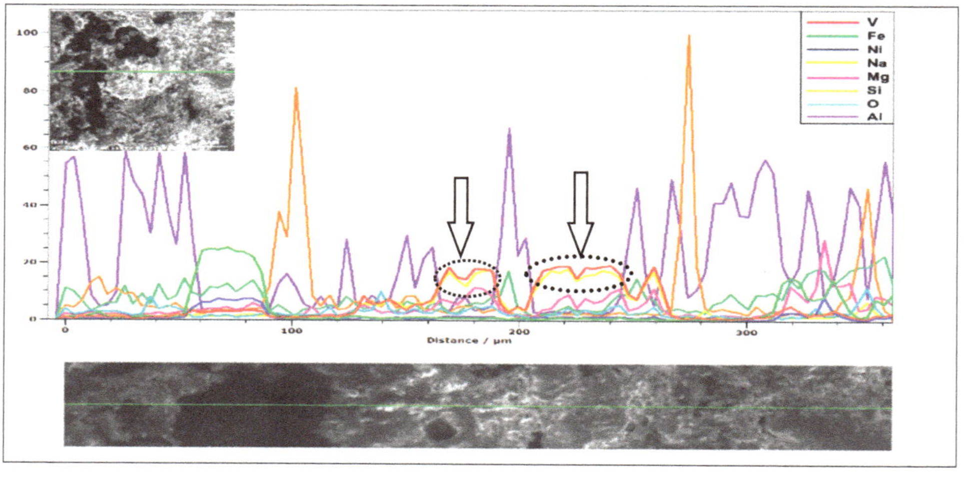

Figure 21. Line scan and distribution elements of the roasted boiler ash at 850 °C.

Furthermore, there are no signs of unbound sodium present in the mixture, suggesting that it was entirely consumed in the process of forming a water-soluble sodium vanadate compound. The results of the point-scan analysis of three different surface products are listed in Table 11. Meanwhile, the excessive roasting temperature of up to 1000 °C (Figure 22) would allow for the formation of refractory compounds (i.e., sintered from V, Ni, Fe). In addition, the formation of the aluminum silicate ($NaAlSi_2O_6$) phase increases due to the melting of NaCl salt, which leads to the formation of agglomerate that could hinder oxygen transfer during the roasting process and hence decrease the vanadium percent recovery.

Generally, the direct leaching process is poorly selective and has varying leaching efficiencies among different vanadium-containing phases. Consequently, subsequent separation and purification processes for vanadium and other metals in the leaching solution are more complex, with a wide range of reagent types and high consumption rates. On the other hand, the roasting-leaching process enables control over the occurrence form of vanadium through roasting, thereby ensuring efficient vanadium leaching. Specifically, the sodium salt roasting process can be employed to selectively separate vanadium using the water leaching method, which offers advantages such as high leaching efficiency, excellent selectivity, reduced consumption of leaching reagents, and suitability for treating materials with complex forms of vanadium occurrence. Based on the characteristics of V-bearing waste of high vanadium content and complex vanadium occurrence forms, NaCl has been used, a low-cost and abundantly available salt compared to other salts in the roasting-water leaching process, to study the extraction process of vanadium and focus on the transition law and mechanism of the vanadium-containing phase during sodium salt roasting so as to provide the theoretical basis for efficient extraction of vanadium. A summary of the comprehensive research carried out on various vanadium resources using sodium-salt roasting and leaching processes is presented in Table 12.

Table 11. Spots EDS analysis of roasted boiler ash at 850 °C and 1000 °C for 2 h and 20 wt.% of NaCl.

	Spots	Composition (wt., %)								
		V	Ni	Fe	O	Na	Mg	Ca	Al	Si
850 °C	S_1	17.86	1.45	5.29	38.23	20.45	7.19	1.02	3.96	4.55
	S_2	12.62	1.74	4.83	35.01	25.11	7.05	0.98	7.58	5.08
	S_3	10.06	1.41	3.27	37.54	27.04	6.89	1.08	7.26	5.45
	S_4	14.08	1.97	2.46	33.93	25.06	8.87	1.5	5.97	6.16
	M_1	0.92	22.11	38.77	27.87	1.15	5.45	1.01	1.63	1.09
	M_2	0.81	4.28	7.24	35.3	15.42	3.95	1.92	14.57	16.51
	M_3	1.07	19.95	36.81	25.76	2.08	9.03	1.02	1.45	2.83
1000 °C	S_1	19.35	16.76	23.21	25.61	1.03	9.5	2.02	1.51	1.01
	S_2	18.87	15.98	21.28	27.84	1.31	10.46	1.98	1.36	0.92
	S_3	19.06	19.19	20.14	26.75	1.56	9.69	1.04	1.47	1.1
	M	-	-	-	47.18	3.14	32.21	1.5	-	15.97
	N_1	-	-	-	30.65	46.04	-	3.51	9.7	10.1
	N_2	-	-	-	28.71	39.45	-	2.08	19.84	9.92

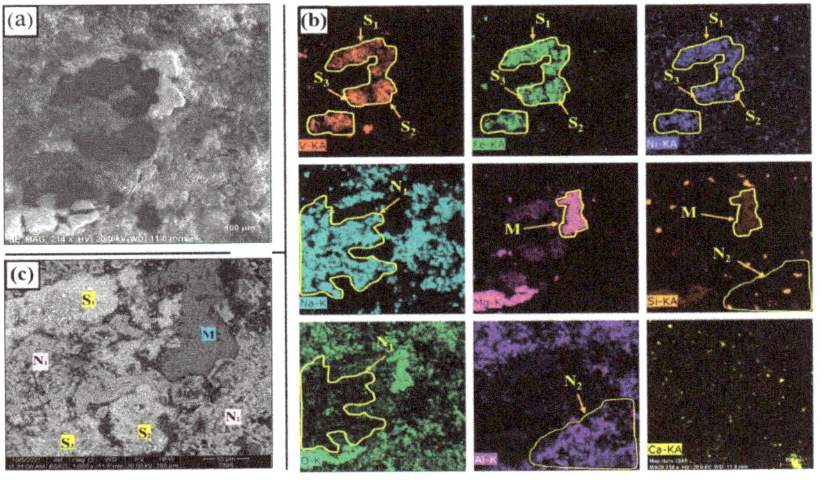

Figure 22. SEM images of roasted boiler ash sample: (**a**) at 1000 °C for 2 h, (**b**) elemental mapping images of the polished surface, and (**c**) back-scattered morphology image.

Table 12. Results of different studies on vanadium's sodium-salt roasting and leaching from various V sources.

Ref.	Roasting: Sodium-Salt Roasting	Roasting: Conditions	Leaching: Lixiviant	Leaching: Conditions	Results
[59,165]	NaCl	1st stage Roasting at 850 °C for 150 min. 2nd stage Water leaching solid residue (WLSR)	Water H_2SO_4	10 (mL/g) L/S ratio at 25 °C for 90 min. −63 µm, 8% H_2SO_4 Conc., in 15 (mL/g) L/S ratio at 85 °C; for 240 min.	1st stage: • V = 95.5% 2nd stage: • Ni = 95.02% • Zn = 90.13%
[43]	$Na_2CO_3 + O_2$	1st stage (V-roasting) 3.3 Na/V molar ratio at 800 °C for 120 min. 2nd stage (Cr-roasting) 2.86 Na/(V,Cr) molar ratio at 950 °C for 120 min.	Water	1st stage: 3 (mL/g) L/S ratio at 25 °C; for 20 min. 2nd stage: 3 (mL/g) L/S ratio at 25 °C; for 20 min.	1st stage: • V = 87.9% • Cr = 6.3% 2nd stage: • Cr = 96.4% • V = 90.7%
[166]	Na_2CO_3	41/9 Na_2CO_3/solid ratio at 700 °C for 150 min.	Water	5 (mL/g) L/S ratio at 90 °C for 30 min.	• V = 90%
[167]	Na_2CO_3	0.1 Na_2CO_3/solid ratio at 0.2 1000 °C for 45 min.	Na_2CO_3 + NaOH	45 + 10 (g/L) of Na_2CO_3 + NaOH Conc., in 20 L/S at 80 °C for 60 min.	• V = 80%
[41]	Na_2CO_3	1.0 Na_2CO_3/solid ratio, at 2.0 1000 °C for 120 min.	H_2SO_4	3.0 M H_2SO_4 Conc., in 15 (mL/g) L/S ratio at 70 °C; for 150 min.	• V = 95%
[161]	Na_2CO_3	2.5 Na_2CO_3/solid ratio at 850 °C for 60 min.	$(NH_4)_2SO_4$	30.0 (g/L) $(NH_4)_2SO_4$ Conc., in 3 (mL/g) L/S ratio at 20 °C for 60 min.	• V = 94.6% • Cr = 96.5%
[153]	NaOH + O_2	7.67 Na/V molar ratio at 700 °C for 120 min.	Water	70 (mL/g) L/S ratio at 25 °C for 60 min.	• V = 99%
[168]	NaOH	0.3 NaOH/solid ratio at 800 °C for 180 min.	Water	10 (mL/g) L/S ratio at 30 °C for 60 min.	• V = 94.9% • Cr = 80.5%
[169]	Na_2CO_3 + NaCl	3:2:5 Na_2CO_3/NaCl/solid ratio at 700 °C for 120 min.	Water	4 (mL/g) L/S ratio at 95 °C for 180 min.	• V = 96% • Cr = 91%
[170]	NaOH	0.3 NaOH/solid ratio at 850 °C for 120 min.	NaOH pressure leaching	5 (mL/g) L/S ratio; 0–5 MPa O_2 partial pressure at 210 °C for 120 min.	• V = 98%

Likewise, Ibrahim [165] studied the cost-effective extraction of precious metals, such as Ni and Zn, by hydro-metallurgical processing of the water-leaching solid residue (as shown in Figure 23) after vanadium extraction from salt-roasting Egyptian boiler ash [59]. Under the most favorable leaching conditions of 8% (vol%) H_2SO_4 concentration, 85 °C leaching temperature, and 1/15 S/L ratio, a maximum extraction of 95.02% Ni and 90.13% Zn was achieved after 240 min of leaching, while the iron impurity was removed by Fe_2O_3 as a nucleating agent and the magnesium impurity was effectively removed by oxalic acid precipitant. After the removal of Fe and Mg, Ni and Zn in the purified solution were precipitated at a pH of 10 as Ni-Zn hydroxide ($Ni(OH)_2$ and $Zn(OH)_2$), which was subsequently transformed into NiO-ZnO by its calcining at 450 °C for 2 h. The precipitation efficiency of Ni and Zn was 95.25% and 89.51%, respectively, and the final calcined product was composed mainly of 37% Ni and 23% Zn. Based on a kinetic analysis, it was discovered that the process of leaching nickel is primarily controlled by diffusion through the solid product layer and chemical reactions. The diffusion process through the solid product layer is the main contributor, with an activation energy of 20.26 kJ/mol. The kinetic of zinc dissolving is governed by the diffusion that occurs through a layer of a solid product, and this diffusion has an activation energy of 11.67 kJ/mol.

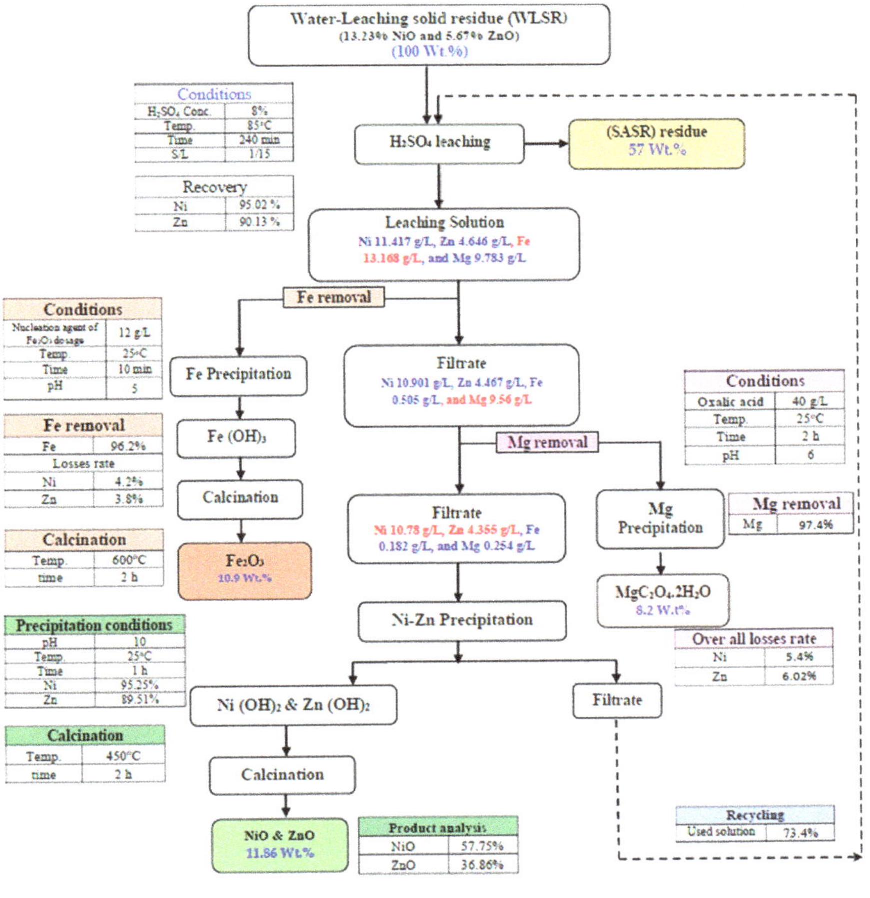

Figure 23. The quantitative scheme for the recovery of Ni, Zn, and other by-products from water leaching solid residue obtained from the salt roasting of Egyptian boiler ash.

3.2.4. Promising Modification Methods

Continuous efforts have been made to develop alternative processes, including sub-molten salts and supercritical fluid for recovering vanadium from various sources. Sub-molten salt and novel leaching methods can extract vanadium from various sources (vanadium-bearing slag, stone coal, spent catalysts, etc.) and should be discussed in this review. The method of sub-molten salts ($NaOH/KOH/NaOH$-$NaNO_3$ = 75% (*w*/*w*) at high pressure) can decrease the cost of energy due to the lower reaction temperature [171,172]. According to a report by Liu [171], a new study has discovered a unique approach to breaking down vanadium slag using KOH as a sub-molten salt under regular pressure. When the reaction conditions are optimal, including a temperature of 180 °C, an initial ratio of 4:1 potassium hydroxide to ore mass, a stirring speed of 700 rpm, a gas flow rate of 1 L/min, and a reaction time of 300 min, the extraction rates of vanadium and chromium can attain up to 95% and 90%, respectively. Still, the requirement of reactors with high corrosion resistance, the large consumption of alkaline, and the high dissolution rate of silica phases limit its wide application. The problems of inefficient vanadium separation and serious environmental pollution greatly hinder the application of these V extraction processes [158,159]. Recently, novel leaching methods have been reported for vanadium recovery using urea, specific chelating agents, or supercritical fluid, resulting in low environmental impact [173–176].

A leaching method has also been reported by adjusting the vanadium particles' surface wettability to intensify the solid–liquid contact using certain surfactants [177]. Though tested in a few cases, as shown in Table 13, such conditions have resulted in considerably lower vanadium recovery yields (60~70%); however, they can still be considered a "green option" to the traditional processes. Accordingly, the broadly available theoretical underpinnings and technological background of such approaches make them worth investigating in the future. Once proven promising, they may eventually be considered for exploitation and optimization.

Table 13. Details of some novel leaching methods for vanadium recovery from various sources.

Source	Method	Conditions	Results	Principle	Highlight	Ref.
Vanadium spent catalyst	Urea leaching	Time, 1 h; temperature, 20 °C; urea conc., 20%; pH = 14	V = 80%	• Considering the mutual solubility effect of urea to vanadium solubility in water	• High kinetic leaching of vanadium even at low temperatures is an advantage	[175]
Vanadium-bearing shale, vanadium-bearing coal	Chelating leaching	Time, 4 h; temperature, 95 °C; chelator conc., 6.0 M; additive (CaF_2), 5% (wt.)	87% V	• The principle is similar to that of bioleaching, which is to obtain a product of a complex metal-chelator. • Chelators used are citric acid, oxalic acid, and ethylenediaminetetraacetic acid (EDTA).	• The process offers the advantage of an eco-friendly approach compared to conventional leaching using mineral acids. • Even so, slower kinetic leaching becomes the main problem of this approach, being the limitation for its wider application.	[3,173]
Vanadium-bearing slag	Surface wettability control	Leaching conditions: not specified Surfactant added: 0.25–1.0% (wt.)	69% V (with surfactant), 51% V (without surfactant)	• To intensify the interaction between the solid particle and leaching solution by adding surfactants like polydimethylsiloxane/PDMS and sodium dodecyl sulfate/SDS. • The surfactants play roles in reducing the surface tension of the solution and changing the surface properties of the solid particles	• The pioneered study showed the capacity of this approach to enhance vanadium extraction. • Further studies in this area are worth encouraging while assessing the economic and environmental impact of the added surfactants.	[177]
Vanadium spent catalyst	Supercritical leaching	Fluid, acetylacetone; temperature, 190 °C; time, 7 h	V = 60%	• Characterized by the use of fluids where they are in their critical point; for instance, CO_2 (critical pressure/P_c, 7.38 MPa; critical temperature/T_c, 31.1 °C), ethanol (P_c, 6.3 MPa; T_c, 241 °C), methanol (P_c, 7.85 MPa; T_c, 240 °C), acetylacetone (P_c, 4.05 MPa; T_c, 339 °C) and other alcohol groups.	• Research on this area is still green. • The advantages in the environment, energy, and chemical aspects are worth considering while preliminary studies show an unsatisfactory result. • The development of novel supercritical solvents and co-solvents is believed to be the main factor attributing to the greater interest in applying this technology in the extractive metallurgy field, including extracting vanadium.	[178]

4. Utilizing By-Products for Industrial Applications

The world is dealing with water scarcity issues due to the discharge of partially treated or untreated wastewater from industries and groundwater pollution. The demand for clean water has increased to meet the needs of a sustainable society. The need for cost-effective and efficient water treatment methods has become a growing concern as we look towards the future. This has led to an increased focus on recycling industrial waste to produce low-cost adsorbents and reduce environmental pollution [179]. Thus, developing an affordable, accessible, and highly effective adsorbent for removing hazardous metal ions presented in wastewater has become necessary. The utilization of solid industrial wastes by reusing them in the manufacturing of useful materials used for environmental purposes has been the subject of several recent studies, and it is one of the most crucial issues for maintaining sustainable development [180,181].

The solid wastes containing silica (SiO_2) and alumina (Al_2O_3), including industrial waste forms, i.e., fly ash, steel slag, waste perlite, waste porcelain, lithium slag, waste metallic residues, paper sludge, cupola slag, kaolin waste, windshield waste, asbestos wastes, etc., could be converted into ecological zeolite-based adsorbents [182–191]. To date, several physicochemical and solvothermal techniques, including the hydrothermal approach [192–194], alkali-fusion processes [195–197], sol–gel processes [198–200], microwave processes [201–204], and alkali-leaching method [205–208] have been adopted and developed to produce synthetic adsorbents. Figure 24 shows the major preparation processes and methods for synthesizing zeolite from industrial solid waste. So, in this review, we focused on the synthesis of low-cost adsorbent from industrial waste.

However, many researchers have studied the synthesis of zeolites from solid wastes to be used as an adsorbent for removing heavy metal ions from wastewater. Zeolites are crystal formations built on stiff anionic alumino-silicate structures with distinct pores or channels that link at cavities or cages [209]. These materials are favorable for adsorption processes due to their high cation exchange capacity (CEC), large surface area, good thermal stability, porosity, surface active functional groups, and nontoxicity [210–212]. New zeolite materials have been processed from fly ash, successfully eliminating heavy metal ions from aqueous solutions [213]. Chen [214] synthesized zeolite X by an alkaline fusion-hydrothermal reaction using industrial lithium slag containing SiO_2:Al_2O_3 in a weight ratio of 71.73:25.16 (wt. %). The results highlight the performance similarity between Zeolite NaX-1 and commercial Zeolite X.

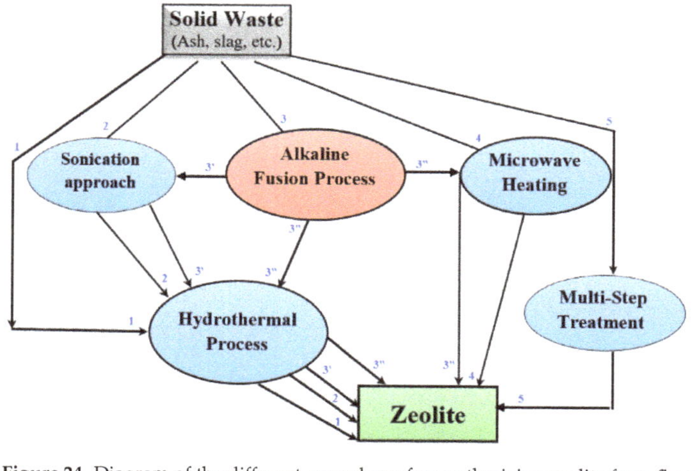

Figure 24. Diagram of the different procedures for synthesizing zeolite from fly ash [215].

R. Anuwattana [216] has reported successfully synthesizing Na-A type zeolite using industrial waste materials such as the solid by-product of cupola slag, which contains SiO_2/Al_2O_3 in a weight ratio of 47:12 (wt.%), and aluminum sludge from an aluminum plating plant. In their investigation, two preparation methods were carried out using the same starting material compositions. To begin the process, alkaline fusion was utilized, and then the material was subjected to hydrothermal treatment to produce sodium aluminosilicate. This substance was subsequently crystallized in a NaOH solution at a temperature of 90 °C for 1~9 h with varying H_2O/SiO_2 ratios. The results indicated that a more excellent ratio of H_2O/SiO_2 resulted in an increased crystallization rate. The highest level of crystallization was observed for Na-A after 3 h. The second method involved alkaline hydrothermal treatment without fusion, using the same conditions as the first procedure. However, this approach did not yield any Na-A zeolite. It is worth considering blast furnace slag (BFS) as a possible source of raw materials for zeolite synthesis. This material primarily comprises CaO, SiO_2, Al_2O_3, and MgO, with minor Fe, Ti, and Mn quantities. Several groups have reported that zeolite can be made from industrial waste.

However, prior attempts to use BFS as a chemical source of zeolite were unsuccessful due to its complex composition, specifically its high Ca content [192]. Hence, further study is needed to produce a variety of zeolites for possible uses. In the same context, Kuwahara [217,218] discovered a new way to synthesize a hydroxyapatite-zeolite micro composite from BFS in a recent study. The process involves using affordable chemical reagents such as H_3PO_4 and NaOH and suitable preparation techniques. This innovative approach provides practical solutions for waste management. Furthermore, using the alkali fusion process, Takaaki Wajima [197] demonstrated how to create zeolite-A, zeolite-X, and hydroxysodalite from the BFS. A more recent approach (Figure 25) for synthesizing a high-crystallin and affordable zeolite from a combination of SASR-kaolin as a readily accessible and low-cost raw material using an alkaline fusion hydrothermal process [219]. The weight ratio of the SASR- kaolin mixture of 1:1.5 gives the best composition and properties of the synthesized zeolite. The optimal conditions for the synthesized zeolite are a fusion temperature of 600 °C, a 1:3 wt. ratio (SASR-kaolin)-NaOH, a 1:4 solid–liquid ratio, an 80 °C crystallization temperature, and a 24 h crystallization time.

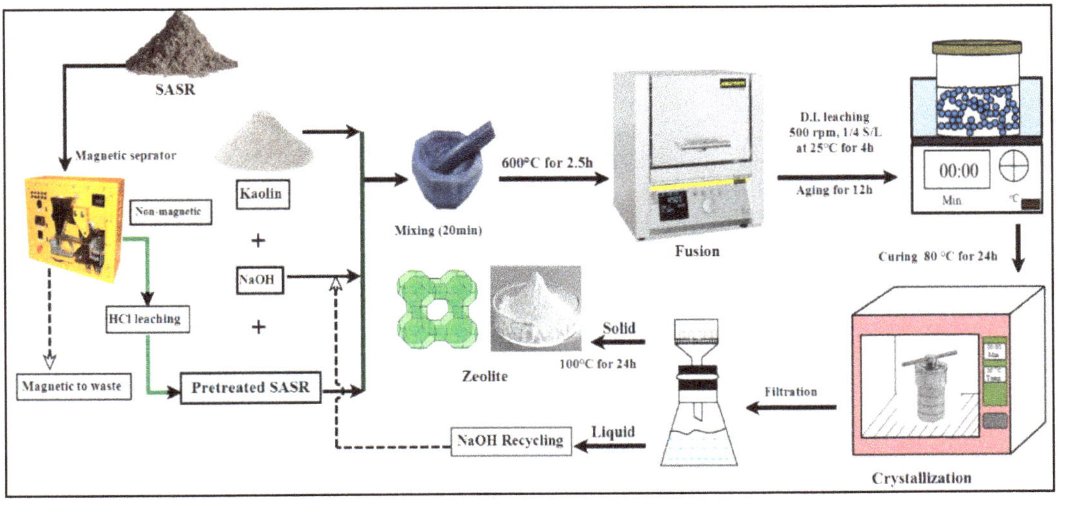

Figure 25. Schematic procedures of synthesized zeolite based on Egyptian boiler ash residues [219].

Figure 26 illustrates the mineralogical structure of the produced synthesized zeolite by mixing pretreated SASR and kaolin mixtures at different weight ratios in the presence of NaOH (Mixture-NaOH mass ratio of 1:1.3). As shown in Figure 26A, the Faujasite zeolite ($Na_2\ Al_2Si_{3.8}O_{11.63}{\cdot}8H_2O$) has been formed as the major mineral phase from SASR-Kaolin weight ratios of 1:0, 1:2, 1:1.5, 1:1, and 2:1.

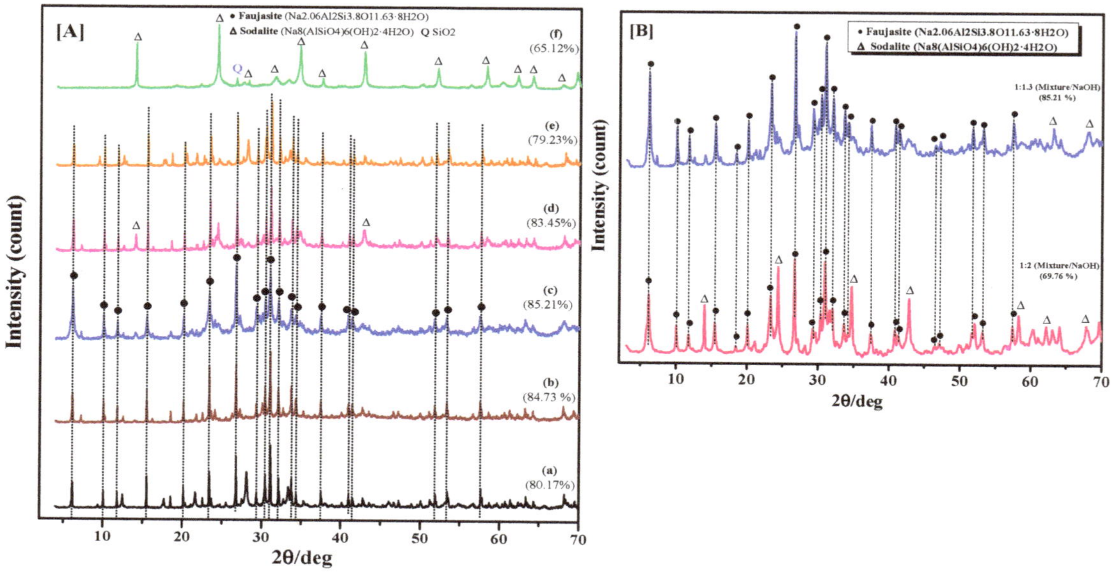

Figure 26. XRD Patterns of synthesized zeolite produced from mixture of SASR-Kaolin at weight ratios of (a) 1:0, (b) 1:2, (c) 1:1.5, (d) 1:1, (e) 2:1, and (f) 0:1. (**A**) XRD Patterns of synthesized products with different mass ratio of mixture-NaOH of 1:1.3 and 1:2. (**B**) where •: Faujasite, and Δ: Sodalite zeolite [219].

However, the intensity of its peaks increased with increases in the mass of kaolin, and a high rate of crystallinity (85.21%) can be obtained when the ratio of SASR to kaolin is 1:1.5, due to the stability of the zeolite composition. A percentage of sodalite zeolite ($Na_8(AlSiO_4)_6(OH)_2.4H_2O$) groups have also existed at an SASR-Kaolin weight ratio of 0:1. Figure 26B shows that the crystallinity of the synthesized zeolite decreased from 85.21% to 69.76% at 1:2 mixture-NaOH weight ratio. However, when the system's alkalinity is too high, a decline in crystalline is caused. This can be attributed to one of the following two reasons: either the created zeolite spontaneously transforms into hydroxyl sodalite, which has more excellent thermodynamic stability [220], or the zeolite will dissolve in the hot alkali solution because of its metastable state and exposure to disturbances [180,221]. Table 14 illustrates how various industrial wastes can be used to create synthetic zeolites, each with its own advantages and disadvantages.

Additionally, in the same context of research [219], synthesized zeolite was an effective adsorbent for removing Zn^{2+}, Pb^{2+}, Cu^{2+}, and Cd^{2+} heavy metal ions from industrial wastewater based on readily accessible and low-cost raw materials. It was noted that the adsorption process was exothermic, pH-dependent, and spontaneous in nature. Significant adsorption occurs when the pH is 7.0 for Zn and Cd ions and 6.0 for Pb and Cu ions. The adsorption process was found to follow the pseudo second-order kinetic model and Langmuir isotherm model.

Table 14. Synthetic zeolites from different industrial wastes and their advantages and limitations.

Raw Material	Chemical Composition %	Advantages	Limitations	Synthesis Route	Zeolite Type	Ref.
BFS	CaO (40.1), SiO_2 (34.58), Al_2O_3 (14.78), MgO (5.29)	Availability, low cost, convenient preparation steps.	• Higher CaO content.	Hydrothermal	A	[217,218]
Natural obsidian	80.04% SiO_2, 12.27% Al_2O_3, 0.16% TiO_2, 0.84% FeO, 0.18% MgO, 1.10% CaO, 3.14% Na_2O and 3.04% K_2O	Higher silica content, availability.	• Still, limited studies did not mention any specific application. • Due to the wide range of behavior changes concerning synthetic conditions.	Hydrothermal	Organic template-free EMT-type, natrolite, Za-gmelinite	[222,223]
RHA (Residue is rich in amorphous silica)	80% silica, Al_2O_3, iron oxide, CaO, MgO, sodium and potassium oxides, and others	Low cost, ultrafine size, highly porous, and chemically reactive.	• Pre-treatment of RHA and waste glasses increase the cost.	Hydrothermal	ZSM5, T, Na-Y	[224,225]
Waste glass materials	SiO_2 (63%), Al_2O_3 (18%), B_2O_3 (10%), and alkaline earth oxides	Excellent mechanical and thermal properties, free of harmful elements.	• Problem in handling, need a substrate or base material for commercial use.	Hydrothermal	A	[198,199,226]
SASR	SiO_2 26.57%, Al_2O_3 5.81%, Fe_2O_3 18.93%, TiO_2 0.13%, MgO 0.04%, K_2O 0.13%, Na_2O 0.18%, P_2O_5 3.61%, LOI 17.88%	Availability, low cost, ultrafine size, convenient preparation steps.	• Pre-treatment (Physical and Chemical) needs to be a source of Al.	Hydrothermal, Alkali fusion	Faujasite	[219]
Paper sludge ash	SiO_2 35.9%, Al_2O_3 22.8%, CaO 33.2%, Na_2O 0.6%, MgO 4.5%, Fe_2O_3 0.9%, TiO_2 2.2%	Amorphous and crystalline phases formed by incineration; low temperature required.	• Low abundance of Si and significant Ca content.	Hydrothermal	Na-P1	[227,228]
Waste stone cake	SiO_2 38.9%, Al_2O_3 12.2%, CaO 6.9%, Na_2O 1.4%, K_2O 1.8%, MgO 2.7%, Fe_2O_3 3.1%, CO_2 32.6%, SO_3 0.1%, P_2O_5 0.1%	Specific pore sizes and large surface areas, high silica content.	• Most SiO_2 and Al_2O_3 are in crystalline phases; it is difficult to dissolve into an alkali solution, contains impurities.	Hydrothermal, Alkali fusion	zeolite-A, P, X and ZSM-5	[195]
Clay materials (kaolin, smectite)	SiO_2 46.5%, Al_2O_3 41.18%, Fe_2O_3 0.19%, TiO_2 0.13%, MgO 0.04%, K_2O 0.13%, Na_2O 0.18%, ZrO_2 0.01%, SO_3 0.15%, P_2O_5 0.03%, LOI 16.25%	Availability, convenient source for producing low silica zeolites like Y, and use of kaolin waste for zeolite synthesis reduces the cost of reagents.	• High energy consumption processes like grinding, calcination, and fusion are required. • Raw materials mining destroys the natural landscape.	Hydrothermal, Alkali fusion	NaA, mordenite, Faujasite, and, NaP	[229]

Table 14. *Cont.*

Raw Material	Chemical Composition %	Advantages	Limitations	Synthesis Route	Zeolite Type	Ref.
Coal fly ash	SiO_2 38.3%, Al_2O_3 34.8%, CaO 11.0%, Fe_2O_3 8.1%, Others 7.8%	The main constituents are silica and alumina, which offer the potential to convert it to zeolite, producing low-price zeolite with high purity; and no harmful effect.	• Effect of the impeller type and agitation during the hydrothermal treatment stage of the process.	Hydrothermal, microwave-assisted hydrothermal method, and fusion methods.	X, Na-P1, A, Y	[183,206,230]
Lithium slag	SiO_2 70.67%, Al_2O_3 27.24%, Fe_2O_3 0.52%, SO_3 0.45%, CaO 0.29%, K_2O 0.22%, MgO 0.16%, Na_2O 0.13%, P_2O_5 0.12%, Others < 0.1%	High silicon aluminum ratio.	• Very few experiments have been conducted in this field.	Hydrothermal	X, FAU/LTA	[214,231]
Waste of iron mine tailings/iron ore tailing	SiO_2 67.58%, Al_2O_3 8.70%, Fe_2O_3 7.42%, CaO 5.78%, MgO 4.37%, K_2O 2.32%, Na_2O 2.15%, Cl 0.69%, TiO_2 0.33%, P_2O_5 0.26%, SO_3 0.23%, MnO 0.10%, SrO 0.06%	Have economic and environmental aspects.	• Very few experiments have been conducted in this field, which contains hazardous impurities that decrease the quality of zeolites.	Hydrothermal	A, ZSM-5	[232]

The maximum monolayer adsorption capacities of Zn^{2+}, Pb^{2+}, Cu^{2+}, and Cd^{2+} ions onto zeolite at 20 °C were 12.025, 15.96, 12.247, and 16.17 $mg.g^{-1}$, respectively, showing a significantly greater removal efficiency, which agrees with the previously obtained results. Ion exchange was the principal mechanism controlling the removal of Zn^{2+} and Cd^{2+} from aqueous solution by synthesized zeolite. In contrast, the main mechanism controlling the process of the removal of Pb^{2+} and Cu^{2+} ions is proposed to be either surface adsorption or precipitation. The proposed adsorption mechanism in Figure 27 shows that the adsorption of Zn^{2+}, Pb^{2+}, Cu^{2+}, and Cd^{2+} on zeolite likely can be divided into three major phenomena: precipitation, ion exchange, and surface adsorption [233,234]. The cations exchange is the crucial mechanism for absorbing Zn^{2+} and Cd^{2+} from an aqueous solution. The following equation can represent this process:

$$(Zeolite\text{-}SASR)\ Na^{2+} + Me^{2+} \rightarrow (Zeolite\text{-}SASR)\ Me^{2+} + Na^{2+} \quad (43)$$

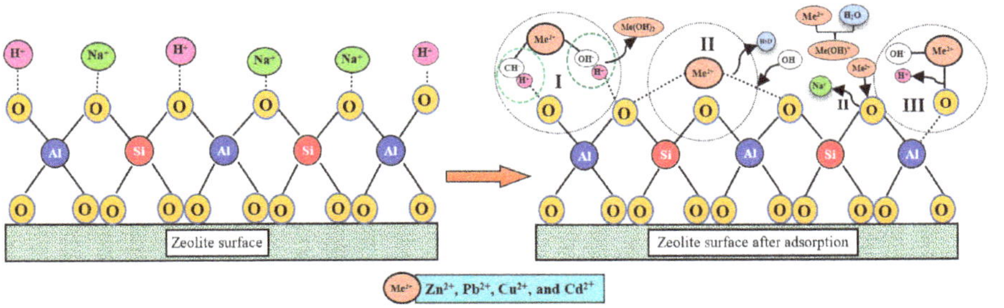

I. Precipitation ($Me(OH_2)\downarrow$). II. Ion exchange. III. Surface complexation ($(Me(OH)^+)$).

Figure 27. The proposed adsorption mechanism of heavy metal ions on synthesized zeolite.

As a result of industrially applying a synthetic zeolite derived from the SASR by-product, it was demonstrated that the synthesized zeolite effectively removed heavy metal ions from the wastewater sample collected from the Egyptian General Petroleum Corporation (EGPC) (Eastern Desert, Egypt). The targeted levels of heavy metal ion concentrations before and during adsorption using synthesized zeolite are shown in Table 15. However, in the fourth adsorption cycle, the residual heavy metal ion concentrations follow WHO guidelines for wastewater disposal in the marine environment with high removal efficiency [235]. When the number of circulations increases (5th and 6th cycles), the amount of heavy metal ions in the effluent also goes up. As a result, when the pollutants are enriched to a specific level, the adsorbent needs to be cleaned to achieve closed recycling. Finally, the synthesized zeolite can be utilized to safeguard industrial wastewater drainage systems.

Table 15. Applying the synthetic zeolites based on SASR-kaolin mixture for removing heavy metal ions from EGPC industrial wastewater sample.

	Metals	Cd	Cr	Cu	Mn	Pb	Fe	Ni	Zn	V
	$C_0(mg.L^{-1})$	6.127	7.016	10.294	8.152	11.493	8.219	0.002	17.051	0.01
	MLD_{WHO}	0.01	0.01	1	0.1	0.01	1.5	0.1	1	0.002
Number of Cycles	1	U. D	U. D	U. D	2.1	U. D	3.1	U. D	U. D	U. D
	2	U. D	U. D	U. D	0.6	U. D	1.1	U. D	U. D	U. D
	3	U. D	U. D	U. D	0.05	U. D	U. D	U. D	U. D	U. D
	4	U. D	U. D	U. D	U. D	U. D	U. D	U. D	U. D	U. D
	5	0.08	0.05	1.3	2.12	0.51	2.5	0.054	1.66	0.027
	6	0.45	2.01	1.7	7.12	1.07	4.5	0.098	3.87	0.089

U. D: Under detection limit; MLD_{WHO}: Maximum limits for wastewater disposal according to WHO.

It was noted that zeolites synthesized based on oil shale ash by an alkaline hydrothermal process were effectively used to extract Pb^{2+} and Cd^{2+} metal ions [236]. Synthesized Na-A zeolite from class F fly ash (FA) and modified oil shale ash (MOSA) by alkaline fusion followed by refluxing was used as an adsorbent for lead, zinc, and chrome [237]. Fly ash hydrothermally modified with NaOH solution was utilized to synthesize zeolites for Cd^{2+} adsorption, which showed effective removal of Cd^{2+} from the wastewater source [238]. Applying a low-temperature roasting process, the novel adsorbent was prepared from fly ash and solid alkali (NaOH). This adsorbent was manufactured under the following conditions: the pristine fly ash and NaOH mass ratio was 5:8, the calcination temperature was from 300 to 350 °C, and the reaction time was 3 h. The obtained results showed a better adsorption capacity of Cd^{2+} on modified fly ash (MFA) [239]. A zeolitic material that was prepared from coal fly ash (CFA) through NaOH fusion treatment, followed by hydrothermal processing, was applied to remove heavy metal ions such as Ni^{2+}, Cu^{2+}, Cd^{2+}, and Pb^{2+} from the wastewater source [240]. The adsorption capacity of the synthesized Na-A zeolite based on the fusion and hydrothermal treatment of oil shale was evaluated by measuring the maximum removal efficiency of Cu^{2+}, Ni^{2+}, Pb^{2+}, and Cd^{2+} from aqueous solutions [241]. A new composite material was formulated based on fly ash (FA), meta-kaolin (MK), and TiO_2 to form a micro-porous zeolitic material with enhanced photocatalytic properties for the adsorption of methylene blue (MB) dye [242]. A modified geopolymer based on fly ash was used to study the effectiveness of adsorbing Cd^{2+} ions from aqueous solutions, as previously described by Javadian [243]. The alkaline fusion-hydrothermal process was used to synthesize zeolites from Brazil oil shale ash. It was noted that the synthesized zeolites are composed of likewise mixed phases (Na-A zeolite, Na-X zeolite, hydroxy sodalite, and quartz) [244].

Yu [245] investigated the efficiency of zeolite-based fly ash to adsorb and remove Ni^{2+} ions from aqueous solutions. This study examined the adsorption capacities of synthetic and commercial zeolite (4A), indicating that zeolite 4A had a lower adsorption capacity than fly ash-based zeolite (75.6 mg/g). Chen [246] developed four different fly ash-based zeolites to adsorb Ni^{2+} ions from aqueous solutions and achieved removal efficiencies ranging from 92.5% to 96.2%. A specific kind of geopolymer made from fly ash showed a significant effect in removing Pb^{2+} ions from aqueous solutions. The maximum removal efficiency of 90.66% has been achieved at 5 pH and a contact time of 2 h [247]. In order to remove Pb^{2+} ions from aqueous solutions, fly ash was employed to create the hydroxysodalite zeolite. The maximum adsorption efficiency of synthetic zeolite was 98.1%, demonstrating the adsorbent's excellent efficiency [181].

Multi-cation wastewater containing Pb^{2+}, Zn^{2+}, and Cd^{2+} was treated using a type of zeolite synthesized from fly ash. Pb^{2+} and Zn^{2+} ions were efficiently and selectively removed with 100% and 70% adsorption efficiencies, respectively, compared to Cd^{2+} ions with a 60% adsorption efficiency [248]. Shyam [249] used fly ash adsorbent modified with $CaCO_3$ to minimize the concentration of Pb^{2+}, Ni^{2+}, and Cr^{6+} ions from aqueous solutions. This study found that 1:10 $CaCO_3$/FA eliminated Pb^{2+} up to 90%, Ni^{2+} 50%, and Cr^{6+} 30% under ideal conditions. In the same way, the adsorption of several heavy metal ions from aqueous solutions was also studied using FA coated with chitosan. It was noted that within 3 h, the produced adsorbent exhibited satisfactory performance. The adsorption capacities of different metal ions have been reported as follows: 36.22, 28.65, 55.52, and 19.10 mg/g for $Cr^{3+,\,6+}$, Cu^{2+}, Zn^{2+}, and As^{5+} ions, respectively [250]. The highest Cd^{2+} ions removal efficiency of 84% was obtained under the best-recommended conditions. The isotherm and kinetic studies indicated that the Langmuir and pseudo second-order models were in good agreement with the adsorption data. Also, they employed FA to synthesize zeolite through the fusion process and showed effective adsorbent to remove Cr^{6+} ions from aqueous solutions. More than 80% of the Cr^{6+} ions were eliminated, and both the Freundlich and Langmuir isotherm models provided an excellent fit to the equilibrium data. The thermodynamic studies revealed that the process was spontaneous and endothermic, and the adsorption kinetics exhibited pseudo second-order behavior [251]. In conclusion, fly ash has been converted into various ceramic products, zeolite materials, and geopolymers, which in turn are used for wastewater treatment [252]. Table 16 summarizes the adsorbents synthesized and their metal ion removal efficiency using various preparation methods.

Table 16. Adsorption conditions of various heavy metals by zeolite-based adsorbents [252].

Heavy Metals and Nutrients	Adsorbent	Removal Efficiency	Ref.
Cd^{2+}	ZFA-600	84%	[243]
	Zeolite X	100%	[253]
	TiO_2/FA)	80%	[254]
	FA-Z	60%	[248]
	MG-Z and MT-Z	~98% and 75%	[255]
Co^{2+}	MCM-41	~90%	[256]
Cr^{3+}		~90%	
Cr^{6+}	ZFA	>80%	[257]
	CFA-FeOOH	84.9%	[258]
	MSFA/PPy		[259]
	ZFA-Na-A		[260]
	Zeolite X		[261]
	Chitosan/CFA		[262]
Cu^{2+}	FA-MS	98%	[263]
	FA-IOT-Geo	98.3%	[264]
	CFA-Geo	93.9%	[265]
	MPF		[266]
	TiO_2/FA	90%	[254]
	MG-Z and MT-Z	100%	[255]
	Ag-Fe_3O_4/FA		[267]
Fe^{3+}	Zeolite NaeP1	100%	[268]
Hg^{2+}	Zeolite LTA	94%	[269]
	ZFA	91.27%	[270]
	HMAS zeolite	~95%	[271]
Mn^{2+}	CFA Zeolite	100%	[272]
	Zeolite	100%	[273]
Ni^{2+}	Zeolite		[245]
	FA-Na-P/TEA	96.2%	[246]
	FA-Na-X	95.5%	
	FA-Na-P/Na-Br	95%	
	FA-Na-P	92.5%	
	Zeolite X	95%	[253]
	MG-Z and MT-Z	~50% and 52%	[255]
	PB/FA—SA-FA	100%	[274]
Pb^{2+}	ZCFA	98.1%	[181]
	Geopolymer	90.66%	[247]
	FA-Z	100%	[248]
	1:10 FA	>90%	[249]
	MG-Z and MT-Z	100%	[255]
	PB/FA—SA-FA	100%	[274]

Table 16. *Cont.*

Heavy Metals and Nutrients	Adsorbent	Removal Efficiency	Ref.
Zn^{2+}	FA-Z	70%	[248]
	FAICS		[250]
	FA		[275]
	FA		[276]
NH_4^+	PB/FA—SA-FA	100%	[274]
	Zeolite NaeP1	~61%	[277]
	Z-P1	65.2%	[278]
PO_4^{3+}	ZHLO	60%	[279]
	ZFA	91%	[280]
	Z-P1	92.3%	[278]

5. Future Recommendations

The synthesizing and application of zeolite from useless by-product waste need further investigation.

To enhance the sustainability and efficiency of salt-roasting processes, alternative methods must be explored for recycling released gasses and barren solutions, along with extracting valuable metals from a secondary vanadium source. Additionally, producing chemical compounds for further use in the relevant process needs further studies.

Other synthetic pathways for producing zeolite materials with high surface area, adjustable molecular dimensions, and adsorption capacities are needed for industrial by-product application. Further research is needed to identify low-cost raw materials for the synthesis of zeolites.

In addition, to reduce the production costs, it is necessary to develop accessible synthetic routes. Kinetic studies and energy calculations are needed to investigate how nucleation occurs during zeolite synthesis. Further studies on the regeneration of synthetic zeolite are required to increase the cost-effectiveness of the process.

6. Conclusions

The shortage of vanadium primary resources from high-grade ores makes it necessary to find a suitable alternative that covers this shortage. Also, the world vision concerning the solution to climate change and environmental pollution problems is based on decreasing or completely getting rid of the amounts of the produced waste. One of these promising alternatives is the utilization of different wastes containing vanadium to decrease their environmental adverse effects and, at the same time, enhance their economic value by converting them into valuable products.

The current study aims to present and analyze the worldwide available alternative vanadium-bearing waste resources, proprieties, reserves, and processing technologies. The heavy oil fly ash "HOFA" and vanadium slag showed a promising alternative for the production of vanadium metal due to their high content of vanadium and the cost-effectiveness of their processing to extract vanadium metal compared to processing their corresponding low-grade ores.

Vanadium can be extracted from these wastes using pyro- or hydro-metallurgical methods or a combination. The hydro-metallurgical extraction can be carried out using one of the vital mineral acids such as HCl, H_2SO_4, or HNO_3, and previous studies reported that HCl and H_2SO_4 have comparable leaching efficiencies. H_2SO_4 is preferred due to its significantly lower cost and its wide application in the leaching of many metals. In addition to V and Ni, H_2SO_4 effectively dissolves almost all metals in HOFA and other vanadium-bearing wastes, such as Mo, Mg, Mn, and others. In addition to acidic leaching reagents, alkaline leaching reagents like NaOH, Na_2CO_3, and NH_4OH can be used for leaching vanadium from their containing wastes. The previous study showed that NaOH has the highest selectivity and leaching efficiency for V compared to other alkaline leaching reagents.

Two-stage leaching can be used for more selectivity and productivity of vanadium from their containing wastes. The first stage includes the alkaline leaching of high-contented and alkaline leachable vanadium using NaOH, followed by the second acidic leaching of acidic leachable elements using H_2SO_4. A combination of pyro- and hydro-metallurgical processes is carried out for more effective extraction of vanadium from their containing wastes. Firstly, pyro-metallurgical processes like roasting are carried out to convert low-leachable vanadium to high-leachable phase. Alkaline salt roasting using NaCl and Na_2CO_3 converts contained vanadium into water-leachable phase $NaVO_3$. Previous studies showed that the roasting temperature is the main controlling parameter on the efficiency of vanadium extraction. The hydro-metallurgical processes include leaching the modified (roasted) wastes using water or a very low-concentration acidic leaching process. Recent studies confirm the cost-effectiveness of the salt roasting process followed by the water leaching process for treating vanadium-bearing wastes, which showed high percent recovery and purity of the extracted vanadium. After leaching, V can recover from the leaching solution by chemical precipitation, solvent extraction, or ion exchange processes.

Complex processing of vanadium-bearing wastes, including firstly the extraction of valuable metals followed by the complete utilization of the resulting residue in the production of adsorbent, i.e., zeolite with different composition and properties to be used in wastewater treatment, is very promising and applied alternative for complex processing of these wastes.

Author Contributions: Conceptualization, X.L. and A.H.I.; methodology, A.H.I., X.L., H.E.S. and A.B.E.; software, A.H.I., H.E.S. and A.B.E.; validation, X.L. and A.B.E.; formal analysis, A.H.I., H.E.S. and A.B.E.; investigation, A.H.I. and A.B.E.; data curation, A.H.I. and A.B.E.; writing—original draft preparation, A.H.I.; writing—review and editing, X.L., H.E.S. and A.B.E.; visualization, A.H.I.; supervision, X.L.; project administration, X.L.; funding acquisition, X.L. All authors have read and agreed to the published version of the manuscript.

Funding: This work was supported by Qingdao West Coast New Area 2020 Annual Science and Technology Project (3-2-2020), the National Natural Science Foundation of China (No. 51674161), and the Natural Science Foundation of Shandong Province (No. ZR2017ZC0735).

Institutional Review Board Statement: Not applicable.

Informed Consent Statement: Not applicable.

Data Availability Statement: No new data were created or analyzed in this study. Data sharing is not applicable to this article.

Acknowledgments: All the authors would like to express their sincere gratitude and appreciation to Shandong University of Science and Technology, Qingdao, China for providing the required funding.

Conflicts of Interest: The authors declare no conflicts of interest.

References

1. Perles, T. *Vanadium Market Analysis*; TTP Squared Inc.: Upper St. Clair, PA, USA, 2021; pp. 23–38.
2. Simandl, G.J.; Paradis, S. Vanadium as a Critical Material: Economic Geology with Emphasis on Market and the Main Deposit Types. *Appl. Earth Sci. Trans. Inst. Min. Metall.* **2022**, *131*, 218–236. [CrossRef]
3. Hu, P.; Zhang, Y.; Huang, J.; Liu, T.; Yuan, Y.; Xue, N. Eco-Friendly Leaching and Separation of Vanadium over Iron Impurity from Vanadium-Bearing Shale Using Oxalic Acid as a Leachant. *ACS Sustain. Chem. Eng.* **2018**, *6*, 1900–1908. [CrossRef]

4. Hu, P.; Zhang, Y.; Liu, T.; Yuan, Y.; Xue, N. Source Separation of Vanadium over Iron from Roasted Vanadium-Bearing Shale during Acid Leaching via Ferric Fluoride Surface Coating. *J. Clean. Prod.* **2018**, *181*, 399–407. [CrossRef]
5. Yan, B.; Wang, D.; Wu, L.; Dong, Y. A Novel Approach for Pre-Concentrating Vanadium from Stone Coal Ore. *Miner. Eng.* **2018**, *125*, 231–238. [CrossRef]
6. Wenk, H.; Bulakh, A. *Minerals, Their Constitution and Origin*; Cambridge University Press: Cambridge, UK, 2004.
7. Schindler, M.; Hawthorne, F.C.; Baur, W.H. A Crystal-Chemical Approach to the Composition and Occurrence of Vanadium Minerals. *Can. Mineral.* **2000**, *38*, 1443–1456. [CrossRef]
8. Boni, M.; Bouabdellah, M.; Boukirou, W.; Putzolu, F.; Mondillo, N. Vanadium Ore Resources of the African Continent: State of the Art. *Ore Geol. Rev.* **2023**, *157*, 105423. [CrossRef]
9. Judd, J.C.; Sandberg, R.G.; Huiatt, J.L. *Recovery of Vanadium, Uranium and Phosphate From Idaho Phosphorite Ores*; Bureau of Mines, Report of Investigations 9025; United States Department of the Interior: Washington, DC, USA, 1986.
10. U.S. Geological Survey. *Mineral Commodity Summaries 2023*; U.S. Geological Survey: Reston, NA, USA, 2023; ISBN 9781411345041.
11. Reznitsky, L.Z.; Sklyarov, E.V.; Suvorova, L.F.; Karmanov, N.S.; Ushchapovskaya, Z.F. The Chromite–Coulsonite–Magnetite Solid Solution: The First Find of a Rare Variety of Spinel in Terrestrial Rocks. *Dokl. Earth Sci.* **2005**, *404*, 1121–1125.
12. Xu, C.; Zhang, Y.; Liu, T.; Huang, J. Characterization and Pre-Concentration of Low-Grade Vanadium-Titanium Magnetite Ore. *Minerals* **2017**, *7*, 137. [CrossRef]
13. Lauf, R.J. *Mineralogy of Uranium and Thorium*; Schiffer Publishing: Atglen, PA, USA, 2016; ISBN 0764351133.
14. Merritt, R.C. *Extractive Metallurgy of Uranium*; Colorado School of Mines Research Institute: Golden, CO, USA, 1971.
15. Henning Stonehenge Metals Corporate Presentation. In Proceedings of the Australian-Uranium-Rare-Earths-Conference, Fremantle, Australia, 16–17 July 2013.
16. Liu, S.; Jaireth, S. Exploring for Calcrete–Hosted Uranium Deposits in the Paterson Region, Western Australia. *Ausgeo News*, 2011; pp. 1–5.
17. Wang, M.; Xian, P.; Wang, X.; Li, B. Extraction of Vanadium from Stone Coal by Microwave Assisted Sulfation Roasting. *JOM* **2015**, *67*, 369–374. [CrossRef]
18. Lewis, S.E.; Henderson, R.A.; Dickens, G.R.; Shields, G.A.; Coxhell, S. The Geochemistry of Primary and Weathered Oil Shale and Coquina across the Julia Creek Vanadium Deposit (Queensland, Australia). *Miner. Depos.* **2010**, *45*, 599–620. [CrossRef]
19. Dechaine, G.P.; Gray, M.R. Chemistry and Association of Vanadium Compounds in Heavy Oil and Bitumen, and Implications for Their Selective Removal. *Energy Fuels* **2010**, *24*, 2795–2808. [CrossRef]
20. Martínez-Palou, R.; Cerón-Camacho, R.; Chávez, B.; Vallejo, A.A.; Villanueva-Negrete, D.; Castellanos, J.; Karamath, J.; Reyes, J.; Aburto, J. Demulsification of Heavy Crude Oil-in-Water Emulsions: A Comparative Study between Microwave and Thermal Heating. *Fuel* **2013**, *113*, 407–414. [CrossRef]
21. Mofarrah, A.; Husain, T. Evaluation of Environmental Pollution and Possible Management Options of Heavy Oil Fly Ash. *J. Mater. Cycles Waste Manag.* **2013**, *15*, 73–81. [CrossRef]
22. Mohammed, H.; Sadeek, S.; Mahmoud, A.R.; Zaky, D. Comparison of AAS, EDXRF, ICP-MS and INAA Performance for Determination of Selected Heavy Metals in HFO Ashes. *Microchem. J.* **2016**, *128*, 1–6. [CrossRef]
23. Gupta, C.K.; Krishnamurthy, N. *Extractive Metallurgy of Vanadium*; Elsevier Science Publishing: Amsterdam, The Netherlands, 1992.
24. Brough, C.; Bowell, R.J.; Larkin, J. The Geology of Vanadium Deposits. In *An Introduction to Vanadium*; Nova Science Publishers, Inc.: New York, NY, USA, 2019; pp. 87–117.
25. Abd El-Hamid, A.A.M.; Abu Khoziem, H.A. Physical and Chemical Characterization of El Kriymat Boiler Ash to Optimize the Leachability of Some Valuable Elements. *J. Environ. Chem. Eng.* **2019**, *7*, 103362. [CrossRef]
26. Vitolo, S.; Seggiani, M.; Filippi, S.; Brocchini, C. Recovery of Vanadium from Heavy Oil and Orimulsion Fly Ashes. *Hydrometallurgy* **2000**, *57*, 141–149. [CrossRef]
27. Holloway, P.C.; Etsell, T.H. Salt Roasting of Suncor Oil Sands Fly Ash. *Can. Metall. Q.* **2004**, *43*, 535–544. [CrossRef]
28. White, D.; Levy, L. Vanadium: Environmental Hazard or Environmental Opportunity? A Perspective on Some Key Research Needs. *Environ. Sci. Process. Impacts* **2021**, *23*, 527–534. [CrossRef]
29. International Energy Agency. *Electricity Market Report*; IEA: Paris, France, 2020.
30. Egypt Electricity Holding Company (EEHC). *Egypt Electricity Holding Company Annual Report*; EEHC: Cairo, Egypt, 2021.
31. Mowafa taib. The Mineral Industry of Egypt. In *U.S. Geological Survey Minerals Yearbook*; United States Geological Survey: Reston, VA, USA, 2018; pp. 1–17.
32. Survey of Energy Resources. *World Energy Resources*; World Energy Council Registered in England and Wales: London, UK, 2013; pp. 1–56.
33. Statistical Review of World Energy. *Country Analysis Executive Summary: China*; U.S. Energy Information Administration (EIA): Washington, DC, USA, 2022.

34. Statistical Review of World Energy. *Country Analysis Executive Summary: Egypt*; U.S. Energy Information Administration (EIA): Washington, DC, USA, 2022.
35. Bakkar, A.; Seleman, M.M.E.; Ahmed, M.M.Z.; Harb, S.; Goren, S.; Howsawi, E. Recovery of Vanadium and Nickel from Heavy Oil Fly Ash (HOFA): A Critical Review. *RSC Adv.* **2023**, *13*, 6327–6345. [CrossRef]
36. Li, X.; Xie, B. Extraction of Vanadium from High Calcium Vanadium Slag Using Direct Roasting and Soda Leaching. *Int. J. Miner. Metall. Mater.* **2012**, *19*, 595–601. [CrossRef]
37. Song, W.C.; Li, K.; Zheng, Q.; Li, H. A Novel Process of Vanadium Extraction from Molten Vanadium Bearing Slag. *Waste Biomass Valorization* **2014**, *5*, 327–332. [CrossRef]
38. Xiao, Q.; Chen, Y.; Gao, Y.; Xu, H.; Zhang, Y. Leaching of Silica from Vanadium-Bearing Steel Slag in Sodium Hydroxide Solution. *Hydrometallurgy* **2010**, *104*, 216–221. [CrossRef]
39. Wu, K.; Wang, Y.; Wang, X.; Wang, S.; Liu, B.; Zhang, Y.; Du, H. Co-Extraction of Vanadium and Chromium from High Chromium Containing Vanadium Slag by Low-Pressure Liquid Phase Oxidation Method. *J. Clean. Prod.* **2018**, *203*, 873–884. [CrossRef]
40. Li, X.; Yu, H.; Xue, X. Extraction of Iron from Vanadium Slag Using Pressure Acid Leaching. *Procedia Environ. Sci.* **2016**, *31*, 582–588. [CrossRef]
41. Aarabi-Karasgani, M.; Rashchi, F.; Mostoufi, N.; Vahidi, E. Leaching of Vanadium from LD Converter Slag Using Sulfuric Acid. *Hydrometallurgy* **2010**, *102*, 14–21. [CrossRef]
42. Zhang, J.; Zhang, W.; Zhang, L.; Gu, S. Mechanism of Vanadium Slag Roasting with Calcium Oxide. *Int. J. Miner. Process.* **2015**, *138*, 20–29. [CrossRef]
43. Li, H.Y.; Fang, H.X.; Wang, K.; Zhou, W.; Yang, Z.; Yan, X.M.; Ge, W.S.; Li, Q.W.; Xie, B. Asynchronous Extraction of Vanadium and Chromium from Vanadium Slag by Stepwise Sodium Roasting-Water Leaching. *Hydrometallurgy* **2015**, *156*, 124–135. [CrossRef]
44. Zhang, X.; Fang, D.; Song, S.; Cheng, G.; Xue, X. Selective Leaching of Vanadium over Iron from Vanadium Slag. *J. Hazard. Mater.* **2019**, *368*, 300–307. [CrossRef]
45. Hobson, A.J.; Stewart, D.I.; Mortimer, R.J.G.; Mayes, W.M.; Rogerson, M.; Burke, I.T. Leaching Behaviour of Co-Disposed Steel Making Wastes: Effects of Aeration on Leachate Chemistry and Vanadium Mobilisation. *Waste Manag.* **2018**, *81*, 1–10. [CrossRef]
46. Speight, J.G. *The Chemistry and Technology of Coal*, 3rd ed.; CRC Press: Boca Raton, FL, USA, 2013; ISBN 9781439836484.
47. Hsieh, Y.-M.; Tsai, M.-S. An Investigation of the Characteristics of Unburned Carbon in Oil Fly Ash. In *Environmental Challenges and Greenhouse Gas Control for Fossil Fuel Utilization in the 21st Century*; Springer: Boston, MA, USA, 2002; Volume 21, pp. 387–401, ISBN 9789896540821.
48. Tsygankova, M.V.; Bukin, V.I.; Lysakova, E.I.; Smirnova, A.G.; Reznik, A.M. The Recovery of Vanadium from Ash Obtained during the Combustion of Fuel Oil at Thermal Power Stations. *Russ. J. Non-Ferrous Met.* **2011**, *52*, 19–23. [CrossRef]
49. Kim, J.Y.; Mukherjee, S.; Ngo, L.; Christiani, D.C. Urinary 8-Hydroxy-2′-Deoxyguanosine as a Biomarker of Oxidative DNA Damage in Workers Exposed to Fine Particulates. *Environ. Health Perspect.* **2004**, *112*, 666–671. [CrossRef] [PubMed]
50. Di Pietro, A.; Visalli, G.; Munaò, F.; Baluce, B.; La Maestra, S.; Primerano, P.; Corigliano, F.; De Flora, S. Oxidative Damage in Human Epithelial Alveolar Cells Exposed in Vitro to Oil Fly Ash Transition Metals. *Int. J. Hyg. Environ. Health* **2009**, *212*, 196–208. [CrossRef] [PubMed]
51. Pattanaik, S.; Huggins, F.E.; Huffman, G.P.; Linak, W.P.; Miller, C.A. XAFS Studies of Nickel and Sulfur Speciation in Residual Oil Fly-Ash Particulate Matters. *Environ. Sci. Technol.* **2007**, *41*, 1104–1110. [CrossRef] [PubMed]
52. Gomez-Bueno, C.O.; Spink, D.R.; Rempel, G.L. Extraction of Vanadium and Nickel From Athabasca Oil Sands Fly Ash. *Metall. Trans.* **1984**, *12*, 341–352. [CrossRef]
53. Al-Malack, M.H.; Bukhari, A.A.; Al-Amoudi, O.S.; Al-Muhanna, H.H.; Zaidi, T.H. Characteristics of Fly Ash Produced at Power and Water Desalination Plants Firing Fuel Oil. *Int. J. Environ. Res.* **2013**, *7*, 455–466.
54. Mofarrah, A.; Husain, T. Use of Heavy Oil Fly Ash as a Color Ingredient in Cement Mortar. *Int. J. Concr. Struct. Mater.* **2013**, *7*, 111–117. [CrossRef]
55. Stas, J.; Dahdouh, A.; Al-chayah, O. Recovery of Vanadium, Nickel and Molybdenum from Fly Ash of Heavy Oil-Fired Electrical Power Station. *Period. Polytech. Chem. Eng.* **2007**, *51*, 67–70. [CrossRef]
56. Vitolo, S.; Seggiani, M.; Falaschi, F. Recovery of Vanadium from a Previously Burned Heavy Oil Fly Ash. *Hydrometallurgy* **2001**, *62*, 145–150. [CrossRef]
57. Aslam, Z.; Hussein, I.A.; Shawabkeh, R.A.; Parvez, M.A.; Ahmad, W. Ihsanullah Adsorption Kinetics and Modeling of H2S by Treated Waste Oil Fly Ash. *J. Air Waste Manag. Assoc.* **2019**, *69*, 246–257. [CrossRef]
58. Tokuyama, H.; Nii, S.; Kawaizumi, F.; Takahashi, K. Process Development for Recovery of Vanadium and Nickel from Heavy Oil Fly Ash by Leaching and Ion Exchange. *Sep. Sci. Technol.* **2003**, *38*, 1329–1344. [CrossRef]
59. Ibrahim, A.H.; Lyu, X.; Atia, B.M.; Gado, M.A.; ElDeeb, A.B. Phase Transformation Mechanism of Boiler Ash Roasted with Sodium Salt for Vanadium Extraction. *J. Mater. Cycles Waste Manag.* **2022**, *25*, 86–102. [CrossRef]
60. Yildirim, I.Z.; Prezzi, M. Chemical, Mineralogical, and Morphological Properties of Steel Slag. *Adv. Civ. Eng.* **2011**, *2011*, 463638. [CrossRef]

61. Li, W.; Zhang, Y.; Liu, T.; Huang, J.; Wang, Y. Comparison of Ion Exchange and Solvent Extraction in Recovering Vanadium from Sulfuric Acid Leach Solutions of Stone Coal. *Hydrometallurgy* **2013**, *131–132*, 1–7. [CrossRef]
62. Zhang, X.; Xie, B.; Diao, J.; Li, X.J. Nucleation and Growth Kinetics of Spinel Crystals in Vanadium Slag. *Ironmak. Steelmak.* **2012**, *39*, 147–154. [CrossRef]
63. Dabbagh, A.; Heidary Moghadam, A.; Naderi, S.; Hamdi, M. Erratum: A Study on the Effect of Coke Particle Size on the Thermal Profile of the Sinters Produced in Esfahan Steel Company (ESCO) (SAIMM Journal 113:12 (941–945)). *J. South. Afr. Inst. Min. Metall.* **2014**, *114*, 418.
64. Dondi, M.; Ercolani, G.; Guarini, G.; Raimondo, M. Orimulsion Fly Ash in Clay Bricks—Part 1: Composition and Thermal Behaviour of Ash. *J. Eur. Ceram. Soc.* **2002**, *22*, 1729–1735. [CrossRef]
65. Miller, C.A.; Srivastava, R.K. Combustion of Orimulsion and Its Generation of Air Pollutants. *Prog. Energy Combust. Sci.* **2000**, *26*, 131–160. [CrossRef]
66. Xiao, Y.; Mambote, C.R.; Jalkanen, H.; Yang, Y.; Boom, R. Vanadium Recovery as FeV from Petroleum Fly Ash. In Proceedings of the 12th International Ferroalloys Congress Sustainable Future, Helsinki, Finland, 6–9 June 2010; pp. 179–188.
67. Li, J.; Chen, Z.; Shen, B.; Xu, Z.; Zhang, Y. The Extraction of Valuable Metals and Phase Transformation and Formation Mechanism in Roasting-Water Leaching Process of Laterite with Ammonium Sulfate. *J. Clean. Prod.* **2017**, *140*, 1148–1155. [CrossRef]
68. Mu, W.; Huang, Z.; Xin, H.; Luo, S.; Zhai, Y.; Xu, Q. Extraction of Copper and Nickel from Low-Grade Nickel Sulfide Ore by Low-Temperature Roasting, Selective Decomposition and Water-Leaching Process. *JOM* **2019**, *71*, 4647–4658. [CrossRef]
69. Li, J.; Li, D.; Xu, Z.; Liao, C.; Liu, Y.; Zhong, B. Selective Leaching of Valuable Metals from Laterite Nickel Ore with Ammonium Chloride-Hydrochloric Acid Solution. *J. Clean. Prod.* **2018**, *179*, 24–30. [CrossRef]
70. Huaiwei, Z.; Xin, H. An Overview for the Utilization of Wastes from Stainless Steel Industries. *Resour. Conserv. Recycl.* **2011**, *55*, 745–754. [CrossRef]
71. Alex, P.; Mishra, P.; Suri, A.K. Studies on Processing of an Alnico Scrap. *Miner. Process. Extr. Metall. Rev.* **2001**, *22*, 547–565. [CrossRef]
72. Shen, Y.; Xue, W.; Niu, W. Recovery of Co(II) and Ni(II) from Hydrochloric Acid Solution of Alloy Scrap. *Trans. Nonferrous Met. Soc. China* **2008**, *18*, 1262–1268. [CrossRef]
73. Goel, S.; Pant, K.K.; Nigam, K.D.P. Extraction of Nickel from Spent Catalyst Using Fresh and Recovered EDTA. *J. Hazard. Mater.* **2009**, *171*, 253–261. [CrossRef]
74. Vuyyuru, K.R.; Pant, K.K.; Krishnan, V.V.; Nigam, K.D.P. Recovery of Nickel from Spent Industrial Catalysts Using Chelating Agents. *Ind. Eng. Chem. Res.* **2010**, *49*, 2014–2024. [CrossRef]
75. Sahu, K.K.; Agarwal, A.; Pandey, B.D. Nickel Recovery from Spent Nickel Catalyst. *Waste Manag. Res. J. A Sustain. Circ. Econ.* **2005**, *23*, 148–154. [CrossRef]
76. Idris, J.; Musa, M.; Yin, C.-Y.; Hamid, K.H.K. Recovery of Nickel from Spent Catalyst from Palm Oil Hydrogenation Process Using Acidic Solutions. *J. Ind. Eng. Chem.* **2010**, *16*, 251–255. [CrossRef]
77. Marafi, M.; Stanislaus, A. Waste Catalyst Utilization: Extraction of Valuable Metals from Spent Hydroprocessing Catalysts by Ultrasonic-Assisted Leaching with Acids. *Ind. Eng. Chem. Res.* **2011**, *50*, 9495–9501. [CrossRef]
78. Oza, R.; Shah, N.; Patel, S. Recovery of Nickel from Spent Catalysts Using Ultrasonication-Assisted Leaching. *J. Chem. Technol. Biotechnol.* **2011**, *86*, 1276–1281. [CrossRef]
79. Ognyanova, A.; Ozturk, A.T.; De Michelis, I.; Ferella, F.; Taglieri, G.; Akcil, A.; Vegliò, F. Metal Extraction from Spent Sulfuric Acid Catalyst through Alkaline and Acidic Leaching. *Hydrometallurgy* **2009**, *100*, 20–28. [CrossRef]
80. Ferella, F.; Ognyanova, A.; De Michelis, I.; Taglieri, G.; Vegliò, F. Extraction of Metals from Spent Hydrotreating Catalysts: Physico-Mechanical Pre-Treatments and Leaching Stage. *J. Hazard. Mater.* **2011**, *192*, 176–185. [CrossRef] [PubMed]
81. Yang, Q.Z.; Ng, R.S.; Qi, G.J.; Low, H.C.; Zhang, Y.P. Economic Viability of Nickel Recovery from Waste Catalyst. *Key Eng. Mater.* **2010**, *447–448*, 765–769. [CrossRef]
82. Yang, Q.Z.; Qi, G.J.; Low, H.C.; Song, B. Sustainable Recovery of Nickel from Spent Hydrogenation Catalyst: Economics, Emissions and Wastes Assessment. *J. Clean. Prod.* **2011**, *19*, 365–375. [CrossRef]
83. Abd El-Hamid, A.M.; Zahran, M.A.; Khalid, F.M.; Mahmoud, A.H. Leaching of Hafnium, Zirconium, Uranium and Other Nuclear Economic Elements from Petroleum Ash. *RSC Adv.* **2014**, *4*, 12506. [CrossRef]
84. Deng, Z.G.; Wei, C.; Fan, G.; Li, M.T.; Li, C.X.; Li, X. Bin Extracting Vanadium from Stone-Coal by Oxygen Pressure Acid Leaching and Solvent Extraction. *Trans. Nonferrous Met. Soc. China (Engl. Ed.)* **2010**, *20*, s118–s122. [CrossRef]
85. Barik, S.P.; Park, K.H.; Nam, C.W. Process Development for Recovery of Vanadium and Nickel from an Industrial Solid Waste by a Leaching-Solvent Extraction Technique. *J. Environ. Manag.* **2014**, *146*, 22–28. [CrossRef]
86. Akcil, A.; Vegliò, F.; Ferella, F.; Okudan, M.D.; Tuncuk, A. A Review of Metal Recovery from Spent Petroleum Catalysts and Ash. *Waste Manag.* **2015**, *45*, 420–433. [CrossRef]
87. Nazari, E.; Rashchi, F.; Saba, M.; Mirazimi, S.M.J. Simultaneous Recovery of Vanadium and Nickel from Power Plant Fly-Ash: Optimization of Parameters Using Response Surface Methodology. *Waste Manag.* **2014**, *34*, 2687–2696. [CrossRef]

88. Navarro, R.; Guzman, J.; Saucedo, I.; Revilla, J.; Guibal, E. Vanadium Recovery from Oil Fly Ash by Leaching, Precipitation and Solvent Extraction Processes. *Waste Manag.* **2007**, *27*, 425–438. [CrossRef]
89. Aburizaiza, A. Sequential Leaching of Vanadium from Heavy Fuel Oil Fly Ash Generated from Saudi Arabia Thermal Power Plants. *Curr. J. Appl. Sci. Technol.* **2019**, *32*, 1–17. [CrossRef]
90. Khalafalla, M.S.; Abdellah, W.M.; Khoziem, H.A.A.; Abd El-Hamid, A.A.M. Ecological Treatment of El Kriymat Boiler Ash for Recovering Vanadium, Nickel and Zinc from Sulfate Leach Liquor. *J. Mater. Cycles Waste Manag.* **2023**, *25*, 441–455. [CrossRef]
91. Weshahy, A.R.; Gouda, A.A.; Atia, B.M.; Sakr, A.K.; Al-Otaibi, J.S.; Almuqrin, A.; Hanfi, M.Y.; Sayyed, M.I.; El Sheikh, R.; Radwan, H.A.; et al. Efficient Recovery of Rare Earth Elements and Zinc from Spent Ni–Metal Hydride Batteries: Statistical Studies. *Nanomaterials* **2022**, *12*, 2305. [CrossRef] [PubMed]
92. Rahimi, G.; Rastegar, S.O.; Rahmani Chianeh, F.; Gu, T. Ultrasound-Assisted Leaching of Vanadium from Fly Ash Using Lemon Juice Organic Acids. *RSC Adv.* **2020**, *10*, 1685–1696. [CrossRef] [PubMed]
93. Mu, W.; Zhang, T.; Dou, Z.; Lü, G.; Hu, L.; Yu, B.; Liu, Y. Oxygen Pressure Acid Leaching of Vanadium Slag. *Appl. Mech. Mater.* **2011**, *79*, 242–247. [CrossRef]
94. Zhang, G.; Zhang, T.; Lü, G.; Zhang, Y.; Liu, Y.; Liu, Z. Extraction of Vanadium from Vanadium Slag by High Pressure Oxidative Acid Leaching. *Int. J. Miner. Metall. Mater.* **2015**, *22*, 21–26. [CrossRef]
95. Guoquan, Z.; Ting'an, Z.; Guozhi, L.; Ying, Z.; Yan, L.; Gang, X. Extraction of Vanadium from LD Converter Slag by Pressure Leaching Process with Titanium White Waste Acid. *Rare Met. Mater. Eng.* **2015**, *44*, 1894–1898. [CrossRef]
96. Zhang, G.-Q.; Zhang, T.-A.; Zhang, Y.; Lv, G.-Z.; Liu, Y.; Liu, Z.-L. Pressure Leaching of Converter Vanadium Slag with Waste Titanium Dioxide. *Rare Met.* **2016**, *35*, 576–580. [CrossRef]
97. Zhou, X.; Wei, C.; Xia, W.; Li, M.; Li, C.; Deng, Z.; Xu, H. Dissolution Kinetics and Thermodynamic Analysis of Vanadium Trioxide during Pressure Oxidation. *Rare Met.* **2012**, *31*, 296–302. [CrossRef]
98. Amer, A.M. Processing of Egyptian Boiler-Ash for Extraction of Vanadium and Nickel. *Waste Manag.* **2002**, *22*, 515–520. [CrossRef]
99. Tsai, S.-L.; Tsai, M.-S. A Study of the Extraction of Vanadium and Nickel in Oil-Fired Fly Ash. *Resour. Conserv. Recycl.* **1998**, *22*, 163–176. [CrossRef]
100. Al-Zuhairi, M.A.H. Vanadium Extraction from Residual of Fired Crude Oil in Power Plants. *Iraqi J. Mech. Mater. Eng.* **2014**, *14*, 423–431.
101. Hakimi, M.; Kiani, P.; Alikhani, M.; Feizi, N.; Bajestani, A.M.; Alimard, P. Reducing Environmental Pollution of Fuel Fly Ash by Extraction and Removal Vanadium Pentoxide. *Solid Fuel Chem.* **2020**, *54*, 337–342. [CrossRef]
102. Akita, S.; Maeda, T.; Takeuchi, H. Recovery of Vanadium and Nickel in Fly Ash from Heavy Oil. *J. Chem. Technol. Biotechnol.* **1995**, *62*, 345–350. [CrossRef]
103. Al-Ghouti, M.A.; Al-Degs, Y.S.; Ghrair, A.; Khoury, H.; Ziedan, M. Extraction and Separation of Vanadium and Nickel from Fly Ash Produced in Heavy Fuel Power Plants. *Chem. Eng. J.* **2011**, *173*, 191–197. [CrossRef]
104. Shahnazi, A.; Rashchi, F.; Vahidi, E. A Kinetics Study on the Hydrometallurgical Recovery of Vanadium from LD Converter Slag in Alkaline Media. In Proceedings of the TMS 2012 Annual Meeting, Orlando, FL, USA, 11–15 March 2012; pp. 425–433. [CrossRef]
105. Qiu, S.; Wei, C.; Li, M.; Zhou, X.; Li, C.; Deng, Z. Dissolution Kinetics of Vanadium Trioxide at High Pressure in Sodium Hydroxide-Oxygen Systems. *Hydrometallurgy* **2011**, *105*, 350–354. [CrossRef]
106. Liu, B.; Du, H.; Wang, S.N.; Zhang, Y.; Zheng, S.L.; Li, L.J.; Chen, D.H. A Novel Method to Extract Vanadium and Chromium from Vanadium Slag Using Molten NaOH-NaNO3 Binary System. *AIChE J.* **2013**, *59*, 541–552. [CrossRef]
107. Mirazimi, S.M.J.; Rashchi, F.; Saba, M. A New Approach for Direct Leaching of Vanadium from LD Converter Slag. *Chem. Eng. Res. Des.* **2015**, *94*, 131–140. [CrossRef]
108. Zhang, G.; Chen, D.; Zhao, W.; Zhao, H.; Wang, L.; Li, D.; Qi, T. A Novel Synergistic Extraction Method for Recovering Vanadium (V) from High-Acidity Chloride Leaching Liquor. *Sep. Purif. Technol.* **2016**, *165*, 166–172. [CrossRef]
109. Li, M.; Zheng, S.; Liu, B.; Wang, S.; Dreisinger, D.B.; Zhang, Y.; Du, H.; Zhang, Y. A Clean and Efficient Method for Recovery of Vanadium from Vanadium Slag: Nonsalt Roasting and Ammonium Carbonate Leaching Processes. *Miner. Process. Extr. Metall. Rev.* **2017**, *38*, 228–237. [CrossRef]
110. Wang, Z.; Chen, L.; Aldahrib, T.; Li, C.; Liu, W.; Zhang, G.; Yang, Y.; Luo, D. Direct Recovery of Low Valence Vanadium from Vanadium Slag- Effect of Roasting on Vanadium Leaching. *Hydrometallurgy* **2020**, *191*, 105156. [CrossRef]
111. Li, D.-Q.; Yang, Y.; Li, H.-Y.; Xie, B. Study on Vanadium Phase Evolution Law in Vanadium Slag During the Interface Reaction Process of Sodium Roasting. In *Rare Metal Technology 2020*; Minerals, Metals and Materials Series; Springer International Publishing: New York, NY, USA, 2020; pp. 253–264, ISBN 9783030367572.
112. Zhang, G.; Zhang, Y.; Bao, S.; Yuan, Y.; Jian, X.; Li, R. Selective Vanadium Extraction from Vanadium Bearing Ferro-Phosphorus via Roasting and Pressure Hydrogen Reduction. *Sep. Purif. Technol.* **2019**, *220*, 293–299. [CrossRef]
113. Ettler, V.; Kvapil, J.; Šebek, O.; Johan, Z.; Mihaljevič, M.; Ratié, G.; Garnier, J.; Quantin, C. Leaching Behaviour of Slag and Fly Ash from Laterite Nickel Ore Smelting (Niquelândia, Brazil). *Appl. Geochem.* **2016**, *64*, 118–127. [CrossRef]

114. Hu, P.; Zhang, Y.; Liu, T.; Huang, J.; Yuan, Y.; Zheng, Q. Highly Selective Separation of Vanadium over Iron from Stone Coal by Oxalic Acid Leaching. *J. Ind. Eng. Chem.* **2017**, *45*, 241–247. [CrossRef]
115. Liu, Z.; Huang, J.; Zhang, Y.; Liu, T.; Hu, P.; Liu, H.; Zheng, Q. Separation and Recovery of Iron Impurities from a Complex Oxalic Acid Solution Containing Vanadium by K3Fe(C2O4)3·3H2O Crystallization. *Sep. Purif. Technol.* **2020**, *232*, 115970. [CrossRef]
116. Mirazimi, S.M.J.; Rashchi, F. Optimization of Bioleaching of a Vanadium Containing Slag Using RSM. In Proceedings of the 7th International Chemical Engineering Congress & Exihibition, Kish, Iran, 21–24 November 2011.
117. Mirazimi, S.M.J.; Abbasalipour, Z.; Rashchi, F. Vanadium Removal from LD Converter Slag Using Bacteria and Fungi. *J. Environ. Manag.* **2015**, *153*, 144–151. [CrossRef]
118. Ntita, J.; Nheta, W. Investigation on the Mechanisms of Bio-Processing Vanadium Slags. In Proceedings of the 5th International Slag Volarisation Symposium, Leuven, Belgia, 3–5 April 2017; pp. 173–176.
119. Ntita, J.; Nheta, W.; Staden, P.V. Selective Leaching of Vanadium from Vanadium Slag Using Organic Acids. In Proceedings of the 9th International Conference on Advances in Science, Engineering, Technology & Waste Management (ASETWM-17), Parys, South Africa, 27–28 November 2017.
120. Thallner, S.; Hemmelmair, C.; Martinek, S.; Schnitzhofer, W. Bioleaching for Removal of Chromium and Associated Metals from LD Slag. *Solid State Phenom.* **2017**, *262*, 79–83. [CrossRef]
121. Lee, J.C.; Pandey, B.D. Bio-Processing of Solid Wastes and Secondary Resources for Metal Extraction—A Review. *Waste Manag.* **2012**, *32*, 3–18. [CrossRef]
122. Zeng, L.; Cheng, C.Y. A Literature Review of the Recovery of Molybdenum and Vanadium from Spent Hydrodesulphurisation Catalysts. Part I: Metallurgical Processes. *Hydrometallurgy* **2009**, *98*, 1–9. [CrossRef]
123. Gomes, H.I.; Funari, V.; Mayes, W.M.; Rogerson, M.; Prior, T.J. Recovery of Al, Cr and V from Steel Slag by Bioleaching: Batch and Column Experiments. *J. Environ. Manag.* **2018**, *222*, 30–36. [CrossRef]
124. Liu, Z.; Nueraihemaiti, A.; Chen, M.; Du, J.; Fan, X.; Tao, C. Hydrometallurgical Leaching Process Intensified by an Electric Field for Converter Vanadium Slag. *Hydrometallurgy* **2015**, *155*, 56–60. [CrossRef]
125. Peng, H.; Liu, Z.; Tao, C. Selective Leaching of Vanadium from Chromium Residue Intensified by Electric Field. *J. Environ. Chem. Eng.* **2015**, *3*, 1252–1257. [CrossRef]
126. Liu, Z.; Li, Y.; Chen, M.; Nueraihemaiti, A.; Du, J.; Fan, X.; Tao, C.Y. Enhanced Leaching of Vanadium Slag in Acidic Solution by Electro-Oxidation. *Hydrometallurgy* **2016**, *159*, 1–5. [CrossRef]
127. Lee, J.; Kurniawan; Kim, E.; Chung, K.W.; Kim, R.; Jeon, H.-S. A Review on the Metallurgical Recycling of Vanadium from Slags: Towards a Sustainable Vanadium Production. *J. Mater. Res. Technol.* **2021**, *12*, 343–364. [CrossRef]
128. Peng, H. A Literature Review on Leaching and Recovery of Vanadium. *J. Environ. Chem. Eng.* **2019**, *7*, 103313. [CrossRef]
129. Ye, P.; Wang, X.; Wang, M.; Fan, Y.; Xiang, X. Recovery of Vanadium from Stone Coal Acid Leaching Solution by Coprecipitation, Alkaline Roasting and Water Leaching. *Hydrometallurgy* **2012**, *117–118*, 108–115. [CrossRef]
130. Gao, H.; Jiang, T.; Zhou, M.; Wen, J.; Li, X.; Wang, Y.; Xue, X. Effect of Microwave Irradiation and Conventional Calcification Roasting with Calcium Hydroxide on the Extraction of Vanadium and Chromium from High-Chromium Vanadium Slag. *Miner. Eng.* **2020**, *145*, 106056. [CrossRef]
131. Chen, L.; Wang, Z.; Qin, Z.; Zhang, G.; Yue, H.; Liang, B.; Luo, D. Investigation of the Selective Oxidation Roasting of Vanadium-iron Spinel. *Powder Technol.* **2021**, *387*, 434–443. [CrossRef]
132. Zeng, L.; Yong Cheng, C. A Literature Review of the Recovery of Molybdenum and Vanadium from Spent Hydrodesulphurisation Catalysts. Part II: Separation and Purification. *Hydrometallurgy* **2009**, *98*, 10–20. [CrossRef]
133. Li, M.; Du, H.; Zheng, S.; Wang, S.; Zhang, Y.; Liu, B.; Dreisinger, D.B.; Zhang, Y. Extraction of Vanadium from Vanadium Slag Via Non-Salt Roasting and Ammonium Oxalate Leaching. *JOM* **2017**, *69*, 1970–1975. [CrossRef]
134. Li, M.; Liu, B.; Zheng, S.; Wang, S.; Du, H.; Dreisinger, D.B.; Zhang, Y. A Cleaner Vanadium Extraction Method Featuring Non-Salt Roasting and Ammonium Bicarbonate Leaching. *J. Clean. Prod.* **2017**, *149*, 206–217. [CrossRef]
135. Zhang, X.; Liu, F.; Xue, X.; Jiang, T. Effects of Microwave and Conventional Blank Roasting on Oxidation Behavior, Microstructure and Surface Morphology of Vanadium Slag with High Chromium Content. *J. Alloys Compd.* **2016**, *686*, 356–365. [CrossRef]
136. Wang, M.; Xiang, X.; Zhang, L.; Xiao, L. Effect of Vanadium Occurrence State on the Choice of Extracting Vanadium Technology from Stone Coal. *Rare Met.* **2008**, *27*, 112–115. [CrossRef]
137. Zhang, Y.; Zhang, T.A.; Dreisinger, D.; Lv, C.; Lv, G.; Zhang, W. Recovery of Vanadium from Calcification Roasted-Acid Leaching Tailing by Enhanced Acid Leaching. *J. Hazard. Mater.* **2019**, *369*, 632–641. [CrossRef]
138. Zhang, J.; Zhang, W.; Xue, Z. Oxidation Kinetics of Vanadium Slag Roasting in the Presence of Calcium Oxide. *Miner. Process. Extr. Metall. Rev.* **2017**, *38*, 265–273. [CrossRef]
139. Wen, J.; Jiang, T.; Zheng, X.; Wang, J.; Cao, J.; Zhou, M. Efficient Separation of Chromium and Vanadium by Calcification Roasting–Sodium Carbonate Leaching from High Chromium Vanadium Slag and V2O5 Preparation. *Sep. Purif. Technol.* **2020**, *230*, 115881. [CrossRef]

140. Zhang, J.; Zhang, W.; Xue, Z. An Environment-Friendly Process Featuring Calcified Roasting and Precipitation Purification to Prepare Vanadium Pentoxide from the Converter Vanadium Slag. *Metals* **2019**, *9*, 21. [CrossRef]
141. Peng, H.; Guo, J.; Zheng, X.; Liu, Z.; Tao, C. Leaching Kinetics of Vanadium from Calcification Roasting Converter Vanadium Slag in Acidic Medium. *J. Environ. Chem. Eng.* **2018**, *6*, 5119–5124. [CrossRef]
142. Xiang, J.; Huang, Q.; Lv, X.; Bai, C. Extraction of Vanadium from Converter Slag by Two-Step Sulfuric Acid Leaching Process. *J. Clean. Prod.* **2018**, *170*, 1089–1101. [CrossRef]
143. Li, H.-Y.; Wang, K.; Hua, W.-H.; Yang, Z.; Zhou, W.; Xie, B. Selective Leaching of Vanadium in Calcification-Roasted Vanadium Slag by Ammonium Carbonate. *Hydrometallurgy* **2016**, *160*, 18–25. [CrossRef]
144. Xu, Y.; Cai, Y.; Lu, X.; Ma, P. Study on Recovery of Vanadium from Low-Vanadium Slag by Calcified Roasting Methods. *Adv. Mater. Res.* **2011**, *284–286*, 2127–2130. [CrossRef]
145. Wen, J.; Jiang, T.; Zhou, M.; Gao, H.Y.; Liu, J.Y.; Xue, X.X. Roasting and Leaching Behaviors of Vanadium and Chromium in Calcification Roasting–Acid Leaching of High-Chromium Vanadium Slag. *Int. J. Miner. Metall. Mater.* **2018**, *25*, 515–526. [CrossRef]
146. Xiang, J.; Huang, Q.; Lv, X.; Bai, C. Effect of Mechanical Activation Treatment on the Recovery of Vanadium from Converter Slag. *Metall. Mater. Trans. B Process Metall. Mater. Process. Sci.* **2017**, *48*, 2759–2767. [CrossRef]
147. Yang, Z.; Li, H.-Y.; Yin, X.-C.; Yan, Z.-M.; Yan, X.-M.; Xie, B. Leaching Kinetics of Calcification Roasted Vanadium Slag with High CaO Content by Sulfuric Acid. *Int. J. Miner. Process.* **2014**, *133*, 105–111. [CrossRef]
148. Wang, G.; Lin, M.M.; Diao, J.; Li, H.Y.; Xie, B.; Li, G. A Novel Strategy for Green Comprehensive Utilization of Vanadium Slag with High-Content Chromium. *ACS Sustain. Chem. Eng.* **2019**, *7*, 18133–18141. [CrossRef]
149. Wen, J.; Jiang, T.; Zhou, W.; Gao, H.; Xue, X. A Cleaner and Efficient Process for Extraction of Vanadium from High Chromium Vanadium Slag: Leaching in $(NH_4)_2SO_4$-H_2SO_4 Synergistic System and NH4+ Recycle. *Sep. Purif. Technol.* **2019**, *216*, 126–135. [CrossRef]
150. Xiang, J.; Pei, G.; Huang, Q.; Lv, W.; Yang, M.; Hu, K.; Lv, X. *A Multi-Step Process for the Cleaner Utilization of Vanadium-Bearing Converter Slag*; Springer International Publishing: New York, NY, USA, 2019; ISBN 9783030057398.
151. Wen, J.; Jiang, T.; Xu, Y.; Liu, J.; Xue, X. Efficient Separation and Extraction of Vanadium and Chromium in High Chromium Vanadium Slag by Selective Two-Stage Roasting–Leaching. *Metall. Mater. Trans. B Process Metall. Mater. Process. Sci.* **2018**, *49*, 1471–1481. [CrossRef]
152. Cai, Z.; Zhang, Y. Phase Transformations of Vanadium Recovery from Refractory Stone Coal by Novel NaOH Molten Roasting and Water Leaching Technology. *RSC Adv.* **2017**, *7*, 36917–36922. [CrossRef]
153. Ji, Y.; Shen, S.; Liu, J.; Xue, Y. Cleaner and Effective Process for Extracting Vanadium from Vanadium Slag by Using an Innovative Three-Phase Roasting Reaction. *J. Clean. Prod.* **2017**, *149*, 1068–1078. [CrossRef]
154. Nkosi, S.; Dire, P.; Nyambeni, N.; Goso, X.C. A Comparative Study of Vanadium Recovery from Titaniferous Magnetite Using Salt, Sulphate, and Soda Ash Roast-Leach Processes. In Proceedings of the 3rd Young Professionals Conference, Pretoria, South Africa, 9–10 March 2017; pp. 191–200.
155. Geng, C.; Sun, T.; Yang, H.; Ma, Y.; Gao, E.; Xu, C. Effect of Na2SO4 on the Embedding Direct Reduction of Beach Titanomagnetite and the Separation of Titanium and Iron by Magnetic Separation. *ISIJ Int.* **2015**, *55*, 2543–2549. [CrossRef]
156. Nugraha, A.; Priyono, B.; Maksum, A.; Juna, H.; Soedarsono, J.W. The Effect of Sodium Sulfate (Na_2SO_4) and Sodium Carbonate (Na_2CO_3) Addition on the Reduction-Roasting Process of Titania Iron Sand. *E3S Web Conf.* **2018**, *67*, 2–7. [CrossRef]
157. Sui, Y.L.; Guo, Y.F.; Travyanov, A.Y.; Jiang, T.; Chen, F.; Qiu, G.Z. Reduction Roasting–Magnetic Separation of Vanadium Tailings in Presence of Sodium Sulfate and Its Mechanisms. *Rare Met.* **2016**, *35*, 954–960. [CrossRef]
158. Ju, Z.-J.; Wang, C.-Y.; Yin, F. Dissolution Kinetics of Vanadium from Black Shale by Activated Sulfuric Acid Leaching in Atmosphere Pressure. *Int. J. Miner. Process.* **2015**, *138*, 1–5. [CrossRef]
159. Huang, K.; Inoue, K.; Harada, H.; Kawakita, H.; Ohto, K. Leaching of Heavy Metals by Citric Acid from Fly Ash Generated in Municipal Waste Incineration Plants. *J. Mater. Cycles Waste Manag.* **2011**, *13*, 118–126. [CrossRef]
160. Pan, B.; Jin, W.; Liu, B.; Zheng, S.; Wang, S.; Du, H.; Zhang, Y. Cleaner Production of Vanadium Oxides by Cation-Exchange Membrane-Assisted Electrolysis of Sodium Vanadate Solution. *Hydrometallurgy* **2017**, *169*, 440–446. [CrossRef]
161. Wen, J.; Jiang, T.; Xu, Y.; Cao, J.; Xue, X. Efficient Extraction and Separation of Vanadium and Chromium in High Chromium Vanadium Slag by Sodium Salt Roasting-(NH4)2SO4 Leaching. *J. Ind. Eng. Chem.* **2019**, *71*, 327–335. [CrossRef]
162. Zhao, Y.; Wang, W.; Zhang, Y.; Song, S.; Bao, S. In-Situ Investigation on Mineral Phase Transition during Roasting of Vanadium-Bearing Stone Coal. *Adv. Powder Technol.* **2017**, *28*, 1103–1107. [CrossRef]
163. Jung, M.; Mishra, B. Vanadium Recovery from Oil Fly Ash by Carbon Removal and Roast-Leach Process. *JOM* **2018**, *70*, 168–172. [CrossRef]
164. Yuan, J.; Cao, Y.; Fan, G.; Du, H.; Dreisinger, D.; Han, G.; Li, M. Study on the Mechanisms for Vanadium Phases' Transformation of Vanadium Slag Non-Salt Roasting Process. In *Rare Metal Technology 2020*; Azimi, G., Forsberg, K., Ouchi, T., Kim, H., Alam, S., Baba, A.A., Eds.; Springer International Publishing: Cham, Switzerland, 2020; pp. 235–242, ISBN 978-3-030-36757-2.

165. Ibrahim, A.H.; Lyu, X.; Atia, B.M.; Gado, M.A.; ElDeeb, A.B. Cost-Effective and High Purity Valuable Metals Extraction from Water Leaching Solid Residues Obtained as a By-Product from Processing the Egyptian Boiler Ash. *Minerals* **2022**, *12*, 1084. [CrossRef]
166. Li, X.; Xie, B.; Wang, G.; Li, X. Oxidation Process of Low-Grade Vanadium Slag in Presence of Na_2CO_3. *Trans. Nonferrous Met. Soc. China* **2011**, *21*, 1860–1867. [CrossRef]
167. Mahdavian, A.; Shafyei, A.; Alamdariand, E.K.; Haghshenas, D.F. Recovery of Vanadium from Esfahan Steel Company Steel Slag; Optimizing of Roasting and Leaching Parameters. *Int. J. ISSI* **2006**, *3*, 17–21.
168. Li, H.Y.; Wang, C.; Lin, M.; Guo, Y.; Xie, B. Green One-Step Roasting Method for Efficient Extraction of Vanadium and Chromium from Vanadium-Chromium Slag. *Powder Technol.* **2020**, *360*, 503–508. [CrossRef]
169. Wang, M.; Chen, B.; Huang, S.; Wang, X.; Liu, B.; Ge, Q.; Xie, S. A Novel Technology for Vanadium and Chromium Recovery from V-Cr-Bearing Reducing Slag. *Hydrometallurgy* **2017**, *171*, 116–122. [CrossRef]
170. Liu, S.; Wang, L.; Chou, K. A Novel Process for Simultaneous Extraction of Iron, Vanadium, Manganese, Chromium, and Titanium from Vanadium Slag by Molten Salt Electrolysis. *Ind. Eng. Chem. Res.* **2016**, *55*, 12962–12969. [CrossRef]
171. Liu, H.; Du, H.; Wang, D.W.; Wang, S.N.; Zheng, S.L.; Zhang, Y. Kinetics Analysis of Decomposition of Vanadium Slag by KOH Sub-Molten Salt Method. *Trans. Nonferrous Met. Soc. China (Engl. Ed.)* **2013**, *23*, 1489–1500. [CrossRef]
172. Wang, Z.; Zheng, S.; Wang, S.; Qin, Y.; Du, H.; Zhang, Y. Electrochemical Decomposition of Vanadium Slag in Concentrated NaOH Solution. *Hydrometallurgy* **2015**, *151*, 51–55. [CrossRef]
173. Hu, P.; Zhang, Y.; Liu, T.; Huang, J.; Yuan, Y.; Yang, Y. Separation and Recovery of Iron Impurity from a Vanadium-Bearing Stone Coal via an Oxalic Acid Leaching-Reduction Precipitation Process. *Sep. Purif. Technol.* **2017**, *180*, 99–106. [CrossRef]
174. Mazurek, K. Recovery of Vanadium, Potassium and Iron from a Spent Vanadium Catalyst by Oxalic Acid Solution Leaching, Precipitation and Ion Exchange Processes. *Hydrometallurgy* **2013**, *134–135*, 26–31. [CrossRef]
175. Mazurek, K.; Białowicz, K.; Trypuć, M. Recovery of Vanadium, Potassium and Iron from a Spent Catalyst Using Urea Solution. *Hydrometallurgy* **2010**, *103*, 19–24. [CrossRef]
176. Sánchez-Camargo, A.P.; Mendiola, J.A.; Ibáñez, E.; Herrero, M. Supercritical Fluid Extraction. In *Reference Module in Chemistry, Molecular Sciences and Chemical Engineering*; Elsevier: Amsterdam, The Netherlands, 2014; ISBN 978-0-12-409547-2.
177. Yang, Q.W.; Xie, Z.M.; Peng, H.; Liu, Z.H.; Tao, C.Y. Leaching of Vanadium and Chromium from Converter Vanadium Slag Intensified with Surface Wettability. *J. Cent. South Univ.* **2018**, *25*, 1317–1325. [CrossRef]
178. Olehile, O.D.; van Dyk, L. Gas Phase Extraction of Vanadium from Spent Vanadium Catalyst and from Tantalum Oxide. 2017.
179. Aljerf, L. High-Efficiency Extraction of Bromocresol Purple Dye and Heavy Metals as Chromium from Industrial Effluent by Adsorption onto a Modified Surface of Zeolite: Kinetics and Equilibrium Study. *J. Environ. Manag.* **2018**, *225*, 120–132. [CrossRef] [PubMed]
180. Ojha, K.; Pradhan, N.C.; Samanta, A.N. Zeolite from Fly Ash: Synthesis and Characterization. *Bull. Mater. Sci.* **2004**, *27*, 555–564. [CrossRef]
181. Golbad, S.; Khoshnoud, P.; Abu-Zahra, N. Hydrothermal Synthesis of Hydroxy Sodalite from Fly Ash for the Removal of Lead Ions from Water. *Int. J. Environ. Sci. Technol.* **2017**, *14*, 135–142. [CrossRef]
182. Kang, Y.; Swain, B.; Im, B.; Yoon, J.H.; Park, K.H.; Lee, C.G.; Kim, D.G. Synthesis of Zeolite Using Aluminum Dross and Waste LCD Glass Powder: A Waste to Waste Integration Valorization Process. *Metals* **2019**, *9*, 1240. [CrossRef]
183. Liu, Y.; Yan, C.; Zhao, J.; Zhang, Z.; Wang, H.; Zhou, S.; Wu, L. Synthesis of Zeolite P1 from Fly Ash under Solvent-Free Conditions for Ammonium Removal from Water. *J. Clean. Prod.* **2018**, *202*, 11–22. [CrossRef]
184. Mallapur, V.P.; Oubagaranadin, J.U.K. A Brief Review on the Synthesis of Zeolites from Hazardous Wastes. *Trans. Indian Ceram. Soc.* **2017**, *76*, 1–13. [CrossRef]
185. Vigil de la Villa Mencía, R.; Goiti, E.; Ocejo, M.; Giménez, R.G. Synthesis of Zeolite Type Analcime from Industrial Wastes. *Microporous Mesoporous Mater.* **2020**, *293*, 109817. [CrossRef]
186. Grela, A.; Łach, M.; Bajda, T.; Mikuła, J.; Hebda, M. Characterization of the Products Obtained from Alkaline Conversion of Tuff and Metakaolin. *J. Therm. Anal. Calorim.* **2018**, *133*, 217–226. [CrossRef]
187. Li, X.Y.; Jiang, Y.; Liu, X.Q.; Shi, L.Y.; Zhang, D.Y.; Sun, L.B. Direct Synthesis of Zeolites from a Natural Clay, Attapulgite. *ACS Sustain. Chem. Eng.* **2017**, *5*, 6124–6130. [CrossRef]
188. Pereira, P.M.; Ferreira, B.F.; Oliveira, N.P.; Nassar, E.J.; Ciuffi, K.J.; Vicente, M.A.; Trujillano, R.; Rives, V.; Gil, A.; Korili, S.; et al. Synthesis of Zeolite A from Metakaolin and Its Application in the Adsorption of Cationic Dyes. *Appl. Sci.* **2018**, *8*, 608. [CrossRef]
189. Aytas, S.; Yurtlu, M.; Donat, R. Adsorption Characteristic of U(VI) Ion onto Thermally Activated Bentonite. *J. Hazard. Mater.* **2009**, *172*, 667–674. [CrossRef] [PubMed]
190. Wang, G.; Wang, X.; Chai, X.; Liu, J.; Deng, N. Adsorption of Uranium (VI) from Aqueous Solution on Calcined and Acid-Activated Kaolin. *Appl. Clay Sci.* **2010**, *47*, 448–451. [CrossRef]

191. Khaleque, A.; Alam, M.M.; Hoque, M.; Mondal, S.; Haider, J.B.; Xu, B.; Johir, M.A.H.; Karmakar, A.K.; Zhou, J.L.; Ahmed, M.B.; et al. Zeolite Synthesis from Low-Cost Materials and Environmental Applications: A Review. *Environ. Adv.* **2020**, *2*, 100019. [CrossRef]
192. Sugano, Y.; Sahara, R.; Murakami, T.; Narushima, T.; Iguchi, Y.; Ouchi, C. Hydrothermal Synthesis of Zeolite a Using Blast Furnace Slag. *ISIJ Int.* **2005**, *45*, 937–945. [CrossRef]
193. Johnson, E.B.G.; Arshad, S.E. Hydrothermally Synthesized Zeolites Based on Kaolinite: A Review. *Appl. Clay Sci.* **2014**, *97–98*, 215–221. [CrossRef]
194. Yao, G.; Lei, J.; Zhang, X.; Sun, Z.; Zheng, S. One-Step Hydrothermal Synthesis of Zeolite X Powder from Natural Low-Grade Diatomite. *Materials* **2018**, *11*, 906. [CrossRef]
195. Wajima, T.; Yoshizuka, K.; Hirai, T.; Ikegami, Y. Synthesis of Zeolite X from Waste Sandstone Cake Using Alkali Fusion Method. *Mater. Trans.* **2008**, *49*, 612–618. [CrossRef]
196. Ayele, L.; Pérez-Pariente, J.; Chebude, Y.; Díaz, I. Conventional versus Alkali Fusion Synthesis of Zeolite A from Low Grade Kaolin. *Appl. Clay Sci.* **2016**, *132–133*, 485–490. [CrossRef]
197. Wajima, T. Synthesis of Zeolite from Blast Furnace Slag Using Alkali Fusion with Addition of EDTA. *Adv. Mater. Res.* **2014**, *1044–1045*, 124–127. [CrossRef]
198. Tsujiguchi, M.; Kobashi, T.; Oki, M.; Utsumi, Y.; Kakimori, N.; Nakahira, A. Synthesis and Characterization of Zeolite A from Crushed Particles of Aluminoborosilicate Glass Used in LCD Panels. *J. Asian Ceram. Soc.* **2014**, *2*, 27–32. [CrossRef]
199. Tsujiguchi, M.; Kobashi, T.; Utsumi, Y.; Kakimori, N.; Nakahira, A. Synthesis of Zeolite a from Aluminoborosilicate Glass Used in Glass Substrates of Liquid Crystal Display Panels and Evaluation of Its Cation Exchange Capacity. *J. Am. Ceram. Soc.* **2014**, *97*, 114–119. [CrossRef]
200. Wu, Y.; Ren, X.; Wang, J. Facile Synthesis and Morphology Control of Zeolite MCM-22 via a Two-Step Sol-Gel Route with Tetraethyl Orthosilicate as Silica Source. *Mater. Chem. Phys.* **2009**, *113*, 773–779. [CrossRef]
201. Le, T.; Wang, Q.; Pan, B.; Ravindra, A.V.; Ju, S.; Peng, J. Process Regulation of Microwave Intensified Synthesis of Y-Type Zeolite. *Microporous Mesoporous Mater.* **2019**, *284*, 476–485. [CrossRef]
202. Ou, X.; Xu, S.; Warnett, J.M.; Holmes, S.M.; Zaheer, A.; Garforth, A.A.; Williams, M.A.; Jiao, Y.; Fan, X. Creating Hierarchies Promptly: Microwave-Accelerated Synthesis of ZSM-5 Zeolites on Macrocellular Silicon Carbide (SiC) Foams. *Chem. Eng. J.* **2017**, *312*, 1–9. [CrossRef]
203. Meng, X.; Xiao, F.S. Green Routes for Synthesis of Zeolites. *Chem. Rev.* **2014**, *114*, 1521–1543. [CrossRef]
204. Khalil, U.; Muraza, O. Microwave-Assisted Hydrothermal Synthesis of Mordenite Zeolite: Optimization of Synthesis Parameters. *Microporous Mesoporous Mater.* **2016**, *232*, 211–217. [CrossRef]
205. Shoppert, A.A.; Loginova, I.V.; Chaikin, L.I.; Rogozhnikov, D.A. Alkali Fusion-Leaching Method For Comprehensive Processing Of Fly Ash. *KnE Mater. Sci.* **2017**, *2*, 89. [CrossRef]
206. Yoldi, M.; Fuentes-Ordoñez, E.G.; Korili, S.A.; Gil, A. Zeolite Synthesis from Industrial Wastes. *Microporous Mesoporous Mater.* **2019**, *287*, 183–191. [CrossRef]
207. Wang, M.; Yang, J.; Ma, H.; Shen, J.; Li, J.; Guo, F. Extraction of Aluminum Hydroxide from Coal Fly Ash by Pre-Desilication and Calcination Methods. *Adv. Mater. Res.* **2012**, *396–398*, 706–710. [CrossRef]
208. Yang, J.; Sun, H.; Peng, T.; Zeng, L.; Chao, L. Separation of Alumina from Aluminum-Rich Coal Fly Ash Using NaOH Molten Salt Calcination and Hydrochemical Process. *Clean Technol. Environ. Policy* **2022**, *24*, 1507–1519. [CrossRef]
209. Ghasemi, Z.; Sourinejad, I.; Kazemian, H.; Hadavifar, M.; Rohani, S.; Younesi, H. Kinetics and Thermodynamic Studies of Cr(VI) Adsorption Using Environmental Friendly Multifunctional Zeolites Synthesized from Coal Fly Ash under Mild Conditions. *Chem. Eng. Commun.* **2020**, *207*, 808–825. [CrossRef]
210. Hammed, A.K.; Dewayanto, N.; Du, D.; Ab Rahim, M.H.; Nordin, M.R. Novel Modified ZSM-5 as an Efficient Adsorbent for Methylene Blue Removal. *J. Environ. Chem. Eng.* **2016**, *4*, 2607–2616. [CrossRef]
211. Ji, Y.; Xu, F.; Wei, W.; Gao, H.; Zhang, K.; Zhang, G.; Xu, Y.; Zhang, P. Efficient and Fast Adsorption of Methylene Blue Dye onto a Nanosheet MFI Zeolite. *J. Solid State Chem.* **2021**, *295*, 121917. [CrossRef]
212. Yuan, W.; Yuan, P.; Liu, D.; Yu, W.; Laipan, M.; Deng, L.; Chen, F. In Situ Hydrothermal Synthesis of a Novel Hierarchically Porous TS-1/Modified-Diatomite Composite for Methylene Blue (MB) Removal by the Synergistic Effect of Adsorption and Photocatalysis. *J. Colloid Interface Sci.* **2016**, *462*, 191–199. [CrossRef]
213. Visa, M. Synthesis and Characterization of New Zeolite Materials Obtained from Fly Ash for Heavy Metals Removal in Advanced Wastewater Treatment. *Powder Technol.* **2016**, *294*, 338–347. [CrossRef]
214. Chen, D.; Hu, X.; Shi, L.; Cui, Q.; Wang, H.; Yao, H. Synthesis and Characterization of Zeolite X from Lithium Slag. *Appl. Clay Sci.* **2012**, *59–60*, 148–151. [CrossRef]
215. Belviso, C. State-of-the-Art Applications of Fly Ash from Coal and Biomass: A Focus on Zeolite Synthesis Processes and Issues. *Prog. Energy Combust. Sci.* **2018**, *65*, 109–135. [CrossRef]

216. Anuwattana, R.; Khummongkol, P. Conventional Hydrothermal Synthesis of Na-A Zeolite from Cupola Slag and Aluminum Sludge. *J. Hazard. Mater.* **2009**, *166*, 227–232. [CrossRef]
217. Kuwahara, Y.; Ohmichi, T.; Kamegawa, T.; Mori, K.; Yamashita, H. A Novel Synthetic Route to Hydroxyapatite-Zeolite Composite Material from Steel Slag: Investigation of Synthesis Mechanism and Evaluation of Physicochemical Properties. *J. Mater. Chem.* **2009**, *19*, 7263–7272. [CrossRef]
218. Kuwahara, Y.; Ohmichi, T.; Kamegawa, T.; Mori, K.; Yamashita, H. A Novel Conversion Process for Waste Slag: Synthesis of a Hydrotalcite-like Compound and Zeolite from Blast Furnace Slag and Evaluation of Adsorption Capacities. *J. Mater. Chem.* **2010**, *20*, 5052–5062. [CrossRef]
219. Ibrahim, A.H.; Lyu, X.; ElDeeb, A.B. Synthesized Zeolite Based on Egyptian Boiler Ash Residue and Kaolin for the Effective Removal of Heavy Metal Ions from Industrial Wastewater. *Nanomaterials* **2023**, *13*, 1091. [CrossRef] [PubMed]
220. Wang, P.; Cao, J.; Zhang, Y.; Sun, Q. Controllable Preparation of Cubic Zeolite A and Application of Langmuir Model in Carbon Dioxide Adsorption. *Nanomaterials* **2021**, *11*, 3375. [CrossRef] [PubMed]
221. Bai, S.X.; Zhou, L.M.; Chang, Z.B.; Zhang, C.; Chu, M. Synthesis of Na-X Zeolite from Longkou Oil Shale Ash by Alkaline Fusion Hydrothermal Method. *Carbon Resour. Convers.* **2018**, *1*, 245–250. [CrossRef]
222. Mamedova, G.A. Hydrothermal Synthesis of Natrolite-Type Zeolite in the Natural Halloysite-Obsidian System. *Glas. Phys. Chem.* **2014**, *40*, 380–383. [CrossRef]
223. Mamedova, G.A.; Kamseu, E.; Beleuk à Moungam, L.M.; Cannio, M.; Billong, N.; Chaysuwan, D.; Melo, U.C.; Leonelli, C. Synthesis of Zeolite with a Gmelinite Structure in the Dolomite-Halloysite-Obsidian System. *J. Clean. Prod.* **2016**, *42*, 518–521. [CrossRef]
224. Kamseu, E.; Beleuk à Moungam, L.M.; Cannio, M.; Billong, N.; Chaysuwan, D.; Melo, U.C.; Leonelli, C. Substitution of Sodium Silicate with Rice Husk Ash-NaOH Solution in Metakaolin Based Geopolymer Cement Concerning Reduction in Global Warming. *J. Clean. Prod.* **2017**, *142*, 3050–3060. [CrossRef]
225. Alam, M.M.; Hossain, M.A.; Hossain, M.D.; Johir, M.A.H.; Hossen, J.; Rahman, M.S.; Zhou, J.L.; Hasan, A.T.M.K.; Karmakar, A.K.; Ahmed, M.B. The Potentiality of Rice Husk-Derived Activated Carbon: From Synthesis to Application. *Processes* **2020**, *8*, 203. [CrossRef]
226. Taylor, J.H.; Elmes, V.K.; Hurt, A.P.; Coleman, N.J. Synthesis of Feldspathoids and Zeolite K–F from Waste Amber Container Glass. *Mater. Chem. Phys.* **2020**, *246*, 122805. [CrossRef]
227. Mun, S.P.; Ahn, B.J. Chemical Conversion of Paper Sludge Incineration Ash into Synthetic Zeolite. *J. Ind. Eng. Chem.* **2001**, *7*, 292–298.
228. Wajima, T.; Ikegami, Y. Zeolite Synthesis from Paper Sludge Ash via Acid Leaching. *Chem. Eng. Commun.* **2008**, *195*, 305–315. [CrossRef]
229. Tavasoli, M.; Kazemian, H.; Sadjadi, S.; Tamizifar, M. Synthesis and Characterization of Zeolite Nay Using Kaolin with Different Synthesis Methods. *Clays Clay Miner.* **2014**, *62*, 508–518. [CrossRef]
230. Amoni, B.d.C.; de Freitas, A.D.L.; Loiola, A.R.; Soares, J.B.; Soares, S.d.A. A Method for NaA Zeolite Synthesis from Coal Fly Ash and Its Application in Warm Mix Asphalt. *Road Mater. Pavement Des.* **2019**, *20*, S558–S567. [CrossRef]
231. Lin, G.; Zhuang, Q.; Cui, Q.; Wang, H.; Yao, H. Synthesis and Adsorption Property of Zeolite FAU/LTA from Lithium Slag with Utilization of Mother Liquid. *Chin. J. Chem. Eng.* **2015**, *23*, 1768–1773. [CrossRef]
232. Zhang, C.; Li, S. Utilization of Iron Ore Tailing for the Synthesis of Zeolite A by Hydrothermal Method. *J. Mater. Cycles Waste Manag.* **2018**, *20*, 1605–1614. [CrossRef]
233. Poudel, M.B.; Awasthi, G.P.; Kim, H.J. Novel Insight into the Adsorption of Cr(VI) and Pb(II) Ions by MOF Derived Co-Al Layered Double Hydroxide @hematite Nanorods on 3D Porous Carbon Nanofiber Network. *Chem. Eng. J.* **2021**, *417*, 129312. [CrossRef]
234. Kandel, D.R.; Kim, H.J.; Lim, J.M.; Poudel, M.B.; Cho, M.; Kim, H.W.; Oh, B.T.; Nah, C.; Lee, S.H.; Dahal, B.; et al. Cold Plasma-Assisted Regeneration of Biochar for Dye Adsorption. *Chemosphere* **2022**, *309*, 136638. [CrossRef]
235. WHO (World Health Organization). Guidelines for the Safe Use of Wastewater, Excreta and Greywater. In *Wastewater Use in Agriculture*; World Health Organization: Geneva, Switzerland, 2006; Volume 2.
236. Shawabkeh, R.; Al-Harahsheh, A.; Hami, M.; Khlaifat, A. Conversion of Oil Shale Ash into Zeolite for Cadmium and Lead Removal from Wastewater. *Fuel* **2004**, *83*, 981–985. [CrossRef]
237. Hamadi, A.; Nabih, K. Synthesis of Zeolites Materials Using Fly Ash and Oil Shale Ash and Their Applications in Removing Heavy Metals from Aqueous Solutions. *J. Chem.* **2018**, *2018*, 6207910. [CrossRef]
238. Qiu, R.; Cheng, F.; Huang, H. Removal of Cd^{2+} from Aqueous Solution Using Hydrothermally Modified Circulating Fluidized Bed Fly Ash Resulting from Coal Gangue Power Plant. *J. Clean. Prod.* **2018**, *172*, 1918–1927. [CrossRef]
239. Huang, X.; Zhao, H.; Hu, X.; Liu, F.; Wang, L.; Zhao, X.; Gao, P.; Ji, P. Optimization of Preparation Technology for Modified Coal Fly Ash and Its Adsorption Properties for Cd^{2+}. *J. Hazard. Mater.* **2020**, *392*, 122461. [CrossRef] [PubMed]
240. Jha, V.K.; Nagae, M.; Matsuda, M.; Miyake, M. Zeolite Formation from Coal Fly Ash and Heavy Metal Ion Removal Characteristics of Thus-Obtained Zeolite X in Multi-Metal Systems. *J. Environ. Manag.* **2009**, *90*, 2507–2514. [CrossRef] [PubMed]

241. Bao, W.; Zou, H.; Gan, S.; Xu, X.; Ji, G.; Zheng, K. Adsorption of Heavy Metal Ions from Aqueous Solutions by Zeolite Based on Oil Shale Ash: Kinetic and Equilibrium Studies. *Chem. Res. Chin. Univ.* **2013**, *29*, 126–131. [CrossRef]
242. Setthaya, N.; Pimraksa, K.; Damrongwiriyanupap, N.; Panias, D.; Mekrattanachai, P.; Chindawong, C. Modified Zeolite from Metakaolin and Fly Ash as Efficient Adsorbent for Cationic Methylene Blue Dye Removal. *Chem. Eng. Commun.* **2022**, *210*, 1178–1194. [CrossRef]
243. Javadian, H.; Ghorbani, F.; Tayebi, H.A.; Asl, S.H. Study of the Adsorption of Cd (II) from Aqueous Solution Using Zeolite-Based Geopolymer, Synthesized from Coal Fly Ash; Kinetic, Isotherm and Thermodynamic Studies. *Arab. J. Chem.* **2015**, *8*, 837–849. [CrossRef]
244. Fernandes Machado, N.R.C.; Miotto, D.M.M. Synthesis of Na-A and -X Zeolites from Oil Shale Ash. *Fuel* **2005**, *84*, 2289–2294. [CrossRef]
245. Yu, J.; Yang, Y.; Chen, W.; Xu, D.; Guo, H.; Li, K.; Liu, H. The Synthesis and Application of Zeolitic Material from Fly Ash by One-Pot Method at Low Temperature. *Green Energy Environ.* **2016**, *1*, 166–171. [CrossRef]
246. Chen, Y.; Xu, T.; Xie, C.; Han, H.; Zhao, F.; Zhang, J.; Song, H.; Wang, B. Pure Zeolite Na-P and Na-X Prepared from Coal Fly Ash under the Effect of Steric Hindrance. *J. Chem. Technol. Biotechnol.* **2016**, *91*, 2018–2025. [CrossRef]
247. Al-Zboon, K.; Al-Harahsheh, M.S.; Hani, F.B. Fly Ash-Based Geopolymer for Pb Removal from Aqueous Solution. *J. Hazard. Mater.* **2011**, *188*, 414–421. [CrossRef]
248. Visa, M.; Isac, L.; Duta, A. Fly Ash Adsorbents for Multi-Cation Wastewater Treatment. *Appl. Surf. Sci.* **2012**, *258*, 6345–6352. [CrossRef]
249. Shyam, R.; Puri, J.K.; Kaur, H.; Amutha, R.; Kapila, A. Single and Binary Adsorption of Heavy Metals on Fly Ash Samples from Aqueous Solution. *J. Mol. Liq.* **2013**, *178*, 31–36. [CrossRef]
250. Adamczuk, A.; Kołodyńska, D. Equilibrium, Thermodynamic and Kinetic Studies on Removal of Chromium, Copper, Zinc and Arsenic from Aqueous Solutions onto Fly Ash Coated by Chitosan. *Chem. Eng. J.* **2015**, *274*, 200–212. [CrossRef]
251. Mostafa Hosseini Asl, S.; Ghadi, A.; Sharifzadeh Baei, M.; Javadian, H.; Maghsudi, M.; Kazemian, H. Porous Catalysts Fabricated from Coal Fly Ash as Cost-Effective Alternatives for Industrial Applications: A Review. *Fuel* **2018**, *217*, 320–342. [CrossRef]
252. Asl, S.M.H.; Javadian, H.; Khavarpour, M.; Belviso, C.; Taghavi, M.; Maghsudi, M. Porous Adsorbents Derived from Coal Fly Ash as Cost-Effective and Environmentally-Friendly Sources of Aluminosilicate for Sequestration of Aqueous and Gaseous Pollutants: A Review. *J. Clean. Prod.* **2019**, *208*, 1131–1147. [CrossRef]
253. Wang, H.; Wang, X.; Xu, Z.; Zhang, M. Synthetic Zeolite from Coal Bottom Ash and Its Application in Cadmium and Nickel Removal from Acidic Wastewater. *Desalin. Water Treat.* **2016**, *57*, 26089–26100. [CrossRef]
254. Visa, M.; Duta, A. TiO_2/Fly Ash Novel Substrate for Simultaneous Removal of Heavy Metals and Surfactants. *Chem. Eng. J.* **2013**, *223*, 860–868. [CrossRef]
255. Itskos, G.; Koutsianos, A.; Koukouzas, N.; Vasilatos, C. Zeolite Development from Fly Ash and Utilization in Lignite Mine-Water Treatment. *Int. J. Miner. Process.* **2015**, *139*, 43–50. [CrossRef]
256. Hui, K.S.; Hui, K.N.; Lee, S.K. A Novel and Green Approach to Produce Nano- Porous Materials Zeolite A and MCM-41 from Coal Fly Ash and Their Applications in Environmental Protection. *Int. J. Chem. Biol. Eng.* **2009**, *3*, 165–175.
257. Asl, S.H.; Ahmadi, M.; Ghiasvand, M.; Tardast, A.; Katal, R. Artificial Neural Network (ANN) Approach for Modeling of Cr(VI) Adsorption from Aqueous Solution by Zeolite Prepared from Raw Fly Ash (ZFA). *J. Ind. Eng. Chem.* **2013**, *19*, 1044–1055. [CrossRef]
258. Asl, S.M.H.; Masomi, M.; Hosseini, M.; Javadian, H.; Ruiz, M.; Sastre, A.M. Synthesis of Hydrous Iron Oxide/Aluminum Hydroxide Composite Loaded on Coal Fly Ash as an Effective Mesoporous and Low-Cost Sorbent for Cr(VI) Sorption: Fuzzy Logic Modeling. *Process Saf. Environ. Prot.* **2017**, *107*, 153–167. [CrossRef]
259. Zhou, Q.; Yan, C.; Luo, W. Polypyrrole Coated Secondary Fly Ash-Iron Composites: Novel Floatable Magnetic Adsorbents for the Removal of Chromium (VI) from Wastewater. *Mater. Des.* **2016**, *92*, 701–709. [CrossRef]
260. Xiao, M.; Hu, X.; Gong, Y.; Gao, D.; Zhang, P.; Liu, Q.; Liu, Y.; Wang, M. Solid Transformation Synthesis of Zeolites from Fly Ash. *RSC Adv.* **2015**, *5*, 100743–100749. [CrossRef]
261. Wang, W.X.; Qiao, Y.; Li, T.; Liu, S.; Zhou, J.; Yao, H.; Yang, H.; Xu, M. Improved Removal of Cr(VI) from Aqueous Solution Using Zeolite Synthesized from Coal Fly Ash via Mechano-Chemical Treatment. *Asia-Pac. J. Chem. Eng.* **2017**, *12*, 259–267. [CrossRef]
262. Wen, Y.; Tang, Z.; Chen, Y.; Gu, Y. Adsorption of Cr(VI) from Aqueous Solutions Using Chitosan-Coated Fly Ash Composite as Biosorbent. *Chem. Eng. J.* **2011**, *175*, 110–116. [CrossRef]
263. Pizarro, J.; Castillo, X.; Jara, S.; Ortiz, C.; Navarro, P.; Cid, H.; Rioseco, H.; Barros, D.; Belzile, N. Adsorption of Cu^{2+} on Coal Fly Ash Modified with Functionalized Mesoporous Silica. *Fuel* **2015**, *156*, 96–102. [CrossRef]
264. Duan, P.; Yan, C.; Zhou, W.; Ren, D. Development of Fly Ash and Iron Ore Tailing Based Porous Geopolymer for Removal of Cu(II) from Wastewater. *Ceram. Int.* **2016**, *42*, 13507–13518. [CrossRef]
265. Al-Harahsheh, M.S.; Al Zboon, K.; Al-Makhadmeh, L.; Hararah, M.; Mahasneh, M. Fly Ash Based Geopolymer for Heavy Metal Removal: A Case Study on Copper Removal. *J. Environ. Chem. Eng.* **2015**, *3*, 1669–1677. [CrossRef]

266. Wu, X.W.; Ma, H.W.; Zhang, L.T.; Wang, F.J. Adsorption Properties and Mechanism of Mesoporous Adsorbents Prepared with Fly Ash for Removal of Cu(II) in Aqueous Solution. *Appl. Surf. Sci.* **2012**, *261*, 902–907. [CrossRef]
267. Joshi, M.K.; Pant, H.R.; Liao, N.; Kim, J.H.; Kim, H.J.; Park, C.H.; Kim, C.S. In-Situ Deposition of Silver-Iron Oxide Nanoparticles on the Surface of Fly Ash for Water Purification. *J. Colloid Interface Sci.* **2015**, *453*, 159–168. [CrossRef]
268. Cardoso, A.M.; Paprocki, A.; Ferret, L.S.; Azevedo, C.M.; Pires, M. Synthesis of Zeolite Na-P1 under Mild Conditions Using Brazilian Coal Fly Ash and Its Application in Wastewater Treatment. *Fuel* **2015**, *139*, 59–67. [CrossRef]
269. Attari, M.; Bukhari, S.S.; Kazemian, H.; Rohani, S. A Low-Cost Adsorbent from Coal Fly Ash for Mercury Removal from Industrial Wastewater. *J. Environ. Chem. Eng.* **2017**, *5*, 391–399. [CrossRef]
270. Tauanov, Z.; Shah, D.; Itskos, G.; Inglezakis, V. Optimized Production of Coal Fly Ash Derived Synthetic Zeolites for Mercury Removal from Wastewater. *IOP Conf. Ser. Mater. Sci. Eng.* **2017**, *230*, 012044. [CrossRef]
271. Liu, M.; Hou, L.A.; Xi, B.; Zhao, Y.; Xia, X. Synthesis, Characterization, and Mercury Adsorption Properties of Hybrid Mesoporous Aluminosilicate Sieve Prepared with Fly Ash. *Appl. Surf. Sci.* **2013**, *273*, 706–716. [CrossRef] [PubMed]
272. Belviso, C.; Cavalcante, F.; Di Gennaro, S.; Lettino, A.; Palma, A.; Ragone, P.; Fiore, S. Removal of Mn from Aqueous Solution Using Fly Ash and Its Hydrothermal Synthetic Zeolite. *J. Environ. Manag.* **2014**, *137*, 16–22. [CrossRef]
273. He, K.; Chen, Y.; Tang, Z.; Hu, Y. Removal of Heavy Metal Ions from Aqueous Solution by Zeolite Synthesized from Fly Ash. *Environ. Sci. Pollut. Res.* **2016**, *23*, 2778–2788. [CrossRef]
274. Koukouzas, N.; Vasilatos, C.; Itskos, G.; Mitsis, I.; Moutsatsou, A. Removal of Heavy Metals from Wastewater Using CFB-Coal Fly Ash Zeolitic Materials. *J. Hazard. Mater.* **2010**, *173*, 581–588. [CrossRef]
275. Đolić Maja, K.M.; Đorđe, V.; Veličković, Z.R.-O.V.; Vladimir, P.; Marinković, A. The Removal of Zn^{2+}, Pb^{2+}, and As(V) Ions by Lime Activated Fly Ash and Valorization of the Exhausted Adsorbent. *Waste Manag.* **2018**, *78*, 366–378. [CrossRef]
276. Visa, M.; Andronic, L.; Enesca, A. Behavior of the New Composites Obtained from Fly Ash and Titanium Dioxide in Removing of the Pollutants from Wastewater. *Appl. Surf. Sci.* **2016**, *388*, 359–369. [CrossRef]
277. Cardoso, A.M.; Horn, M.B.; Ferret, L.S.; Azevedo, C.M.; Pires, M. Integrated Synthesis of Zeolites 4A and Na-P1 Using Coal Fly Ash for Application in the Formulation of Detergents and Swine Wastewater Treatment. *J. Hazard. Mater.* **2015**, *287*, 69–77. [CrossRef]
278. He, H.; Xu, S.; Han, R.; Wang, Q. Nutrient Sequestration from Wastewater by Using Zeolite Na-P1 Synthesized from Coal Fly Ash. *Environ. Technol.* **2017**, *38*, 1022–1029. [CrossRef]
279. Wang, Z.; Fan, Y.; Li, Y.; Qu, F.; Wu, D.; Kong, H. Synthesis of Zeolite/Hydrous Lanthanum Oxide Composite from Coal Fly Ash for Efficient Phosphate Removal from Lake Water. *Microporous Mesoporous Mater.* **2016**, *222*, 226–234. [CrossRef]
280. Chen, X.Y.; Wendell, K.; Zhu, J.; Li, J.L.; Yu, X.; Zhang, Z. Synthesis of Nano-Zeolite from Coal Fly Ash and Its Potential for Nutrient Sequestration from Anaerobically Digested Swine Wastewater. *Bioresour. Technol.* **2012**, *110*, 79–85. [CrossRef] [PubMed]

recycling

MDPI

Review

Bioleaching of Gold from Printed Circuit Boards: Potential Sustainability of Thiosulphate

Zahra Ilkhani and Farid Aiouache *

School of Engineering, Lancaster University, Lancaster LA1 4YR, UK; z.ilkhani@lancaster.ac.uk
* Correspondence: f.aiouache@lancaster.ac.uk; Tel.: +44-1524593526

Abstract: The rapid consumption and disposal of electronic waste due to technological innovations and changes in living commodities are causing the development of a significant environmental challenge. Among the components of these wastes, spent printed circuit boards are particularly considered to be among the most valuable owing to their content of precious metals, such as gold first and potentially platinum, which may be available in a lower proportion. Effective methods as part of gold recovery strategies by industries and policymakers are developed and envisioned from economic and environmental perspectives. Currently, cyanidation dominates global gold production from e-waste due to its selectivity for gold. The high toxicity of cyanide, however, poses serious environmental issues, leading thiosulphate leaching to emerge as a non-toxic and promising alternative for gold extraction. Its industrial viability has been demonstrated by Barrick Gold Corporation at the Goldstrike site with the pretreatment of acidic or alkaline pressure oxidation. This review introduces bioleaching as a promising economic and environmentally friendly process for gold extraction. This review explores thiosulphate leaching of gold as an alternative to conventional cyanidation, with a particular focus on biothiosulphate production by adapted microorganisms. The factors that affect the pretreatment, chemical reaction mechanism, and design engineering are discussed. The consumption of thiosulphate was identified as one of the main challenges, restricting the reliability of the process. Various solutions for the reduction of its consumption and relevant process costs were discussed, with a particular examination from the engineering aspect of the process design and scalability to industrially relevant operating conditions by using bioreactors adapted to large pulp density loads of electrical waste.

Keywords: bioleaching; waste printed circuit boards; gold; thiosulphate; metal recycling

1. Introduction

Nowadays, the increasing reliance of the modern lifestyle on electronic devices with relatively short replacement cycles has led to a substantial rise in the accumulation of electrical waste (e-waste) [1]. According to statistics by [2], 40 million tons of e-waste are produced annually, which makes up around 5% of global solid waste. In 2019, Asia produced 46.6% of the world's total e-waste, with America ranking the second highest with 24.4% e-waste generation followed by Europe (22.4%), Africa (5.4%), and Oceania (1.3%) [3]. Ref. [4] reported the production of 52.2 Mt global e-waste in 2021, which is expected to exceed 74.7 Mt by 2030 and 120 Mt by 2050. Despite this rise, only a proportion of 10–15% of generated e-waste is currently recycled, and the rest is stored or deposited in landfills, often with reduced strategies for sustainable recovery of the metals they contain, leading to unprecedented contamination of the environment and risks of diseases such as cancer, kidney and heart malfunctions, and brain swelling [5]. As a result, landfilling is considered the worst option as an e-waste destination, and there is an urgent need to recover metals from e-waste in order to mitigate environmental contamination, promote the economic advantages as a secondary source of precious metals, and save the consumption of natural ores [6].

Among e-waste, printed circuit boards (PCBs) contribute to approximately 3% of the total e-waste weight, while we note that mobile phones and computer PCBs have been widely studied owing to their high disposal volumes and precious metal contents [7]. They are utilised as random-access memory motherboards and network interface cards to facilitate electrical and mechanical connections, and their compositions can vary depending on the technology, year of fabrication, and manufacturer brand; however, they are typically composed of a metallic portion (40%) and non-metallic components such as plastics (30%) and ceramics (30%) [8]. They consist of a lamination layer of copper-clad fiberglass reinforced with epoxy resin material and other essential components such as capacitors, resistors, microchips, and diodes. The hazardous components of PCBs include heavy metals (Pb, Hg, Cd, As, and Cr), polychlorinated biphenyls, polybrominated biphenyls, and epoxy resins, which pose environmental challenges. Generally, PCBs contain about 10–20% Cu, 7% Fe, 5% Al, 3% Sn, 1–3% Ni, 1.5% Pb, 25% organic compounds, and precious metals such as 200–3000 ppm Ag, 20–250 ppm Au, and 10–200 ppm Pt, which are used because of their favourable electrical conductivity, chemical stability, and resistance to oxidation, corrosion, and acids [9]. The worth of Au and Ag in discarded PCBs in 2022 was estimated at USD 54,514.97/kg–USD 65,714.28/kg and USD 592.22/kg–USD 866.13/kg, respectively. Although their total weight is less than 1% of PCBs, they are still precious and financially important [10].

Every year, around 17 million PCBs are discarded, which is equivalent to the disposal of nearly 0.5 million tons of waste PCBs. In computer PCBs, the concentrations of Cu and Au reach 20–40 times and 25–250 times, respectively, more than natural ores, have the potential to be recovered from e-waste with less energy consumption, and are seen as alternative sources to natural ores [11]. As an example, 210 kg of Cu and 1.5 kg of Au can be recovered from one metric ton of waste PCBs, while only around 5 g and 5.25 kg of Au and Cu can be recovered from relevant natural ores, respectively [3]. Thus, waste PCBs have drawn interest from industries and academia as a secondary metal resource to create efficient, sustainable, and economical metal recovery technologies [12]. Au is one of the first metals used by human civilisation, even before 3400 BC, and has captivated attention for its wide-ranging applications (i.e., jewellery, high-tech industries, chemical processes, and medical applications) [13,14]. According to the U.S. Geological Survey, China leads the world in Au production (420 tons), followed closely by Australia and Russia [15]. As a result, the growing demand for Au in different sectors and the depletion of natural resources make it essential to recover Au from alternative sources such as PCBs [14].

Pyrometallurgy and hydrometallurgy technologies are conventional methods that are still employed but face growing challenges due to issues linked to economic and environmental outlooks, including the emission of harmful gases, costs, energy requirements, and chemical contamination. Therefore, it is important to develop energy-efficient technologies, such as those that rely on bioprocessing technology, operating at moderate conditions in terms of chemical and energy consumption [16–18]. Bioleaching, as one of the bioprocess technologies, relies on the ability of microorganisms to produce essential leaching agents such as organic acids (i.e., oxalic acid, citric acid, malic acid, gluconic acid, succinic acid, and formic acid); amino acids (i.e., L-valine, glycine, DL-alanine, and L-histidine); biosurfactants such as rhamnolipids (Rhls) produced from the bacterium *P. aeruginosa CVCM* to remove 11% Fe and 25% Zn at low concentration (0.4 mg/mL) and 19%

Fe and 52% Zn at high concentration (1 mg/mL) [19]; siderophores such as hydroxamate and catecholate mixed-type pyoverdine PyoPpC-3B produced by the bacterium *P. putida PpF1* to selectively extract Zn and Mn from primary and secondary mineral residues [20]; chelating; and complexing agents, and it is becoming a promising technology for recycling e-waste because it is straightforward, requires less skilled labour, reduced capital and operating costs, and can be controlled under mild conditions. This process also reduces the costs of final disposal and residue treatment, creating an environmentally friendly waste stream while ensuring selective metal extraction [21]. Currently, about 15–25% of the world's copper production, 5% of Au production, and smaller proportions of Co, Ni, U, and Zn are produced by the bacteria-assisted leaching method [22]. For over a century, cyanide has been used as a cost-effective and efficient agent for gold leaching. However, due to its high toxicity, thiosulphate has been proposed as a safer alternative. Past review papers have explored the principles of bioleaching, including the methods and mechanisms, types of e-waste and microorganisms, and their role in the extraction of heavy and precious metals [23,24]. This review presents an update on bioleaching methods for the extraction of Au from waste PCBs by revisiting the fundamental concepts and discussions involved in the most recent findings regarding the design methodologies and mechanisms, as well as industrial perspectives and applications, contributing as a starting point to future directions. Thiosulphate-based leaching and associated challenges, the importance of engineering design and scale-up in the bioleaching process, and its impact on cost reductions are discussed.

2. Preprocessing of PCBs for Bioleaching

Typically, preprocessing is the initial step before recovering metals from waste PCBs. Dismantling various electronic components from PCBs is a crucial step in the recycling process. During dismantling, priority is given to the removal of reusable or hazardous components such as batteries and cathode ray tubes in order to make the recovery of metals easier [10]. Selective removal of hazardous components from PCBs helps prevent toxic elements from entering the recycling process. Manual dismantling, oven heating, and open burning are some common techniques for dismantling. However, some of these techniques are hazardous and might change the properties of waste PCBs and pose risks. For example, in the oven-heating technique, certain toxic substances may be generated as a result of the high melting point of lead-free solders (270–280 °C). The development of semi-automatic and automatic machineries to be used instead of manual dismantling methods is increasing the efficiency of the process and reducing the negative environmental impacts compared to oven heating, but the associated expenses remain high owing to the intricate geometries of PCBs of non-uniform structures, which highlights the application of manual dismantling. Ref. [25] estimated the costs for manual, mechanical, and heating dismantling techniques, as illustrated in Table 1. Manual dismantling proved to be the most economical option when the quantity of waste PCBs was below 1000 tons, but as the quantity was increased, mechanical dismantling and heating became more advantageous due to the high labour costs of manual dismantling. When the amount of waste PCBs exceeds 5000 tons, heating dismantling is considered the most cost-effective method [10,24]. A crushing stage is necessary to make the further handling of PCB waste easier. Size reduction follows and is performed by cutting PCBs into small pieces (1–2 cm^2) by means of shredders or granulators. Further reduction of PCBs (5–10 mm) is achieved using ball mills, centrifugal mills, cutting mills, and ring mills [26]. The next step, which follows preprocessing, involves the recycling of waste materials in an economical and environmentally sustainable manner [24]. The production of fine dust during crushing and grinding poses a significant difficulty, and it can be challenging to control this process [27]. After removal of the hazardous components, different mineral processing unit operations, such as shredding, crushing, and grinding, can be used to liberate metals from cladding materials such as resin, fiberglass, and plastics. Various types of hammer crushers, rotary crushers, disc crushers, shredders, and cutters equipped with a bottom sieve are used for liberation. As PCBs are made of reinforced resin, copper wires, and glass fibres (multilayer), conventional crushers may not achieve good liberation. In contrast, shredding or cutting, which work on the principle of shearing, are found to be more useful. Unlike mineral ores, PCBs do not have a particular size fraction for liberation; instead, different types of elements are liberated at different size fractions. Figure 1 outlines the preprocessing steps involved in preparing PCBs for bioleaching.

Table 1. Cost of dismantling methods [25].

Method	Cost (USD) (Below 1 kt)	Cost (USD) (5 kt)	Cost (USD) (10 kt)
Manual	~500–1000	~350,000	~400,000
Mechanical	~50,000	~110,000	~350,000
Heating	~60,000	~90,000	~150,000

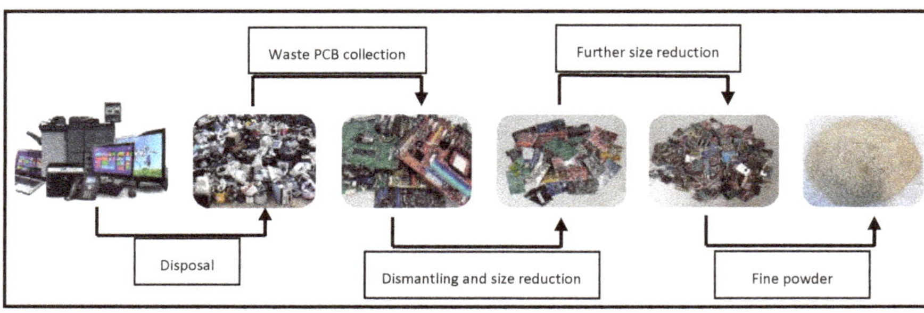

Figure 1. Preprocessing of waste PCBs before bioleaching.

3. Bioleaching

3.1. Bacteria Used for Bioleaching and Mechanism

The primary difference in how autotrophic and heterotrophic microorganisms leach metals lies in their energy sources and metabolic activities. Autotrophic bacteria, commonly known as sulphur-oxidising and iron-oxidising bacteria, which are capable of growing in acidic conditions, are the most prevalent [28]. During the bioleaching process, these microorganisms facilitate the oxidation of ferrous ions to ferric ions and elemental sulphur to sulphuric acid (as shown in Equations (1) and (2)). Biogenic ferric iron and sulphuric acid act as oxidising agents (lixiviants) for the solubilisation of metals from e-wastes (Equations (3) and (4)) [29]. *A. ferrooxidans* and *A. thiooxidans* are the most used bacteria because they are resistant to contamination by other microorganisms, tolerate high concentrations of heavy metals, have a low nutrient requirement, and can grow under extremely acidic pH conditions [5]. These bacteria are able to capture electrons from substrates for their metabolic activities to release heat and solubilise metals without relying on external energy sources to facilitate the process [30]. Heterotrophic microorganisms, including fungi and bacteria, are extensively used in bioleaching due to their ability to metabolise organic compounds as energy sources and their adaptability to high pH levels and complex metal contents. Fungal species, such as *Penicillium*, *Aspergillus*, *Trichoderma*, *Saccharomyces*, and *Phanerochaete*, produce organic acids (i.e., citric acid, gluconic acid, oxalic acid, and formic acid) as leaching agents, which facilitate the mobilisation of metals from e-waste (Equation (5)). Additionally, these organic acids bind with metals to form stable complexes that significantly enhance metal solubility in the leaching solution [31,32]. There are three main mechanisms that occur between microorganisms and e-waste during bioleaching, namely, acidolysis (acid formation), redoxolysis (microbial-driven oxidation and reduction reactions), and complexolysis (biogenic complexing agents). In acidolysis, the oxygen atoms present in a metal oxide form interact with water molecules to facilitate the solubilisation of metals into a leaching solution. During acidolysis, a number of sulphur-oxidising autotrophs, such as *A. thiooxidans*, *A. caldus*, and *S. thermosulfidooxidans*, as well as heterotrophic bacteria species like *Bacillus*, *Pseudomonas*, and *Chromobacterium* and fungi including *Aspergillus* and *Penicillium*, consume nutrients to produce a range of acids, such as sulphuric, gluconic, acetic, malonic, oxalic, lactic, pyruvic, succinic, and formic acids, which facilitate metal solubilisation by maintaining low pH and reducing anion availability. Acidolysis is a fast and effective method commonly used to extract metals like Zn, Ni, and Cu from waste [18,33]. In the redoxolysis mechanism, the metals are dissolved via oxidation–reduction reactions. Through electron transfer, redoxolysis facilitates the supply of energy required for microbial growth [2]. Iron plays a major role as an electron carrier. After the microbial oxidation of Fe^{2+}, Fe^{3+} acts as an oxidising agent that is able to solubilise metals such as Cu into Cu^{2+} and then can be chemically reduced to Fe^{2+} through redox reactions. Subsequently, Fe^{2+} is reoxidised to Fe^{3+} by the metabolic activity of the microorganisms [34]. Fe^{3+} ion is recognised to be an affordable oxidising agent in hydrometallurgical processes and has been proven to be a promising choice for the extraction of various metals from PCBs. Commercial-scale bioleaching commonly uses a combination of redoxolysis by biogenic Fe^{3+} and acidolysis by biogenic sulphuric acid for metal recovery of e-waste. In complexolysis, the target metals interact with ligands like cyanide, organic acids, or siderophores to form stable metal–ligand complexes. Complexolysis is used to leach metals such as Au, Ag, Fe, Al, and Pt by heterotrophic bacteria and fungi like *C. violaceum*, *P. fluorescens*, *B. megaterium*, *P. aeruginosa*, *A. niger*, etc. Common examples of these metal–ligand complexes include citric acid with Mg, oxalic acid with Al and Fe, and cyanide with Au, Ag, and Pt [35].

$$S^0 + \frac{3}{2}O_2 + H_2O \rightarrow H_2SO_4 \qquad \Delta G^\circ = -507\ \mathrm{kJ/mol} \tag{1}$$

$$2Fe^{2+} + O_2 + 2H^+ \rightarrow 2Fe^{3+} + H_2O \qquad \Delta G^\circ = -88.8\ \mathrm{kJ/mol} \tag{2}$$

$$2Fe^{3+} + M \rightarrow M^{2+} + 2Fe^{2+} \tag{3}$$

$$H_2SO_2 + MS \rightarrow H_2O + MSO_4 \quad [M = Cu,\ Al,\ Mn, \ldots] \tag{4}$$

$$M + H^+ + O_2 + H_2O \rightarrow M^{n^+} \tag{5}$$

3.2. Bioleaching Methods

The bioleaching process is carried out through direct and indirect methods. The direct method is classified into one-step and two-step methods, while the indirect method involves the spent medium [36]. The one-step method requires the addition of samples of e-waste and microorganisms directly into a sterile culture medium at the same time, which allows the bacteria to grow in the presence of waste. The intricate composition of e-waste and the presence of toxic metals, however, may hinder microbial activity during the leaching process, decreasing the bioleaching efficiency. A study, for instance, showed that in the one-step bioleaching method, the growth of Frankia spp. decreased as the concentration of waste PCBs increased due to the lack of secondary metabolites and the decreased toxic effects of e-waste [37]. In the two-step method, on the other hand, the microorganisms grow first without any addition of e-waste. In fact, microorganisms are cultivated in a culture medium under adapted environmental conditions suitable for their activity until they reach the logarithmic growth stage [33]. E-waste samples are then introduced into the microbial culture medium. In contrast to the one-step method, the two-step method can notably mitigate the inhibition of microbial growth caused by e-waste and enhance the bioleaching efficiency. Since the microorganisms are still present in the two-step method, the toxicity of e-waste may partially affect the microbial growth and constrain the flexibility of the process compared to the spent-medium method [34]. In the indirect leaching method, the microorganisms are cultivated until they reach the growth phase and produce their metabolites. After the metabolite production, the bacteria are filtered by centrifugation and removed from the culture medium to obtain a cell-free medium, while the e-waste samples are added to the spent medium. This method eliminates the harmful effects of e-waste on microorganisms and provides better control throughout operation conditions [35]. Unlike the one-step and two-step methods, the spent-medium method enables an independent optimisation of the biological and chemical processes, such as shortening the leaching process and increasing the temperature and rotation speed free of the shear limitation on bacteria, which enhances the mass transfer rate and metal recovery [36]. For example, in a study by [32], fungal leaching of metals from a spent lithium-ion phone mobile battery by *A. niger* was conducted under one-step, two-step, and spent-medium methods, and the maximum leaching efficiency of Cu (100%), Li (95%), Mn (70%), Al (65%), Co (45%), and Ni (38%) was obtained at a pulp density of 1% in the spent-medium method. Another study reported that increasing the pulp density led to a decline in the bioleaching efficiency of Au to 11.3% by the bacterium *C. violaceum* at 0.5% pulp density when using the two-step method compared to 18% by the spent-medium method [38].

3.3. *Effective Factors During the Bioleaching Process*

Many studies have shown that factors such as pH, temperature, pulp density, microorganism type, nutrient, and aeration affect bioleaching efficiency. Maintaining an optimal pH in the culture medium is crucial for enhancing microbial activity and metal solubilisation during bioleaching. The pH influences both the effectiveness of metal leaching and the stability of metal ions in solutions. Acidophilic bacteria perform best at a pH level 2.0–2.5. *A. ferrooxidans* can tolerate pH levels below 2.0, but higher pH levels (above 2.5) reduce the bioleaching efficiency due to the precipitation of the ferric iron as jarosite and the bacterial attachment to these compounds, which hinders their activity. Temperature plays a key role in bioleaching, as different microorganisms thrive at specific temperature ranges. Mesophilic acidophiles typically grow at a temperature in the range of 25–30 °C, while thermophiles are grown at 40–45 °C. Fungi generally grow at a temperature within a range of 25–35 °C, with *A. niger* showing optimal performance at 25 °C [18,35]. The microorganisms involved in the bioleaching process are typically aerobic, meaning they require oxygen for their metabolism and growth. In the laboratory, aeration is provided using a shaker incubator, with an agitation speed typically maintained between 120 and 145 rpm for bacterial cultures. An excessive agitation, however, can cause physical stress on the bacteria and negatively affect their activity [23,39]. The effectiveness of microbial leaching is significantly influenced by the composition of the culture medium, as it directly affects the metabolic activities of microorganisms. This medium includes both organic and inorganic nutrients that support microbial growth. For the cultivated cyanogenic bacteria used in leaching valuable metals from e-waste, a nutrient-rich medium containing organic components like peptone, yeast extract, glycine, and amino acids is typically employed. Acidophilic bacteria, however, are grown in a chemically defined medium that supplies essential nutrients such as ammonium sulphate, dipotassium hydrogen phosphate, magnesium sulphate, iron sulphate, and elemental sulphur [40]. The pulp density, which is defined as the weight ratio of a solid material (i.e., e-waste) to a liquid in a solution (i.e., leaching agent), is a key factor affecting the effectiveness of metal leaching. High pulp densities increase toxicity, reduce oxygen transfer, and inhibit microbial growth, leading to a reduced leaching efficiency. Although some microorganisms, like acidophilic bacteria, tolerate heavy metals, their activity decreases at high pulp densities due to the limited presence of oxygen and increased medium viscosity. Several studies identified 1% (*w*/*v*) as an optimal pulp density for e-wastes. For example, ref. [41] found that the consortium of *C. violaceum*, *P. aeruginosa*, and *P. fluorescens* were able to leach 69.3% Au from waste PCBs at a pulp density of 1%, but the leaching rate decreased to 20.28% when the pulp density was increased to 10%. Pulp density is also one of the important parameters that affects reactor design, culture media consumption, and operating cost. When the pulp density is, for instance, increased from 1% to 10%, the volume of the liquid phase and the size of the reactor decrease notably by 90%, leading to a substantial decline in the operation cost [2]. According to [42], there was a significant drop in operating costs when the pulp density was increased from 1% to 5% with pH adjustment, which led to an annual profit of AUD 2749 per ton of PCBs. In spent-medium bioleaching, the optimisation of operating parameters such as temperature, agitation speed, and pulp density are critical. Unlike the one-step and two-step methods, the absence of live microbial culture allows for broader flexibility in the parameter selection. In this context, maximising metal leaching efficiency becomes the primary focus independent of conditions required to sustain microbial viability. For example, high temperatures or agitation speeds may be used to accelerate reaction kinetics without concern for microbial tolerance, enabling greater adaptability and a potential for scale-up to industrial applications.

3.4. Progress in Culture Growth

The toxic metals present in e-waste can negatively impact microbial growth, metabolism, and survival during the leaching process. To maintain a sufficient and active microbial population in the solution, it is often necessary to use microorganisms that are resistant or adapted to these toxic conditions. Therefore, before starting bioleaching, the microorganisms should undergo a gradual adaptation process. This involves serial sub-culturing over time with a gradually increasing pulp density of the waste material, allowing microbes to increase their tolerance to the levels of metal toxicity [17,43]. For instance, Ref. [30] studied the bioleaching of base metals from PCBs using adapted *A. ferrooxidans*. The adaptation process involved stepwise increases in waste concentration from 1 to 15 g/L over 187 days in a flask, followed by further adaptation in a bubble column bioreactor with waste concentrations raised to 40 g/L over 44 days. Under optimal conditions with a solid content of 20 g/L, this approach resulted in 54% of Cu, 75% of Ni, and 55% of Fe recovered after 9 days. Although microorganisms have been widely used across various scientific and industrial fields, most research on bioleaching has focused on pure microbial cultures, which are often lacking in terms of adaptability, efficiency, and the ability to manage complex substrates. Mixed cultures, which consist of two or more microorganisms in a consortium of culture media, offer a more effective approach for bioleaching compared to pure cultures. Species like *A. thiooxidans*, *A. ferrooxidans*, *A. caldus*, *L. ferrooxidans*, and *L. ferriphilum* have been combined for the extraction of metals in bioleaching processes to form biofilms that generate microenvironments, enhancing leaching kinetics. Although the precise mechanisms that illustrate the interaction within these communities remain weakly known, the growing prevalence of mixed cultures in bioleaching is partly due to the presence of diverse metal ions in ores and soils, which cannot be effectively processed by pure cultures. The applications of mixed cultures of microorganisms in bioleaching demonstrated that these cultures can bioleach the concentrate more rapidly and extensively than pure cultures [44,45]. For example, in one study, Cu leaching in a mixed culture of *A. ferrooxidans* and *A. thiooxidans* was 10% higher than in pure cultures [46].

The use of genetically modified organisms (GMOs) has emerged as a promising approach to improving the efficiency of the bioleaching process. Through genetic engineering, microorganisms can be designed to tolerate and adapt to high metal concentrations, enhancing the metabolic activity under optimal conditions and significantly reducing the time needed for metal extraction. Recent advancements in genetic engineering, along with progress in DNA synthesis, sequencing, and integrated -omics technologies (such as genomics, proteomics, transcriptomics, and metabolomics), are providing new opportunities to develop high-performing engineered microorganisms suitable for enhanced leaching performance and resilience in harsh environments [40]. For example, genetic engineering has been successfully used to enhance arsenic bioleaching by *A. ferrooxidans* TFBk [47]. In another study, Tay et al. observed improved Au recovery using two genetically modified strains of *C. violaceum*, named pBAD and pTAC, with a leaching of 30% and 25% of Au, respectively, compared to only an 11% leaching rate by the non-modified microorganisms [48].

4. Precious Metal Leaching

4.1. Cyanide-Based Leaching

The recovery of precious metals like Au and Ag from PCBs has been given much attention, as they are utilised in different industries such as electronic industries owing to their stability and conductivity [49]. Ref. [50] reported a global demand of Au in electronic industries of 254 tons in 2015, which highlights the importance of the recovery of Au through sustainable methods. Cyanidation was introduced by John Stewart MacArthur in the 1880s, and since then, it has been the primary method for Au leaching due to its efficiency for the selective recovery of Au [51]. Cyanide refers to inorganic compounds containing a cyano group (C-N), which are highly toxic, particularly at temperatures above 25.6 °C. The total reaction of precious metal dissolution by cyanide is represented in Equation (6). Although cyanide has been the main leaching agent for Au extraction, it has some drawbacks, such as absence of possible solutions for cyanide regeneration as well as environmental concerns owing to cyanide leakage into groundwater and the overall management of hazards associated with its toxic nature [52]. According to statistics by [53], cyanide leakage from metallurgical plants caused some serious accidents. For instance, in 1995 in Guyana, a tailings dam at the Omai mine collapsed and discharged approximately 2.9 million m^3 of cyanide-laced tailings, which contaminated the nearby Omai River. Another accident happened in 2000 in Baia Mare, Romania, when the collapse of the Aurul S.A. tailings dam released a swage containing up to 100 tons of cyanide, which finally flowed into the Danube River. Gold leaching with a biocyanide solution has emerged as a cost-effective, simple, and efficient alternative to traditional cyanidation methods. Recent studies have shown that bacteria such as *C. violaceum*, Pseudomonas species, *E. coli*, and *B. megaterium* can produce biological cyanide. But still, the high consumption of cyanide by Cu poses challenges. To address this issue, certain autotrophic bacteria have been employed to target base metals and remove them selectively in the first step in order to facilitate the leaching of precious metals [35]. Several studies have reported high Au extraction rates after Cu removal. For instance, in the study by [54], 98.4% of Cu was extracted by *A. ferrivorans* and *A. thiooxidans* under acidic condition (pH 1–1.6) and room temperature after 7 days. In the second step, *P. putida* was employed, and 44% Au was solubilised under alkaline condition (pH 7.3–8.6) after 2 days. In another study, PCBs were pretreated with biooxidation using *A. ferrooxidans*, and over 80% of Cu removal enhanced the Au/Cu ratio in the residual PCBs. Before the pretreatment, the Au leaching rate was only 20.8%, which doubled to 40.1% after the treatment [55]. Table 2 summarises some previous studies on the bioleaching of precious metals from e-waste by different microorganisms. Cyanide-producing microorganisms are the main source for the selective leaching of Au and Ag from waste PCBs.

$$4M + 8CN^- + O_2 + H_2O \rightarrow 4M(CN)_2^- + 4OH^- \quad [M : Ag,\ Au] \tag{6}$$

Table 2. Overview of the previous studies on bioleaching of precious metals under different conditions.

Bacteria	Waste	Leaching (%)	Leaching Condition	Leaching Agent (mg/L)	Reference
P. balearica SAE1	PCBs	Au: 68.5 Ag: 33.8	Pulp density: 1.5%, 30 °C, 7 d	Glycine: 5000	[56]
P. Chlororaphis	PCBs	Au: 8.2 Ag: 12.1	Pulp density: 1.6%, 25 °C, 3 d	Cyanide: 8.71	[57]
C. violaceum	PCBs	Au: 11.3	Pulp density: 1.5%, 30 °C, 8d	Cyanide: 68	[58]
C. violaceum	Landfill of e-waste	Au: 16	Pulp density: 0.5%,30 °C, 8d	Cyanide: 5–15	[59]
P. biofilm	PCBs	Ag: 14.7	Pulp density: 2%, 25 °C, 7 d	Cyanide: 5	[60]
F. casuarinae	PCBs	Au: 75	Pulp density: 0.2%, 28 °C, 30 d	Biomass: 3620	[37]

4.2. Thiosulphate-Based Leaching

Thiosulphate ($S_2O_3^{2-}$), a sulphur oxyanion with a tetrahedral structure, is considered among the most promising alternatives to cyanide, as it has numerous benefits compared to various cyanide and non-cyanide leaching agents, including low corrosivity, stability, rapid leaching kinetics, operability in a safe workplace, and lower risk of environmental pollution [61]. Because of its efficiency, thiosulphate is recognised as a highly effective reagent for extracting Au from waste PCBs [51]. The first report on the use of thiosulphate in precious metal recovery dates back to 1905, when Au was extracted from ores by employing ammonia–thiosulphate leaching [62]. The overall reaction for the thiosulphate leaching of Au is represented by Equation (7). After the formation of the Au(I)–thiosulphate complex, a slightly acidic to highly alkaline condition is needed to prevent the decomposition of the complex and ensure its stability in solution [14]. Thiosulphate-based leaching requires an oxidant to facilitate Au solubilisation. Cu^{2+} is one of the prevalent oxidants, which accelerates Au dissolution by 17–20 times. The high redox potential of Cu^{2+}/Cu^+, however, results in high thiosulphate consumption, which is often moderated by the presence of ammonia to form a cupric ammonia complex ($Cu(NH_3)_4^{2+}$) as a stable catalyst to weaken the interaction between Cu^{2+} and thiosulphate as well as to reduce copper hydroxide precipitation [63–65]. Additionally, ammonia has a non-negligible role in reducing the formation of a passivation layer on the Au surface and enhancing the leaching kinetics [66].

$$4Au + 8S_2O_3^{2-} + O_2 + 2H_2O \rightarrow 4Au(S_2O_3)_2^{3-} + 4OH^- \qquad \Delta G^\circ = -97.9\ \text{kJ/mol} \tag{7}$$

4.2.1. Thiosulphate Leaching Methods

Ammonia-Based Method

There are two main methods of thiosulphate leaching: non-ammonia- and ammonia-based methods. The ammonia-based methods consist of copper ammonia, nickel ammonia, and cobalt ammonia leaching. Ammonia accelerates the dissolution of Au, stabilises thiosulphate, and maintains the pH level [64]. In an ammonia-based thiosulphate system, Cu^{2+} acts as a redox mediator to facilitate Au oxidation through both anodic and cathodic reactions [15]. In the anodic area (Equations (8)–(10)), NH_3 moves toward the Au surface and forms a complex with Au+ ($Au(NH_3)_2^+$). NH_3 is then replaced by $S_2O_3^{2-}$, and a more stable complex is formed ($Au(S_2O_3)_2^{3-}$). Although Au can form different complexes with ammonia and thiosulphate, $Au(S_2O_3)_2^{3-}$ is the most stable complex in an ammonia-based system. In the cathodic area, $Cu(NH_3)_4^{2+}$ is reduced to $Cu(S_2O_3)_3^{5-}$ and rapidly oxidised back to $Cu(NH_3)_4^{2+}$ by dissolved oxygen (Equations (11) and (12)). The values of associated Gibbs free energy offer an overview of the level of spontaneous reactions occurring with reference to the chemical equilibrium. While most values support the forward path, the spontaneous solubilisation of the metal gold, for instance (Equation (8)), is not favoured. In one study, around 98% of Au was solubilised from 1% (*w*/*v*) PCBs by using ammonium thiosulphate as the leaching agent [55]. In another study, 1 M ammonium thiosulphate was utilised to extract around 91% Au from 1% (*w*/*v*) PCBs after 24 h [67]. The authors of [68] compared thiosulphate and thiourea for Au dissolution and reported that approximately 70% of Au was dissolved from PCBs with different concentrations of ammonium thiosulphate (0.08 to 0.12 M) at 20 °C and a pH of 10.5, while only 40% dissolved Au was achieved with thiourea. Table 3 summarises some key information on the ammonia-based thiosulphate leaching of Au reported in previous studies, noting the boundary conditions of typical operations (i.e., thiosulphate 0.1–1 M, ammonia 0.1–4 M, and Cu^{2+} < 0.1 M, with leaching rates ranging from 15% to 99%). Despite the substantial number of lab-scale experiments and advances in understanding the chemical mechanisms and fundamental principles, the development toward an ammonia-based method has been slow, partly due to the high volatility of ammonia and the environmental issues related to the leakage of ammonia during storage or transportation [62]. Table 4 summarises the advantages and disadvantages of employing ammonia-based systems in thiosulphate leaching.

$$Au \rightarrow Au^+ + e^- \quad \Delta G^\circ = 163.2\text{kJ/mol} \tag{8}$$

$$Au^+ + 2NH_3 \rightarrow Au(NH_3)_2^+ \qquad \Delta G^\circ = -74.1 \text{ kJ/mol} \tag{9}$$

$$Au(NH_3)_2^+ + 2S_2O_3^{2-} \rightarrow 2NH_3 + Au(S_2O_3)_2^{3-} \qquad \Delta G^\circ = -74.9 \text{ kJ/mol} \tag{10}$$

$$Cu(NH_3)_4^{2+} + 3S_2O_3^{2-} + e^- \rightarrow Cu(S_2O_3)_3^{5-} + 4NH_3 \quad \Delta G^\circ = -21.9 \text{ kJ/mol} \tag{11}$$

$$2Cu(S_2O_3)_3^{5-} + 16NH_3 + O_2 + 2H_2O \rightarrow 4Cu(NH_3)_4^{2+} + 4OH^- + 12S_2O_3^{2-} \quad \Delta G^\circ = -67.1 \text{ kJ/mol} \tag{12}$$

Table 3. Ammonia-based thiosulphate leaching of Au under typical operating conditions.

Agent (M)	Condition	Au (%)	Waste	References
$S_2O_3^{2-}$ 0.072, NH_3 0.266, Cu^{2+} 0.01	20–25 °C, 400 rpm, 5 min	~50%	Printed circuit boards	[63]
$S_2O_3^{2-}$ 0.2, NH_3 0.2, Cu^{2+} 0.0015	Pulp 30%, 30 °C, aeration (0.2 L/min), 6 h	89%	Pressure oxidised sulphide gold concentrate	[69]
$S_2O_3^{2-}$ 0.2, NH_3 0.4, Cu^{2+} 0.01–0.02	30 °C, 300 rpm, 20%, 24 h, pH 11.5	30%	Gold ore	[70]
$S_2O_3^{2-}$ 0.08–0.12, NH_3 0.1–0.2, Cu^{2+} 0.015 M	20 °C, pH 10.5, 2 h	Up to 70%	Waste PCBs	[68]
$S_2O_3^{2-}$ 0.3, Cu^{2+} 0.05	pH 10, pulp 10%, 15 h	94%	Gold ore	[71]
$S_2O_3^{2-}$ 1, NH_3 1, Cu^{2+} 0.01	Pulp 10%, 24 h	99%	Waste PCBs	[72]

Table 3. *Cont.*

Agent (M)	Condition	Au (%)	Waste	References
$S_2O_3^{2-}$ 0.2, NH_3 1, Co^{2+} 0.03	50 °C, pH 10	70%	Gold ore	[13]
$S_2O_3^{2-}$ 0.3, NH_3 1, Cu^{2+} 0.03	25 °C, pH 10	72.7%	Gold ore	[13]
$S_2O_3^{2-}$ 0.5, NH_3 1, Cu^{2+} 0.04	70 °C, pH 6, 1 h	99%	Silver-bearing ore	[73]
$S_2O_3^{2-}$ 0.8, NH_3 4, Cu^{2+} 0.05	25 °C, pH 10.2, 48 h	90%	Gold-copper sulphide concentrate	[74]
$S_2O_3^{2-}$ 0.1, NH_3 0.2, Cu^{2+} 0.015–0.03	25 °C, pH 9–11	15%	PCB	[75]

Table 4. The advantages and disadvantages of using ammonia in thiosulphate leaching systems [62,66].

Advantages	Disadvantages
Prevents passivation film on Au surface and accelerates dissolution rate	Volatile reagent
Hinders the dissolution of undesirable minerals including silicates, carbonates, and iron oxides	Easy escape from vessel and environmental pollution
Forms stable complexes with Cu and reduces the reactivity of Cu with thiosulphate	Toxic to humans and aquatic animals
	Difficult transport and store

Non-Ammonia Method

Many studies have employed thiosulphate leaching by eliminating ammonia. Alternative non-ammonia methods in thiosulphate leaching include copper thiosulphate, oxygen thiosulphate, ferric ethylenediaminetetraacetic acid (EDTA) thiosulphate, copper EDTA thiosulphate, and ferric oxalate thiosulphate to mitigate the issues associated with ammonia. Ref. [66] employed a copper–citrate–thiosulphate method for Au leaching from a refractory carbonaceous gold concentrate, and the findings demonstrated that substituting citrate with ammonia had a similar extraction capability but reduced the thiosulphate consumption from 0.045 M to 0.025 M. Ref. [76] observed that copper ion–ethanediamine–thiosulphate leaching was more effective than copper ion–ammonia–thiosulphate leaching of Au from gold ore. The use of cetyltrimethyl ammonium bromide (CTAB) was found to increase the extraction rate to 94.3% and to decrease thiosulphate consumption to 1.12 kg/t in a leaching system containing 0.1 M sodium thiosulphate, 0.06 M ethanediamine, 0.005 M Cu^{2+}, and 1.5 kg/t of CTAB. In this process, CTAB broke down into $[CTA^+]$, attracted the negatively charged particles of $[Au(S_2O_3)_2{}^{3-}]{\cdot}nH_2O$ and $[CTA^+][AuBr2^-]{\cdot}nH_2O$, and formed ion pairs that facilitated the stabilisation of $[Au(S2O3)2^{3-}]$. Overall, ethylenediamine can form a more stable complex with Cu^{2+} than ammonia. The cupric–ethylenediamine complex is considered an interesting agent in reducing the catalytic oxidation ability of Cu^{2+} over the cupric/cuprous redox equilibrium potential, which in turn lowers the consumption of thiosulphate. In some of the literature, the effect of external factors on the leaching rate was assessed. For example, in one experiment, ultrasound was applied in a cobalt ammonia thiosulphate leaching system (0.2 M $S_2O_3^{2-}$, 0.03 M Co^{2+}, 1 M NH_3, 50 °C, and 750 W ultrasonic power). The results showed that the leaching rate was 8 times faster and the extraction yield with ultrasound reached 89%, which was found to be 25% higher than the one without ultrasound. Temperature, ultrasonic power, and the reduction in activation energy from 22.65 kJ/mol to 13.86 kJ/mol were the main reasons for the high leaching rate [13].

4.2.2. Challenges with the Thiosulphate-Based Leaching

The primary challenge associated with the use of thiosulphate is the high cost of reagent consumption and its impacts on the generated sulphate, tetrathionate, and trithionate [63]. Addressing this challenge requires the implementation of strategies, including the control of key factors of operations such as temperature, flow rate and mixing speed, oxygen level, Eh, pH, and the concentration of Cu^{2+} and ammonia. Depending on pH and Eh, unstable compounds such as tetrathionate, sulphite, and trithionate may form; therefore, controlling the Eh and pH is essential to minimise thiosulphate loss [77]. Using inorganic additives like phosphate, sulphate, and chloride and organic additives such as EDTA, humic acid (HA), polyamine, CDTA, ethylenediamine (EDA), triethanolamine (TEA), ammonium alcohol polyvinyl phosphate, citric acid, carboxymethyl cellulose (CMC), and amino acids can reduce thiosulphate consumption as they have the potential to compete with thiosulphate anions and form complexes with Cu^{2+} that decrease the interaction between Cu^{2+} and thiosulphate [78]. For example, in a copper–ammonia–thiosulphate leaching system, 1 L of leach solution was applied to 400 g of sulphide ore, with an initial gold concentration of 4.3 mg/kg. The addition of a low concentration of EDTA (2.0 mM) enhanced the efficiency of Au leaching to 100% and decreased the consumption of ammonium thiosulphate from 9.63 kg/t to 3.85 kg/t after 24 h [79]. Ref. [65] employed TEA as an additive for Au leaching in a thiosulphate–copper–ammonia system. The results demonstrated that the dissolution rate increased by 50% while thiosulphate consumption was reduced by around 10%. In a study conducted by [80], ethydiaminedhephen acetic (EDDHA) was employed as a non-toxic, environmentally friendly, and cost-effective organic additive to form a stable complex with Cu^{2+} owing to the numerous coordination sites of carboxylic acid and amine functional groups in it, which led to a stable Cu-EDDHA complex ion and the dissolution of 82.84% Au, which was 56.02% higher than in cyanide leaching. Moreover, EDDHA reduced the thiosulphate consumption from 103.19 kg/t to 10.54 kg/t. Ref. [81] investigated the effect of different additives on Au oxidation and revealed that potassium ethyl xanthate (KEX), imidazole, sulphinic acid, sodium di-ethyl dithiocarbamate, pyridine, thioglycolic acid, and mercaptobenzothiazole completely passivated the Au surface, while a small amount (5 mM) of thiourea and thioacetamide improved the oxidation of Au. Another study found that Au dissolution in a thiosulphate–oxygen–copper medium was only 2% over 24 h and increased to 95% in the presence of activated carbon as an additive [77]. In some experiments, Cu^{2+} was substituted by Co as an oxidising agent. According to [82], in ammoniacal thiosulphate solutions, Co(III)-NH_3 complexes were able to extract around 80% of Au, while they reduced thiosulphate consumption by 44.2% thiosulphate. Table 5 summarises the effects of different additives on Au leaching and the consumption of thiosulphate. As seen, adding additives increased the dissolution of Au by approximately 20% and decreased the thiosulphate consumption by 30%.

Table 5. Effects of additives on Au dissolution and thiosulphate consumption.

Additive (mM)	Au Dissolution (%)	Thiosulphate Consumption (%)	Reference
Humic acid (80)	Before: 72.6 After: 81.4	Before: 42.4 After: 13.2	[83]
EDTA (2.0)	Before: ~80 After: 100	Before: 9.63 After: 3.85	[7]

Table 5. Cont.

Additive (mM)	Au Dissolution (%)	Thiosulphate Consumption (%)	Reference
TEA (12)	Increased by 50	Decreased by 10	[65]
EDDHA (450)	Before: 56.02 After: 82.84	Before: 10.31 After: 1.05	[80]
CTAB (60)	Before: 50 After: 94.3	Before: 1.3 After: 0.11	[76]
EDA (10)	Before: 60.3 After: 80.3	Before: 1.13 After: 0.414	[84]

4.2.3. Biothiosulphate Produced by Microorganisms

Over the past two decades, efforts have been made to minimise the breakdown of thiosulphate during thiosulphate leaching of Au by using different types of additives and controlling the parameters affecting the process. It was discovered that producing thiosulphate by means of microorganisms (biothiosulphate) offers a potential pathway toward the development of a feasible, innovative, and sustainable metal recovery technology [51,85]. The benefits of using biogenic thiosulphate for Au leaching include, for instance, fewer toxic substances, less energy requirements associated with intermediate chemical productions, and scalability combined with e-waste recycling operations. Currently, an industrial thiosulphate leaching process is employed at Nevada Gold Mines that involves thermal treatment of sulphur and calcium hydroxide at 90 °C followed by oxidation of substances under pressure at 550 kPa. Therefore, developing a biogenic process in moderate conditions (e.g., ambient temperature and atmospheric pressure) can be recognised as an innovative alternative to reduce CAPEX and OPEX costs [86]. The two most common species that produce thiosulphate are cyanobacteria and proteobacteria. Table 6 summarises the ability of various microorganisms to produce thiosulphate under different environmental and substrate conditions [87]. The bacterium *Microcoleus chtonoplastes* produces thiosulphate (695 mg/L) using light, sulphide, and hydrogen and consumes carbonate as a carbon source at 22 °C and a pH around 8 [88]. *Methylophaga sulfidovorans* consumes methanol or dimethyl sulphide (DMS) as substrates to produce thiosulphate, as shown in Equation (13), and grows at a pH above 7.5 and a temperature of 22–30 °C. In the thiosulphate production by the bacterium *Thiobacillus thioparus*, sulphide, thiocyanate, elemental sulphur (S0), DMS, and polythionates can act as electron donors. Around 18.5% of thiosulphate production by this bacterium results from sulphide oxidation, which occurs under oxygen-limiting conditions and pH 7 [89]. *Streptomyces fradiae* produced up to 485 mg/L of $Na_2S_2O_3$ as a main byproduct in a mineral medium at pH 7.5, which remained stable without degradation during 20 days of incubation [90]. Ref. [91] discovered that the addition of 1 mM sulphide to the medium containing recombinant E. coli resulted in the production of 0.45 mM sulphite, 0.2 mM thiosulphate, and 0.1 mM sulphur. *Sulphurimonas sp.* was observed to consume sulphide to produce 1 mM thiosulphate after intermediate production of elemental sulphur [89]. In a mixed culture with a neutral pH and temperature range of 20–32 °C, *T. oxidans* and *T. neapolitanus* produced sulphate and thiosulphate through sulphur oxidation. Sulphate was produced as a main product under atmospheric conditions and at pH 8.0; however, during oxygen-limited conditions, especially when the O_2/S_2 consumed ratio was around 0.5, thiosulphate formation (35 mg/L) occurred as a result of the increased sulphide–oxygen ratio [92].

$$(CH_3)_2S + 4O_2 \rightarrow 2CO_2 + H^+ + 0.5S_2O_3^{2-} + 2.5H_2O \quad \Delta G^\circ = -1606.5\ \text{kJ/mol} \tag{13}$$

Table 6. Illustration of thiosulphate production by microorganisms [87].

Microorganism	Substrate	Thiosulphate Production Conditions	Produced Thiosulphate (mg/L)
Microcoleus chtonoplastes	Sulphide	pH 8, temperature 22 °C, anoxic photooxidation	695
Methylophaga sulfidovorans	Ethanethiol, hydrogen sulphide	pH 7.5, temperature 22–30 °C	672
Streptomyces fradiae	L-cystine	pH 7.5, temperature 22 °C, 20 days	485
Pseudomonas sp. *C27*	Sulphide, acetate	pH 7–8, temperature 30 °C, coupled with nitrogen fixation process	451
Calothrix sp. *HI 41*	Sulphide, hydrogen sulphide	Temperature 25 °C, anoxic photooxidation, 10 days	392
Oscillatoria sp. *BO 32*	Sulphide	Temperature 25 °C, anoxic photooxidation	146
Sulfurimonas sp. *CVO*	Sulphur, sulphide	Temperature 30 °C	112
Anabaena cylindrica	Sulphide	Temperature 25 °C	73
E. coli	Sulphide	Temperature 37 °C	45
T. neapolitanus	Sulphide	pH 5–9, temperature 30 °C, limited oxygen	35
D. desulfuricans	Sulphide	pH 1–6, temperature 85 °C	10

Biogenic thiosulphate as an intermediate metabolite is extremely unstable under acidic conditions but can be stabilised through the adjustment of pH and presence of inhibitors [51]. For example, biogenic thiosulphate (500 mg/L) obtained from *A. thiooxidans* was able to leach 65% Au from 0.5% *w*/*v* of PCBs in the presence of 3.25 mg/L sodium azide (NaN_3) as an inhibitor and 1 M ammonia at pH 6–7 after 36 h [93]. Ref. [94] investigated the thiosulphate leaching of Au from ore by the bacterium *Methylophaga sulfidovorans* using sodium sulphide as the substrate. The bacterium was able to produce thiosulphate in a rotating sealed flask of high-salinity medium at pH 7–8 and a temperature of 30 °C. A maximum of 61.9% of Au was leached in 24 h under conditions of 10% pulp density, air flow rate of 0.1 L/min, and 50 °C. The biothiosulphate approach was found to be a viable substitute for chemical thiosulphate, opening the door to further investigation of other thiosulphate-producing microorganisms from inexpensive substrates. This is considered a promising area in bioprocess engineering for metal recovery in an efficient and environmentally friendly way.

5. Engineering Perspective and Scaling up

The phase of scaling up the operation of the laboratory scale has been an important challenge in bioprocessing to achieve representative microbiological and technological enhancements and bring it as close as possible to industrial applications [22]. Currently, industrial-scale metal extraction is performed using methods such as dump bioleaching, heap bioleaching, and tank bioleaching. Among these, dump leaching is the oldest technique used in the industry, primarily for extraction of Cu, Ni, Zn, Au, and U from ores. In this method, a lixiviant generated by bacteria is sprayed onto ores piled in a dump, and the leachate solution is collected in ditches at the bottom of the dump, recovered, and then directed to an oxidation basin, where bacteria are regenerated for reuse in the process. This technique generally has low metal recovery efficiency and requires several days to complete the process. Moreover, there is a risk of leachate seeping into nearby water sources, with a potential risk of water pollution. Appropriate safety measures must therefore be undertaken to control the potential runoff [95]. Heap bioleaching is another method suitable for low-grade ores that are not economically viable for leaching through grinding and smelting and is particularly effective, for instance, for treating complex sulphide ores that have a porous structure [96,97]. Heap bioleaching involves stacking ores on an impermeable surface and applying leaching agents such as acidified solutions for Cu or cyanide solutions for Au over the top of the heap. The leaching occurs through microbial activity and mineral oxidation within the heap, and metal-rich solution is collected at the base and directed to a pond to recover metals [98]. Despite its technical maturity and widespread industrial application, it has drawbacks, such as yielding low leaching rates within a long processing time. Extending the leaching time increases the demand for manpower and material resources, which restricts further development of this technology [96]. Table 7 provides a summary of commercial-scale bioleaching operations for the recovery of Au and Cu from refractory gold concentrates, copper ores, and sulphidic ores across various global mining sites. The table shows details about the mine location, operator, year of operation commencement, production amount, and type of bioleaching technology employed.

Table 7. Commercial-scale bioleaching operations for refractory gold concentrate, copper ores, or sulphide ores [99,100].

Mine	Location	Operator	Year Started	Production	Type
Obuasi	Ghana	Anglo Gold Ashanti	1994	Au: 200,000 ounces	Stirred-tank
Laizhou (BioGold)	China	Sino Gold Mining	2001	Au: 75,000 ounces	Stirred-tank
Olimpiada	Russia	Polyus Gold	2001	Au: 965,000 ounces	Stirred-tank
Suzdal	Kazakhstan	Nordgold	2005	Au: 90,000 ounces	Stirred-tank
Kokpatas	Uzbekistan	Navoi Mining and Metallurgical Combinat	2009	Au: 432,000 ounces	Stirred-tank
Fosterville	Australia	Kirkland Lake Gold	2005	Au: 150,000 ounces	Stirred-tank
Iranian Babak Copper Company	Iran	Iranian Babak Copper Company	2020	Cu: 50,000 t/y	Heap
Zaldivar	Chile	Barrick Gold Corporation	1995	Cu: 150,000 t/y	Heap
Whim Creek and Mons Cupri	Australia	Straits Resources	2006	Cu: 17,000 t/y	Heap
Jinchuan Copper	China	Zijin Mining Group Ltd.	2006	Cu: 10,000 t/y	Heap

To enhance ore leaching efficiency, the process is initially performed in laboratory flasks prior to expanding to larger bioreactors. These bioreactors play a key role in process scale-up and are commonly operated in either batch or continuous modes. The development of an adapted bioreactor is the first step to move from an experimental scale to industrially relevant operating conditions. Bubble column reactors (BCRs) and continuous stirred-tank reactors (CSTRs) have been the widely used bioreactors in the bioleaching process. One of the advantages of these bioreactors compared to others is the ability to control the concentration of product and substrate during the process and to maintain the environmental operating conditions [22,30]. Ref. [101] investigated the extraction of heavy metals from a petroleum spent catalyst using *A. thiooxidans* in a bubble column reactor, and results showed 87% % Mo, 37% Ni, and 15% Al extraction with a pulp density of 0.9%, particle size of 60.7 μm, and aeration rate of 209 mL/min after 7 days. In another study, ref. [30] employed adapted *A. ferrooxidans* in a bubble column and leached approximately 54% Cu, 75% Ni, and 55% Fe with 2% waste PCBs in 9 days. Figure 2 illustrates a schematic diagram of two types of reactors employed in the bioleaching process. As seen in Figure 2a, the stirring in a BCR is conducted by air bubbling, and the aeration happens via air injection through a sparger, operating at high airflow. A water cooling thermostat is fitted to maintain the temperature. Figure 2b shows continuous bioleaching in a 50 L STR. The mixture of sediment slurry and deionised water is pumped to a reactor filled with sulphur-oxidising microorganisms and supplied with aerator [102]. Figure 2c represents a combination of BCR and STR in the bioleaching of metals by an acidophilic consortium. A continuous cultivation experiment of microorganisms carried out in a BCR and bioleaching in batch and continuous modes was performed in an STR [6,103]. CSTRs are more advantageous in comparison with BCRs because of the presence of high mass transfer rates, homogeneous mixing, and better control over parameters such as pH, temperature, and aeration [22]. A typical design of CSTRs involves a circuit of continuously flowing aerated tanks set up in series, parallel, or a combination of both, equipped with agitated impellers that keep the grounded minerals suspended in the solution and ensure an effective transfer of oxygen, which is necessary for both the dissolution of the metals and the growth of aerobic microorganisms. In CSTRs, the feed is added to the first tank and overflows from tank to tank in co-current flow with the microorganisms. When microorganisms are injected at the start-up of an operation, a batch culture is maintained until the microorganisms reach a certain stage close to the middle of the logarithmic phase, at which point fresh substrate is supplied continuously from the feed tank to the reactor. The continuous flow of the substrate through the tanks is provided to ensure the optimal growth of microorganisms for a high metal dissolution rate at the steady-state phase [104]. Since its introduction in 1986, CSTR-related technologies have undergone significant advancement, such as high-temperature performance in bioleaching. During the first decade of this century, the Beaconsfield plant in Australia and the Laizhou plant in China, utilising Mintek BacTech technology, have conducted bioleaching operations at moderate to high temperature, ranging between 45–55 °C. The BioCop™ process, pioneered by the BHP Billiton biotechnology group based in Johannesburg in South Africa, employs thermophilic bacteria thriving at a temperature between 70–80 °C to leach chalcopyrite [105].

In a study conducted by [95], bioleaching with thermophilic bacteria, *L. ferriphilum* and *A. caldus*, was employed at a high pulp density of 8% PCBs using a 3 L stirred-tank reactor, resulting in a recovery of 85.23% Zn, 76.59% Cu, and 70.16% Al in 7 days. Ref. [106] employed a 12 L aerated rotating-drum reactor operated at 50 °C for the bioleaching of 76% Cu with a pulp density of 2.5% by the thermophilic bacterium *S. thermosulfidooxidans* and Fe^{2+} concentration of 5 g/l in 8 days. In another study, ref. [107] achieved a 99% leaching rate of Cu within 72 h using a 3 L CSTR containing 5% of low-grade PCBs and mixed acidophilic culture supplemented with 7.8 g/L of Fe^{2+}. However, large-scale bioleaching has not been sufficiently investigated owing to factors such as reaction conditions, the complexity of the mechanism, and weakly efficient mass transfer rates [108]. More comprehensive investigation is crucial, as the findings are found to be not linearly scalable when the size of equipment is increased. While controlling conditions for a complete reaction at the laboratory scale is straightforward and flexible, maintaining microbial populations along with the aeration rate is challenging at the industrial scale, which affects the steady-state operations over long times of streams [24].

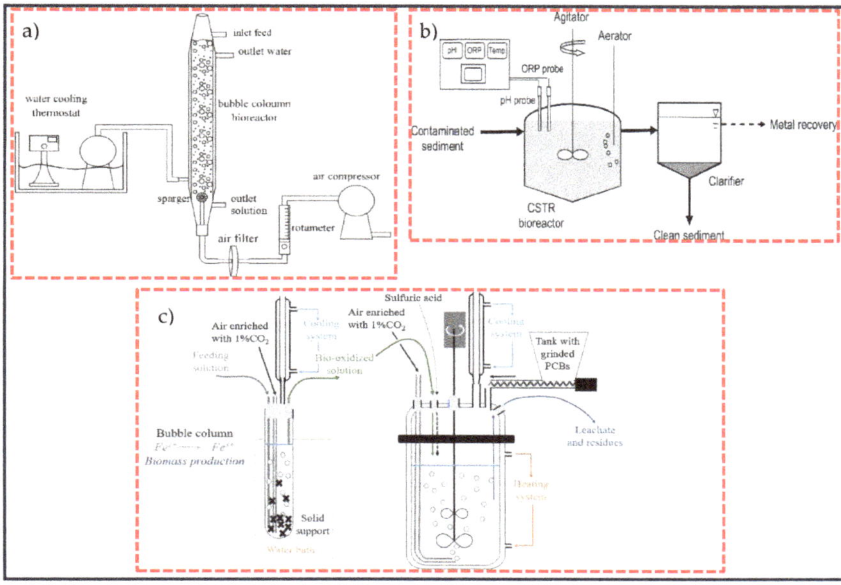

Figure 2. A schematic diagram of two types of reactors in bioleaching process (**a**) bubble column reactor (BCR), (**b**) continuous stirred tank reactor (CSTR) and (**c**) a combination of bubble column and stirred tank reactors. Copyright © 2020, Elsevier (**a**), copyright © 2021, Elsevier (**b**), copyright © 2022, Elsevier (**c**) [6,30,102].

In recent years, several projects in Europe, such as BIORECOVER, RAWMINA, RUBICON, and BIOCriticalMetals, have employed bioleaching for the recovery of various metals from e-wastes. Among the industrial contributors, the UK-based IT lifecycle services provider N2S is applying bioleaching for the extraction of precious metals from printed circuit boards and aligning with the UK's environmental and cybersecurity regulations [109]. In Estonia, BiotaTec (formerly BiotaP), which initially focused on environmental monitoring using metagenomics, has since expanded their activities to the development of microorganism-driven bioleaching and biomining solutions for the recovery of metals from low-grade ores and various waste streams. Another notable company is Ekolive, a Slovak start-up which offers an EU/ETV-certified eco-innovative bioleaching method (InnoBioTech®) for processing minerals and waste using heterotrophic bacteria. Their large-scale pilot project, established in Slovenia in 2019, demonstrated the market maturity of this approach for industrial mineral recovery from mining waste [110].

6. Process Economics

6.1. Cost Estimation

Estimating both capital expenditure (CAPEX) and operational expenditure (OPEX) is essential to assess the economic feasibility of the bioleaching process and give insight into the opportunities to improve, optimise, and conduct further research if needed. CAPEX includes the cost of equipment, installation, construction, piping, electrical systems, service facilities, and land, and OPEX covers ongoing costs like electricity, water, materials, labour, and maintenance [111]. Estimating capital costs helps identify the key factors that affect the economic viability of the process. For instance, while lower pulp densities often yield higher metal leaching rates, they are associated with higher capital costs. In contrast, higher pulp densities may reduce leaching efficiency but result in lower CAPEX, making the process more economically viable. In one study, leaching of base metals from PCBs at 1% pulp density led to high operating costs and an annual loss of AUD 1007 per tonne of PCBs. However, increasing the pulp density to 5% significantly lowered the costs, resulting in an annual profit of AUD 2749 per tonne. Although the raw material costs can change over time, the process would be profitable. Even with a 40% increase in raw material costs, the study showed that an annual profit of AUD 2463 per tonne of PCBs was still achievable [42]. Labour costs for the preprocessing of e-waste, such as collecting, dismantling, and sorting, are the highest operating expenses, making it difficult to expand e-waste recycling in regions such as Australia, New Zealand, and other parts of Oceania. A techno-economic study assessed the use of biooxidation and cyanidation to extract gold from refractory ore. Over a 25-year mine life, processing 1200 tonnes of ore daily, the project required an AUD 220 million investment and had operating costs of AUD 58.27 million annually. With an expected annual revenue of AUD 78.48 million and a net present value (NPV) of AUD 34.4 million at a 7% interest rate, the process was found to be economically viable [112]. Another study compared two bioleaching technologies, including an aerated bioreactor and an aerated and stirred bioreactor, for Cu recovery from goethite using *A. ferrooxidans*. The aerated and stirred bioreactor proved to be financially viable at a pulp density of 10%, achieving an NPV of AUD 1.275 billion and an internal rate of return (IRR) of 65% over 20 years. With a CAPEX of AUD 119.8 million and annual OPEX of AUD 5.9 million, the plant is expected to become profitable within one year [39].

6.2. Cost Minimisation Strategies

In the bioleaching process, the initial investment required for setting up a reactor plant may seem to be high, but the ongoing operational expenses can be reduced through the optimisation of process parameters and the utilisation of cost-effective substances. For example, the use of commercial iron, sulphur, and specific additives contributes to the overall cost [28] but can be reduced by using low-grade pyrite as an alternative source [113]. In the case of heterotrophic bacteria, glucose is the main nutrient carbon source for bacterial growth and accounts for approximately 44% of the total initial costs. Using alternative affordable substrates such as molasses is an option to reduce the cost of the process [24]. Ref. [114] reported that using potato wastewater reduced the cost by 17% compared to glucose, and replacing corn stover with potato wastewater reduced the costs by 22% in comparison with potato wastewater. The formation of iron hydroxides and jarosite precipitations during bioleaching, which leads to a decrease in diffusion rates of oxidants to the surface of PCBs and results in non-economic losses in metal yields, can be avoided by maintaining low pH and using effective sulphur agents that oxidise sulphur to sulphuric acid. The costs of the preprocessing of e-waste (dismantling, separation, and grinding) and operation (labour, raw materials, and utilities) can render the bioleaching technique economically impractical without careful management of the design and process automation. Cost reduction in the grinding step can be achieved by maintaining a particle size that does not significantly impact both the mass transfer rate and the bioleaching yield [28,108]. For example, a study reported that using PCB chips instead of powdered PCBs resulted in high metal leaching efficiency and significantly lowered operational costs, as it avoided the energy-intensive milling process and reduced the risk of metal loss during preprocessing [115].

7. Conclusions

Metal recycling from printed circuit boards (PCBs), as the major component of e-waste, is a beneficial option to protect the environment, conserve natural resources, and meet the need for metals, particularly the valuable ones that are critical for current industries. The bioleaching process as a green technology for metal extraction is typically conducted by direct (one-step and two-step) and indirect (spent-medium) methods, and relevant chemical mechanisms fundamentally include acidolysis, complexolysis, and redoxolysis. The latter is the primary mechanism associated with iron-oxidising bacteria. For over a century, Au extraction has relied on the cyanidation process, but environmental concerns limit its expansion, and the thiosulphate-based process is offered as a safe alternative for Au recovery. High thiosulphate consumption remains, however, one of the main challenges associated with this method, and various strategies such as the use of additives, control of operation conditions, and employment of biothiosulphate are options being considered. Upscaling bioleaching to an industrially feasible process in cost-effective operations remains a challenge, but the selection of a suitable microorganism and adapted design of bioreactor are contributing as closely as possible to the success of the process's feasibility from economic and environmental perspectives. These studies indicate that the biological extraction of metals from PCBs presents a promising viable alternative to physico-chemical methods and that an improvement in handling large pulp density of PCBs within bioreactors is necessary for the practical achievement of a cost-effective and competitively upscaled process.

Author Contributions: Conceptualisation, Z.I. and F.A.; methodology, Z.I. and F.A.; formal analysis, Z.I. and F.A., investigation, Z.I.; resources, F.A.; data curation, Z.I.; writing—original draft preparation, Z.I.; writing—review and editing, Z.I. and F.A.; supervision, F.A.; project administration, F.A.; funding acquisition, F.A. All authors have read and agreed to the published version of the manuscript.

Funding: This research was funded by Innovate UK, KTP012187, and ICT Reverse Asset Management Ltd., A109952 for the financial supports of this work.

Acknowledgments: The authors acknowledge the support of Innovate UK and ICT Reverse Ltd. for financial supports of this work.

Conflicts of Interest: The authors declare no conflicts of interest.

References

1. Vardanyan, A.; Vardanyan, N.; Abrahamyan, N.; Aatach, M.; Gaydardzhiev, S. Sequential biologically assisted extraction of Cu and Zn from printed circuit boards (PCB). *Int. J. Environ.* **2022**, *81*, 1756–1771. [CrossRef]
2. Baniasadi, M.; Vakilchap, F.; Bahaloo-Horeh, N.; Mousavi, S.M.; Farnaud, S. Advances in bioleaching as a sustainable method for metal recovery from e-waste: A review. *J. Ind. Eng. Chem.* **2019**, *76*, 75–90. [CrossRef]
3. Shahabuddin, M.; Uddin, M.N.; Chowdhury, J.I.; Ahmed, S.F.; Uddin, M.N.; Mofijur, M.; Uddin, M.A. A review of the recent development, challenges, and opportunities of electronic waste (e-waste). *Int. J. Environ. Sci. Technol.* **2023**, *20*, 4513–4520. [CrossRef]
4. Zhao, W.; Xu, J.; Fei, W.; Liu, Z.; He, W.; Li, G. The reuse of electronic components from waste printed circuit boards: A critical review. *Environ. Sci. Adv.* **2023**, *2*, 196–214. [CrossRef]
5. Arshadi, M.; Yaghmaei, S. Advances in bioleaching of copper and nickel from electronic waste using *Acidithiobacillus ferrooxidans*: Evaluating daily pH adjustment. *Chem. Pap.* **2020**, *74*, 2211–2227. [CrossRef]
6. Hubau, A.; Minier, M.; Chagnes, A.; Joulian, C.; Silvente, C.; Guezennec, A.G. Recovery of metals in a double-stage continuous bioreactor for acidic bioleaching of printed circuit boards (PCBs). *Sep. Purif. Technol.* **2020**, *238*, 116481. [CrossRef]
7. Arshadi, M.; Mousavi, S.M. Multi-objective optimization of heavy metals bioleaching from discarded mobile phone PCBs: Simultaneous Cu and Ni recovery using *Acidithiobacillus ferrooxidans*. *Separ. Purif. Technol.* **2015**, *147*, 210–219. [CrossRef]
8. Baniasadi, M.; Graves, J.E.; Ray, D.A.; De Silva, A.L.; Renshaw, D.; Farnaud, S. Closed-loop recycling of copper from waste printed circuit boards using bioleaching and electrowinning processes. *Waste Biomass Valorization* **2021**, *12*, 3125–3136. [CrossRef]
9. Hao, J.; Wang, Y.; Wu, Y.; Guo, F. Metal recovery from waste printed circuit boards: A review for current status and perspectives. *Resour. Conserv. Recycl.* **2020**, *157*, 104787. [CrossRef]
10. Deng, S.; Xiao, Z.; Zhang, W.; Noble, A.; Das, S.; Yih, Y.; Sutherland, J.W. Economic analysis of precious metal recovery from electronic waste through gas-assisted microflow extraction. *Resour. Conserv. Recycl.* **2023**, *190*, 106810. [CrossRef]
11. Garg, H.; Nagar, N.; Ellamparuthy, G.; Angadi, S.I.; Gahan, C.S. Bench scale microbial catalysed leaching of mobile phone PCBs with an increasing pulp density. *Heliyon* **2019**, *5*, e02883. [CrossRef]

12. Ahamed, A.; Ge, L.; Zhao, K.; Veksha, A.; Bobacka, J.; Lisak, G. Environmental footprint of voltammetric sensors based on screen-printed electrodes: An assessment towards "green" sensor manufacturing. *Chemosphere* **2021**, *278*, 130462. [CrossRef]
13. Gui, Q.; Khan, M.I.; Wang, S.; Zhang, L. The ultrasound leaching kinetics of gold in the thiosulphate leaching process catalysed by cobalt ammonia. *Hydrometallurgy* **2020**, *196*, 105426. [CrossRef]
14. Syed, S. Recovery of gold from secondary sources—A review. *Hydrometallurgy* **2012**, *115*, 30–51. [CrossRef]
15. Wang, J.; Faraji, F.; Ramsay, J.; Ghahreman, A. A review of biocyanidation as a sustainable route for gold recovery from primary and secondary low-grade resources. *J. Clean. Prod.* **2021**, *296*, 126457. [CrossRef]
16. Ilkhani, Z.; Vakilchap, F.; Sadeghi, N.; Mousavi, S.M. Base metals (Fe, Al, Ti) and rare earth elements (Ce, La, Pr) leaching from red mud through an efficient chemical-biological hybrid approach. *Miner. Eng.* **2024**, *208*, 108603. [CrossRef]
17. Zhang, L.; Zhou, W.; Liu, Y.; Jia, H.; Zhou, J.; Wei, P.; Zhou, H. Bioleaching of dewatered electroplating sludge for the extraction of base metals using an adapted microbial consortium: Process optimization and kinetics. *Hydrometallurgy* **2020**, *191*, 105227. [CrossRef]
18. Dong, Y.; Zan, J.; Lin, H. Bioleaching of heavy metals from metal tailings utilizing bacteria and fungi: Mechanisms, strengthen measures, and development prospect. *J. Environ. Manag.* **2023**, *344*, 118511. [CrossRef]
19. Diaz, M.A.; De Ranson, I.U.; Dorta, B.; Banat, I.M.; Blazquez, M.L.; Gonzalez, F.; Ballester, A. Metal removal from contaminated soils through bioleaching with oxidizing bacteria and rhamnolipid biosurfactants. *Soil Sediment Contam. Int. J.* **2015**, *24*, 16–29. [CrossRef]
20. Williamson, A.J.; Folens, K.; Matthijs, S.; Cortes, Y.P.; Varia, J.; Du Laing, G.; Hennebel, T. Selective metal extraction by biologically produced siderophores during bioleaching from low-grade primary and secondary mineral resources. *Miner. Eng.* **2021**, *163*, 106774. [CrossRef]
21. Sadeghi, N.; Vakilchap, F.; Ilkhani, Z.; Mousavi, S.M. Assessment of the visible light effect on one-step bacterial leaching of metals from spent lithium-ion batteries cathode pretreated by a bio-chemical lixiviant. *J. Clean. Prod.* **2024**, *436*, 140432. [CrossRef]
22. Mahmoud, A.; Cézac, P.; Hoadley, A.F.; Contamine, F.; d'Hugues, P. A review of sulphide minerals microbially assisted leaching in stirred tank reactors. *Int. Biodeter. Biodegr.* **2017**, *119*, 118–146. [CrossRef]
23. Roy, J.J.; Cao, B.; Madhavi, S. A review on the recycling of spent lithium-ion batteries (LIBs) by the bioleaching approach. *Chemosphere* **2021**, *282*, 130944. [CrossRef]
24. Yaashikaa, P.R.; Priyanka, B.; Kumar, P.S.; Karishma, S.; Jeevanantham, S.; Indraganti, S. A review on recent advancements in recovery of valuable and toxic metals from e-waste using bioleaching approach. *Chemosphere* **2022**, *287*, 132230. [CrossRef]
25. Zeng, X.; Li, J.; Xie, H.; Liu, L. A novel dismantling process of waste printed circuit boards using water-soluble ionic liquid. *Chemosphere* **2013**, *93*, 1288–1294. [CrossRef]
26. Li, J.; Duan, H.; Yu, K.; Liu, L.; Wang, S. Characteristic of low-temperature pyrolysis of printed circuit boards subjected to various atmosphere. *Resour. Conserv. Recycl.* **2010**, *54*, 810–815. [CrossRef]
27. Ghosh, B.; Ghosh, M.K.; Parhi, P.; Mukherjee, P.S.; Mishra, B.K. Waste printed circuit boards recycling: An extensive assessment of current status. *J. Clean. Prod.* **2015**, *94*, 5–19. [CrossRef]
28. Srichandan, H.; Mohapatra, R.K.; Parhi, P.K.; Mishra, S. Bioleaching approach for extraction of metal values from secondary solid wastes: A critical review. *Hydrometallurgy* **2019**, *189*, 105122. [CrossRef]
29. Pathak, A.; Kothari, R.; Dastidar, M.G.; Sreekrishnan, T.R.; Kim, D.J. Comparison of bioleaching of heavy metals from municipal sludge using indigenous sulphur and iron-oxidizing microorganisms: Continuous stirred tank reactor studies. *J. Environ. Sci. Health A* **2014**, *49*, 93–100. [CrossRef] [PubMed]
30. Arshadi, M.; Pourhossein, F.; Mousavi, S.M.; Yaghmaei, S. Green recovery of Cu-Ni-Fe from a mixture of spent PCBs using adapted *A. ferrooxidans* in a bubble column bioreactor. *Sep. Purif. Technol.* **2021**, *272*, 118701. [CrossRef]
31. Naderi, A.; Vakilchap, F.; Motamedian, E.; Mousavi, S.M. Regulatory-systemic approach in *Aspergillus niger* for bioleaching improvement by controlling precipitation. *Appl. Microbiol. Biotechnol.* **2023**, *107*, 7331–7346. [CrossRef] [PubMed]
32. Horeh, N.B.; Mousavi, S.M.; Shojaosadati, S.A. Bioleaching of valuable metals from spent lithium-ion mobile phone batteries using *Aspergillus niger*. *J. Power Sources* **2016**, *320*, 257–266. [CrossRef]
33. Vardanyan, A.; Vardanyan, N.; Aâtach, M.; Malavasi, P.; Gaydardzhiev, S. Bio-Assisted Leaching of Non-Ferrous Metals from Waste Printed Circuit Boards—Importance of Process Parameters. *Metals* **2022**, *12*, 2092. [CrossRef]
34. Kaksonen, A.H.; Deng, X.; Bohu, T.; Zea, L.; Khaleque, H.N.; Gumulya, Y.; Cheng, K.Y. Prospective directions for biohydrometallurgy. *Hydrometallurgy* **2020**, *195*, 105376. [CrossRef]
35. Dong, Y.; Zan, J.; Lin, H. Recovery of precious metals from waste printed circuit boards though bioleaching route: A review of the recent progress and perspective. *J. Environ. Manag.* **2023**, *348*, 119354. [CrossRef] [PubMed]
36. Jiang, T.; Shi, Q.; Wei, Z.; Shah, K.; Efstathiadis, H.; Meng, X.; Liang, Y. Leaching of valuable metals from cathode active materials in spent lithium-ion batteries by levulinic acid and biological approaches. *Heliyon* **2023**, *9*, e15788. [CrossRef]

37. Marappa, N.; Ramachandran, L.; Dharumadurai, D.; Nooruddin, T. Recovery of gold and other precious metal resources from environmental polluted E-waste printed circuit board by bioleaching *Frankia*. *Int. J. Environ. Res.* **2020**, *14*, 165–176. [CrossRef]
38. Natarajan, G.; Ting, Y.P. Gold biorecovery from e-waste: An improved strategy through spent medium leaching with pH modification. *Chemosphere* **2015**, *136*, 232–238. [CrossRef] [PubMed]
39. Kara, I.T.; Wagland, S.T.; Coulon, F. Techno-economic assessment of bioleaching for metallurgical by-products. *J. Environ. Manag.* **2024**, *358*, 120904. [CrossRef]
40. Adetunji, A.I.; Oberholster, P.J.; Erasmus, M. Bioleaching of metals from e-waste using microorganisms: A re-view. *Minerals* **2023**, *13*, 828. [CrossRef]
41. Pradhan, J.K.; Kumar, S. Metals bioleaching from electronic waste by *Chromobacterium violaceum* and *Pseudomonads* sp. *Waste Manag. Res.* **2012**, *30*, 1151–1159. [CrossRef] [PubMed]
42. Van Yken, J.; Boxall, N.J.; Cheng, K.Y.; Nikoloski, A.N.; Moheimani, N.; Kaksonen, A.H. Techno-economic analysis of an integrated bio-and hydrometallurgical process for base and precious metal recovery from waste printed circuit boards. *Hydrometallurgy* **2023**, *222*, 106193. [CrossRef]
43. Kim, B.J.; Koh, Y.K.; Kwon, J.S. Bioleaching of pyrrhotite with bacterial adaptation and biological oxidation for iron recovery. *Metals* **2021**, *11*, 295. [CrossRef]
44. Gu, T.; Rastegar, S.O.; Mousavi, S.M.; Li, M.; Zhou, M. Advances in bioleaching for recovery of metals and bioremediation of fuel ash and sewage sludge. *Bioresour. Technol.* **2018**, *261*, 428–440. [CrossRef]
45. Li, X.; Feng, C.; Lei, M.; Luo, K.; Wang, L.; Liu, R.; Hu, Y. Bioremediation of organic/heavy metal contaminants by mixed cultures of microorganisms: A review. *Open Chem.* **2022**, *20*, 793–807. [CrossRef]
46. Qiu, M.Q.; Xiong, S.Y.; Zhang, W.M.; Wang, G.X. A comparison of bioleaching of chalcopyrite using pure culture or a mixed culture. *Miner. Eng.* **2005**, *18*, 987–990. [CrossRef]
47. Yachkula, A.; Rozova, O.; Abashina, T.; Vainshtein, M.; Grouzdev, D.; Bulaev, A. Attempts to stimulate leaching activity of *Acidithiobacillus ferrooxidans* strain TFBk. *Minerals* **2022**, *12*, 1051. [CrossRef]
48. Tay, S.B.; Natarajan, G.; Rahim, M.N.B.A.; Tan, H.T.; Chung, M.C.M.; Ting, Y.P.; Yew, W.S. Enhancing gold recovery from electronic waste via lixiviant metabolic engineering in *Chromobacterium violaceum*. *Sci. Rep.* **2013**, *3*, 2236. [CrossRef]
49. Lu, Y.; Xu, Z. Precious metals recovery from waste printed circuit boards: A review for current status and perspective. *Resour. Conserv. Recycl.* **2016**, *113*, 28–39. [CrossRef]
50. Panda, R.; Dinkar, O.S.; Jha, M.K.; Pathak, D.D. Recycling of gold from waste electronic components of devices. *Korean. J. Chem. Eng.* **2020**, *37*, 111–119. [CrossRef]
51. Pourhossein, F.; Mousavi, S.M. A novel rapid and selective microbially thiosulphate bioleaching of precious metals from discarded telecommunication printed circuited boards (TPCBs). *Resour. Conserv. Recycl.* **2022**, *187*, 106599. [CrossRef]
52. Jorjani, E.; Sabzkoohi, H.A. Gold leaching from ores using biogenic lixiviants—A review. *Curr. Res. Biotechnol.* **2022**, *4*, 10–20. [CrossRef]
53. Hilson, G.; Monhemius, A.J. Alternatives to cyanide in the gold mining industry: What prospects for the future? *J. Clean. Prod.* **2006**, *14*, 1158–1167. [CrossRef]
54. Işıldar, A.; van de Vossenberg, J.; Rene, E.R.; van Hullebusch, E.D.; Lens, P.N. Two-step bioleaching of copper and gold from discarded printed circuit boards (PCB). *J. Waste Manag.* **2016**, *57*, 149–157. [CrossRef] [PubMed]
55. Li, J.; Liang, C.; Ma, C. Bioleaching of gold from waste printed circuit boards by *Chromobacterium violaceum*. *J. Mater. Cycles Waste Manag.* **2015**, *17*, 529–539. [CrossRef]
56. Kumar, A.; Saini, H.S.; Kumar, S. Bioleaching of gold and silver from waste printed circuit boards by *Pseudomonas balearica* SAE1 isolated from an e-waste recycling facility. *Curr. Microbiol.* **2018**, *75*, 194–201. [CrossRef] [PubMed]
57. Ruan, J.; Zhu, X.; Qian, Y.; Hu, J. A new strain for recovering precious metals from waste printed circuit boards. *Waste Manag.* **2014**, *34*, 901–907. [CrossRef] [PubMed]
58. Chi, T.D.; Lee, J.C.; Pandey, B.D.; Yoo, K.; Jeong, J. Bioleaching of gold and copper from waste mobile phone PCBs by using a cyanogenic bacterium. *Miner. Eng.* **2011**, *24*, 1219–1222. [CrossRef]
59. Natarajan, G.; Ting, Y.P. Pretreatment of e-waste and mutation of alkali-tolerant cyanogenic bacteria promote gold biorecovery. *Bioresour. Technol.* **2014**, *152*, 80–85. [CrossRef]
60. Hu, J.; Tang, Y.; Ai, F.; Lin, M.; Ruan, J. Biofilm for leaching precious metals from waste printed circuit boards using biocyanidation technology. *J. Hazard. Mater.* **2021**, *403*, 123586. [CrossRef]
61. Kaksonen, A.H.; Mudunuru, B.M.; Hackl, R. The role of microorganisms in gold processing and recovery—A review. *Hydrometallurgy* **2014**, *142*, 70–83. [CrossRef]
62. Zhang, X.M.; Senanayake, G. A review of ammoniacal thiosulphate leaching of gold: An update useful for further research in non-cyanide gold lixiviants. *Miner. Process. Extr. Metall. Rev.* **2016**, *37*, 385–411. [CrossRef]
63. Ha, V.H.; Lee, J.C.; Jeong, J.; Hai, H.T.; Jha, M.K. Thiosulphate leaching of gold from waste mobile phones. *J. Hazard. Mater.* **2010**, *178*, 1115–1119. [CrossRef] [PubMed]

64. Xie, F.; Chen, J.N.; Wang, J.; Wang, W. Review of gold leaching in thiosulphate-based solutions. *Trans. Nonferrous Met. Soc. China* **2021**, *31*, 3506–3529. [CrossRef]
65. Zhao, H.F.; Yang, H.Y.; Chen, X.; Chen, G.B.; Tong, L.L.; Jin, Z.N. Effect of Triethanolamine as a New and Efficient Additive on Thiosulphate-Copper-Ammonia System Leaching of Gold. *Jom* **2020**, *72*, 946–952. [CrossRef]
66. Wang, J.; Xie, F.; Wang, W.; Bai, Y.; Fu, Y.; Dreisinger, D. Eco-friendly leaching of gold from a carbonaceous gold concentrate in copper-citrate-thiosulphate solutions. *Hydrometallurgy* **2020**, *191*, 105204. [CrossRef]
67. Xu, B.; Kong, W.; Li, Q.; Yang, Y.; Jiang, T.; Liu, X. A review of thiosulphate leaching of gold: Focus on thiosulphate consumption and gold recovery from pregnant solution. *Metals* **2017**, *7*, 222. [CrossRef]
68. Camelino, S.; Rao, J.; Padilla, R.L.; Lucci, R. Initial studies about gold leaching from printed circuit boards (PCB's) of waste cell phones. *Procedia Mater. Sci.* **2015**, *9*, 105–112. [CrossRef]
69. Lampinen, M.; Laari, A.; Turunen, I. Ammoniacal thiosulphate leaching of pressure oxidized sulphide gold concentrate with low reagent consumption. *Hydrometallurgy* **2015**, *151*, 1–9. [CrossRef]
70. Oraby, E.A. Gold leaching in Thiosulphate Solutions and Its Environmental Effects Compared with Cyanide. Doctoral Dissertation, Curtin University, Perth, Australia, 2009.
71. Navarro, P.; Vargas, C.; Villarroel, A.; Alguacil, F.J. On the use of ammoniacal/ammonium thiosulphate for gold extraction from a concentrate. *Hydrometallurgy* **2002**, *65*, 37–42. [CrossRef]
72. Jeon, S.; Tabelin, C.B.; Park, I.; Nagata, Y.; Ito, M.; Hiroyoshi, N. Ammonium thiosulphate extraction of gold from printed circuit boards (PCBs) of end-of-life mobile phones and its recovery from pregnant leach solution by cementation. *Hydrometallurgy* **2020**, *191*, 105214. [CrossRef]
73. Ficeriová, J.; Baláž, P.; Villachica, C.L. Thiosulphate leaching of silver, gold and bismuth from a complex sulphide concentrate. *Hydrometallurgy* **2005**, *77*, 35–39. [CrossRef]
74. Aylmore, M.G.; Muir, D.M. Thiosulphate leaching of gold—A review. *Miner. Eng.* **2001**, *14*, 135–174. [CrossRef]
75. Petter, P.M.H.; Veit, H.M.; Bernardes, A.M. Evaluation of gold and silver leaching from printed circuit board of cellphones. *J. Waste Manag.* **2014**, *34*, 475–482. [CrossRef]
76. Yu, H.; Zi, F.; Hu, X.; Zhong, J.; Nie, Y.; Xiang, P. The copper–ethanediamine–thiosulphate leaching of gold ore containing limonite with cetyltrimethyl ammonium bromide as the synergist. *Hydrometallurgy* **2014**, *150*, 178–183. [CrossRef]
77. Sitando, O.; Senanayake, G.; Dai, X.; Nikoloski, A.N.; Breuer, P. A review of factors affecting gold leaching in non-ammoniacal thiosulphate solutions including degradation and in-situ generation of thiosulphate. *Hydrometallurgy* **2018**, *178*, 151–175. [CrossRef]
78. Breuer, P.L.; Jeffrey, M.I. The reduction of copper(II) and the oxidation of thiosulphate and oxysulphur anions in gold leaching solutions. *Hydrometallurgy* **2003**, *70*, 163–173. [CrossRef]
79. Feng, D.; van Deventer, J.S.J. Thiosulphate leaching of gold in the presence of ethylenediaminetetraacetic acid (EDTA). *Miner. Eng.* **2010**, *23*, 143–150. [CrossRef]
80. Zhang, G.; Hou, L.; Chen, P.; Zhang, Q.; Chen, Y.; Zainiddinovich, N.Z.; Jia, F. Efficient and stable leaching of gold in a novel ethydiaminedhephen acetic-thiosulphate system. *Miner. Eng.* **2024**, *209*, 108639. [CrossRef]
81. Chandra, I.; Jeffrey, M.I. An electrochemical study of the effect of additives and electrolyte on the dissolution of gold in thiosulphate solutions. *Hydrometallurgy* **2004**, *73*, 305–312. [CrossRef]
82. Wu, W.; Liu, X.; Zhang, X.; Zhu, M.; Tan, W. Bioleaching of copper from waste printed circuit boards by bacteria-free cultural supernatant of iron–sulphur-oxidizing bacteria. *Bioresour. Bioprocess* **2018**, *5*, 1–13. [CrossRef]
83. Xu, B.; Yang, Y.; Jiang, T.; Li, Q.; Zhang, X.; Wang, D. Improved thiosulphate leaching of a refractory gold concentrate calcine with additives. *Hydrometallurgy.* **2015**, *152*, 214–222. [CrossRef]
84. Wang, R.; Lin, J.Q.; Liu, X.M.; Pang, X.; Zhang, C.J.; Yang, C.L.; Chen, L.X. Sulphur oxidation in the acidophilic autotrophic *Acidithiobacillus* spp. *Front. Microbiol.* **2019**, *9*, 3290. [CrossRef] [PubMed]
85. Zhang, S.; Yan, L.; Xing, W.; Chen, P.; Zhang, Y.; Wang, W. *Acidithiobacillus ferrooxidans* and its potential application. *Extremophiles* **2018**, *22*, 563–579. [CrossRef]
86. Kara, I.T.; Huntington, V.E.; Simmons, N.; Wagland, S.T.; Coulon, F. Extracting Metal Ions from Basic Oxygen Steelmaking Dust by using Bio-Hydrometallurgy. *Heliyon* **2024**, *10*, e32437. [CrossRef]
87. McNeice, J.; Mahandra, H.; Ghahreman, A. Application of biogenic thiosulfate produced by *methylophaga sulfidovorans* for sustainable gold extraction. *ACS Sustain. Chem. Eng.* **2022**, *10*, 10034–10046. [CrossRef]
88. De Wit, R.; van Gemerden, H. Oxidation of sulphide to thiosulphate by *Microcoleus chtonoplastes*. *FEMS Microbiol. Ecol.* **1987**, *3*, 7–13. [CrossRef]
89. Hutt, L.P.; Huntemann, M.; Clum, A.; Pillay, M.; Palaniappan, K.; Varghese, N.; Boden, R. Permanent draft genome of *Thiobacillus thioparus DSM 505 T*, an obligately chemolithoautotrophic member of the Betaproteobacteria. *Stand. Genom. Sci.* **2017**, *12*, 1–8. [CrossRef]

90. Kunert, J.; Stránský, Z. Thiosulphate production from cystine by the keratinolytic prokaryote Streptomyces fradiae. *Arch. Microbiol.* **1988**, *150*, 600–601. [CrossRef]
91. Xin, Y.; Liu, H.; Cui, F.; Liu, H.; Xun, L. Recombinant *Escherichia coli* with sulphide: Quinone oxidoreductase and persulphide dioxygenase rapidly oxidises sulphide to sulfite and thiosulphate via a new pathway. *Environ. Microbiol.* **2016**, *18*, 5123–5136. [CrossRef]
92. McNeice, J.; Mahandra, H.; Ghahreman, A. Biogenesis of thiosulfate in microorganisms and its applications for sustainable metal extraction. *Rev. Environ. Sci. Biotechnol.* **2022**, *21*, 993–1015. [CrossRef]
93. Pourhossein, F.; Mousavi, S.M. Improvement of gold bioleaching extraction from waste telecommunication printed circuit boards using biogenic thiosulphate by *Acidithiobacillus thiooxidans*. *J. Hazard. Mater.* **2023**, *450*, 131073. [CrossRef] [PubMed]
94. McNeice, J.; Mahandra, H.; Ghahreman, A. Biogenic Production of Thiosulphate from Organic and Inorganic Sulphur Substrates for Application to Gold Leaching. *Sustainability* **2022**, *14*, 16666. [CrossRef]
95. Xia, M.C.; Wang, Y.P.; Peng, T.J.; Shen, L.; Yu, R.L.; Liu, Y.D.; Zeng, W.M. Recycling of metals from pretreated waste printed circuit boards effectively in stirred tank reactor by a moderately thermophilic culture. *J. Biosci. Bioeng.* **2017**, *123*, 714–721. [CrossRef] [PubMed]
96. Li, J.; Yang, H.; Tong, L.; Sand, W. Some aspects of industrial heap bioleaching technology: From basics to practice. *Min. Proc. Ext. Meta. Rev.* **2022**, *43*, 510–528. [CrossRef]
97. Jia, Y.; Ruan, R.; Qu, J.; Tan, Q.; Sun, H.; Niu, X. Multi-Scale and Trans-Disciplinary Research and Technology Developments of Heap Bioleaching. *Minerals* **2024**, *14*, 808. [CrossRef]
98. Devi, A.S.; Ganesh, S. Recent bioleaching approaches employed for the extraction of metals in mining fields for the purpose of utilization and creating the sustainable future. *Environ. Qual. Manag.* **2024**, *33*, 317–333. [CrossRef]
99. Tezyapar Kara, I.; Kremser, K.; Wagland, S.T.; Coulon, F. Bioleaching metal-bearing wastes and by-products for resource recovery: A review. *Environ. Chem. Lett.* **2023**, *21*, 3329–3350. [CrossRef]
100. Roberto, F.F.; Schippers, A. Progress in bioleaching: Part B, applications of microbial processes by the minerals industries. *Appl. Microbiol. Biotechnol.* **2022**, *106*, 5913–5928. [CrossRef]
101. Shahrabi-Farahani, M.; Yaghmaei, S.; Mousavi, S.M.; Amiri, F. Bioleaching of heavy metals from a petroleum spent catalyst using *Acidithiobacillus thiooxidans* in a slurry bubble column bioreactor. *Sep. Purif. Technol.* **2014**, *132*, 41–49. [CrossRef]
102. Chen, S.Y.; Wu, J.Q.; Sung, S. Effects of sulfur dosage on continuous bioleaching of heavy metals from contaminated sediment. *J. Hazard. Mater.* **2022**, *424*, 127257. [CrossRef]
103. Hubau, A.; Minier, M.; Chagnes, A.; Joulian, C.; Perez, C.; Guezennec, A.G. Continuous production of a biogenic ferric iron lixiviant for the bioleaching of printed circuit boards (PCBs). *Hydrometallurgy* **2018**, *180*, 180–191. [CrossRef]
104. Rawlings, D.E.; Johnson, D.B. The microbiology of biomining: Development and optimization of mineral-oxidizing microbial consortia. *Microbiology* **2007**, *153*, 315–324. [CrossRef]
105. Batty, J.D.; Rorke, G.V. Development and commercial demonstration of the BioCOP™ thermophile process. *Hydrometallurgy* **2006**, *83*, 83–89. [CrossRef]
106. Rodrigues, M.L.; Leão, V.A.; Gomes, O.; Lambert, F.; Bastin, D.; Gaydardzhiev, S. Copper extraction from coarsely ground printed circuit boards using moderate thermophilic bacteria in a rotating-drum reactor. *J. Waste Manag.* **2015**, *41*, 148–158. [CrossRef]
107. Mäkinen, J.; Bachér, J.; Kaartinen, T.; Wahlström, M.; Salminen, J. The effect of flotation and parameters for bioleaching of printed circuit boards. *Miner. Eng.* **2015**, *75*, 26–31. [CrossRef]
108. Van Yken, J.; Cheng, K.Y.; Boxall, N.J.; Nikoloski, A.N.; Moheimani, N.; Valix, M.; Kaksonen, A.H. An integrated biohydrometallurgical approach for the extraction of base metals from printed circuit boards. *Hydrometallurgy* **2023**, *216*, 105998. [CrossRef]
109. N2s. Available online: https://www.n2s.co.uk/ (accessed on 31 December 2022).
110. Abdel Azim, A.; Bellini, R.; Vizzarro, A.; Bassani, I.; Pirri, C.F.; Menin, B. Highlighting the role of archaea in urban mine waste exploitation and valorisation. *Recycling* **2023**, *8*, 20. [CrossRef]
111. Thompson, V.S.; Gupta, M.; Jin, H.; Vahidi, E.; Yim, M.; Jindra, M.A.; Reed, D.W. Techno-economic and life cycle analysis for bioleaching rare-earth elements from waste materials. *ACS Sustain. Chem. Eng.* **2018**, *6*, 1602–1609. [CrossRef]
112. Andrianandraina, S.H.; Alamdari, H.; Blais, J.F. Mass balance and techno-economic study of a complete treatment chain of bio-oxidation for the extraction and recovery of precious metals from gold ore. *Miner. Eng.* **2023**, *202*, 108247. [CrossRef]
113. Bryan, C.G.; Watkin, E.L.; McCredden, T.J.; Wong, Z.R.; Harrison, S.T.L.; Kaksonen, A.H. The use of pyrite as a source of lixiviant in the bioleaching of electronic waste. *Hydrometallurgy* **2015**, *152*, 33–43. [CrossRef]

114. Jin, H.; Reed, D.W.; Thompson, V.S.; Fujita, Y.; Jiao, Y.; Crain-Zamora, M.; Fisher, J.; Scalzone, K.; Griffel, M.; Hartley, D.; et al. Sustainable bioleaching of rare earth elements from industrial waste materials using agricultural wastes. *ACS Sustain. Chem. Eng.* **2019**, *7*, 15311–15319. [CrossRef]
115. Maluleke, M.D.; Kotsiopoulos, A.; Govender-Opitz, E.; Harrison, S.T. Bioleaching of printed circuit boards in a two-stage reactor system with enhanced ferric iron regeneration in a re-circulating packed-bed reactor from PCB leaching. *Miner. Eng.* **2024**, *218*, 109000. [CrossRef]

MDPI AG
Grosspeteranlage 5
4052 Basel
Switzerland
Tel.: +41 61 683 77 34

Recycling Editorial Office
E-mail: recycling@mdpi.com
www.mdpi.com/journal/recycling

www.ingramcontent.com/pod-product-compliance
Lightning Source LLC
Chambersburg PA
CBHW040836250926
51850CB00007BA/1330
* 9 7 8 3 7 2 5 8 6 4 2 4 9 *